Advanced Manufacturing Operations Technologies

This book discusses and chronicles various types of manufacturing processes, including casting and molding, machining, joining, shearing, and forming. It refers to repetitive, discrete job shop process manufacturing (continuous) and process manufacturing (batch). It also offers detailed examples from the nuclear, electronic, plastics, adhesives, inks, packaging, chemical, and pharmaceutical industries.

Advanced Manufacturing Operations Technologies: Principles, Applications, and Design Correlations in Chemical Engineering Fields of Practice fills the gap in the connection between production and regulated applications in several industries. It highlights established concepts and provides a new, fresh outlook by concentrating on and creating linkages in the implementation of practices in manufacturing and safe, clean energy systems. Case studies for the overall design, installations, and construction of manufacturing operations in various industries as well as the standard operating procedures are offered. The book also discusses the correlation between design strategies including step-by-step processes to ensure the reliability, safety, and efficacy of products. The fundamentals of controlled techniques, quality by design, risk assessment, and management are covered in support of operations applications and continuous improvement.

This comprehensive book is helpful to all professionals, students, and academicians in many scientific disciplines that utilize fundamental principles of chemical engineering. It is engineering-driven and will be of use to those in industrial and manufacturing, chemical, biochemical, mechanical engineering, and automated control systems fields.

Advanced Manufacturing Operations Technologies

Principles, Applications, and Design
Correlations in Chemical Engineering
Fields of Practice

Sam A. Hout

CRC Press
Taylor & Francis Group
Boca Raton London New York

CRC Press is an imprint of the
Taylor & Francis Group, an **informa** business

Designed cover image: Shutterstock

First edition published 2024
by CRC Press
6000 Broken Sound Parkway NW, Suite 300, Boca Raton, FL 33487–2742

and by CRC Press
4 Park Square, Milton Park, Abingdon, Oxon, OX14 4RN

CRC Press is an imprint of Taylor & Francis Group, LLC

© 2024 Sam A. Hout

Library of Congress Cataloging-in-Publication Data
Names: Hout, Sam A., 1951– author.
Title: Advanced manufacturing operations technologies : principles, applications, and
    design correlations in chemical engineering fields of practice / Sam A. Hout.
Description: First edition. | Boca Raton : CRC Press, [2023] | Includes bibliographical
    references and index.
Identifiers: LCCN 2022061386 (print) | LCCN 2022061387 (ebook) |
    ISBN 9781032469126 (hbk) | ISBN 9781032470115 (pbk) |
    ISBN 9781003384199 (ebk)
Subjects: LCSH: Chemical plants. | Chemical processes. | Operations research.
Classification: LCC TP155.5 .H68 2023 (print) | LCC TP155.5 (ebook) |
    DDC 660—dc23/eng/20230130
LC record available at https://lccn.loc.gov/2022061386
LC ebook record available at https://lccn.loc.gov/2022061387

ISBN: 978-1-032-46912-6 (hbk)
ISBN: 978-1-032-47011-5 (pbk)
ISBN: 978-1-003-38419-9 (ebk)

DOI: 10.1201/9781003384199

Typeset in Times
by Apex CoVantage, LLC

Contents

Preface xi
Acknowledgments xiii
Author xv

Introduction: Heat, Mass, and Momentum Transfer 1

Chapter 1 Instrumentation 3

Connection of Pressure Gauges in Chemical Plants 3
Instrumentation and Control of Distillation Columns 8
Instrumentation of Heat Exchangers 10

Chapter 2 Chemical Plant—Engineering Materials 13

Rheological Behavior 18

Chapter 3 Pressure Vessels 21

Stress = Force/Area Over Which It Acts 21

Chapter 4 Thermodynamics—Mass, Enthalpy, Entropy, Free Energy,
and Internal Energy in Closed and Open Systems 25

Thermodynamic Properties of Fluids 26
The Ideal Gas Model Assumes 27
Prediction of Vapor–Liquid Equilibria of Mixtures at Higher
Pressure 28

Chapter 5 Fluid Dynamics—Crystallization, Sedimentation, and Particle
Dynamics 31

Chapter 6 Fluid Mechanics 33

Fluid Flow 33
Physical Properties of Fluids 33
Compressible Fluid Flow in Pipe 34
Steam 35

Chapter 7   Heat Transmission—Natural Convection, Forced Convection,
                and Single- and Two-Phase (Boiling) Flow Systems 37

                Extend to Horizontal Surface 43

Chapter 8   Transport Phenomena.. 47

                Mass Balance (Steady State).. 47
                T Operates on a Vector to Give a Derivative 49

Chapter 9   Chemical Reaction Engineering.. 51

                Purge... 51
                Reactors 52
                Batch Reactor 52
                Catalytic Processes.. 54
                Reactor Types Are Based on Two Models.. 55

Chapter 10   Fuel and Power.. 57

                Nuclear Power Production—Fission and Nuclear Reactor 58
                Carbon-Based Fuels 61
                Bond Breaking Is a Major Reaction in Cracking 61
                   Distillation Operations.. 62
                   Reforming.. 64

Chapter 11   Electrochemistry/Corrosion.. 67

                Corrosion Testing 71

Chapter 12   Unit Operations 75

                Solvent Extraction 75
                Fluidization... 77
                Sedimentation... 77
                Crystallization 77
                Leaching 78
                Distillation 78

Chapter 13   Process Dynamics.. 89

                Quantifying Flow Sheets 89
                System Stability Determination 92

Chapter 14   Mathematics 95

Chapter 15 Numerical Methods—Computation 97

DO-Loop 97
Least Square Regression 99
Numerical Integration... 100

Chapter 16 Chemical Plant Design 107

Chapter 17 Experimentation 109

Batch Distillation at Constant Reflux Ratio in Packed Column....... 111

Chapter 18 Multi-Phase Flow 113

Chapter 19 Bio-Chemical Engineering... 115

Catalytic Enzymes... 115
Fermentation.. 115

Chapter 20 Industrial Inks, Dyes, and Pigments................................. 117

Composition of Printing Inks 117
Preparation of the Specimens 119
Dissolution... 119
Water Extraction for the Barium Determination 119
Fluorescent Pigments 122
Application Areas... 124
Dispersion.. 125
Heat Stability 125
Additives... 125
Lightfastness.. 125
Plate-Out... 126
Toxicity... 126
UV Curing Offset Inks 126
Ink Manufacturing 128

Chapter 21 Industrial Plastics 129

Plastic Materials 129

Chapter 22 Coating 131

Cylinder and Roller Setting.. 131

Chapter 23 Industrial Adhesives 133

        Packaging Systems 133
        Base Cup Adhesives 133
        Cigarette Adhesives 136
        Foamed Hot Melt 136
        Wood Adhesives 137
           Bag Adhesives 137
        Disposable Products 138
        Pressure-Sensitive Adhesives 138
        Laminating Adhesives 139
        Cohesive Coating Welds 139
        Labeling Adhesives 139

Chapter 24 Electronic Materials Processing 141

        Thick Film Systems 141
        Thin Film Systems 142
           Photo Resolution of Metal Lines on Insulating Materials 142
        Conductive Polymers 143
        Thick Film—Case Study 144
        Effect of Positive Material, Kiln Firing Temperature, and Drying
           on TR and TCR 144
        Effect of Clearance and Squeegee Drive Pressure 144
        Effect of Emulsion Thickness 144
        Conclusions 145

Chapter 25 Nuclear Power Plant—Reactor Heat Removal 153

        Design Criteria 154
        Assumptions 156
        System Description 157
        System Design 157
        Sample Calculation for Determination of RHR
           HX Effectiveness 162
        Estimated RCS Thermal Capacity 163

Chapter 26 Pharmaceutical Manufacturing 169

        Microbiological Limits 175

**Appendix I: Process Flow Diagram—Component Preparation,
Compounding, Fill/Finish, and Packaging of Liquid Aseptic Operation** 179

Appendix II: Master Validation Plan 181

Appendix III: Hand Calculations—System Architecture and Computations for Nuclear Reactor Heat Removal Systems Computer Program 211

Appendix IV: Chemical Reaction Fouling 237

Appendix V: Proposal for New Plant Construction 249

Appendix VI: The Accident at Three Mile Island 265

Appendix VII: Water Hammer in Nuclear Power Plants 269

Appendix VIII: Selection and Design of Industrial Furnace Heaters 271

Appendix IX: Chemical Reaction Fouling in Unit Operations 273

Appendix X: Analysis of Single-Phase Flow in Power Plant Drain Systems 289

Appendix XI: Case Studies 293

Appendix XII: Heavy Metals Testing 297

Bibliography 305

Glossary of Terms 307

Index 319

Preface

Chemical engineering is a highly mathematical discipline. It is my intent for this book to minimize the detailed mathematical calculus, dimensionless numbers, mathematical modeling, differential equations, and several other mathematical analytical and numerical methods and focus on fundamentals. Professional chemical engineers and other professionals who want to understand the natural sciences behind chemical engineering principles and practice for applied implementations can benefit from this book's comprehensive coverage that includes case studies for overall design, installations, and construction of manufacturing operations in various industries including standard operation procedures.

The more complex the chemical and biochemical industry becomes, the more vital the role of a comprehensive understanding of fluid flow and processing is. The transport of fluids in nuclear reactors, petroleum refineries, pharmaceutical aseptic processing, and food processing are all complex fluid transports in pipes and vessels that require modeling for design conditions and conversion utilizing temperatures, pressures, and fluid flow rate calculations. Extensive mathematical calculations and equations based on theory and empirical knowledge allow us to build and construct unit operations for processing, separation, and purification of various compounds and products. This is extended from fuel and power, alternative energy, electronic components and systems manufacturing, and other production systems of chemical and biological conversion processes. The developments include powder systems, inks and dyes, and paints that cover a spectrum of different materials that extend from pure simple solutions to complex emulsions. Cleaning sanitizing agents and sterilants, as well as synthetic detergents all played a major role in the development of chemical engineering.

I conceptualized writing this book on manufacturing design and operation applications after spending over 30 years in drug manufacturing engineering including aseptic liquid and lyophilized parenteral drug vial fill/finish, cartridge filling for injector applications, sterile prefilled syringes, and ophthalmic bottle filling and capping. Quality by design dictates process preparations and compounding of drug formulations are key steps in transferring bulk product batches from beginning to end. These processes all require diligent care and technical know-how in transferring across process steps while maintaining clean and safe operations. Personnel flow and materials flow separations and consideration for movement have rigorous requirements to ensure against cross-contamination and making a safe product. During my career, I also worked on projects and consulted in other chemical engineering-affiliated industries. I wanted to put it all together in one accessible simplified experience-based reference to reinforce knowledge of design, processes, and compliance regulations and provide an easy guide to follow. This book is important in providing a step-by-step understanding of what is required to engineer and manufacture products.

Process flow, product risk-based quality approach, and the significance of process validation are emphasized. Critical step-by-step technologies and systems are described as part of the integrated facilities design. Design calculations based on fundamental principles of chemical engineering are categorized to include the use of partial differential equations and computer interpolation numerical methods. Reliability in batch-to-batch consistency and repeatable production in a compliant manner are discussed. The systematic management of risk to ensure efficacious and safe products and manufacturing systems are explained in terms as it relates to process validation.

The purpose of this book is to help my colleagues in the manufacturing industries have a simple reference book on important issues that we deal with daily and always looking for ways to trouble shoot and solve problems that face us in manufacturing products.

To develop this book, I had to learn more about the subject matter that I am describing and covering, I had to investigate insights and real examples about situations in that I had difficult and complex formulations that required specific innovative ways to process while ensuring compliance. I believe that through learning, training, and asking questions about alternatives, I was able to grow in how to convey this message in a more concise and focused fashion.

The feedback from the reviewers opened my thinking to how others are perceiving the materials and required me to add value in terms of case studies and real examples from experience. Besides writing from my personal experience in the industry, I had to research and follow guidelines among other references that have been reviewed and cited as part of this publication.

I had many communications with colleagues to see how ideas can be explained and described in the desired methods. I have been thinking about how to aid in technical subjects that are pertinent to this field for over 20 years, and I started putting together notes that would help my colleagues in the industry for over 10 years.

The table of contents in this book flows as intended for the reader to explore the subject matter in a continuous fashion or refer to a specific section as needed. Engineering principles, procedures, and calculations are the foundation of this book to cover overall facilities' adherence to SOP and organizational structure, including appropriate protocols and reports covering all aspects of a product submission and process design for pre-approval, namely, exhibit data, and process step-by-step calculations—all presented in an orderly manner with full documentation of real examples from industrial cases.

Acknowledgments

This book is written with the support of my family, my caring loving wife Mona, my empathetic daughter Samantha, and my wise son Owen. Modestly, many colleagues and friends in Lebanon, England, Switzerland, and the United States have played a transformational role in adding to my knowledge and experience to add value to what I learned and how to share in the service of others. My gratitude to colleagues and friends who contributed to my project management, engineering and managerial execution capabilities, manufacturing science, and technology through our many communications over three decades. A special dedication is hereby expressed to my advanced engineering mentor Dr. Barry D. Crittenden, Bath University, United Kingdom, who supervised my work on my Ph.D. in chemical engineering for his guidance and nascent support to my endeavors and growth in the United States. I am thankful to Dr. T. Reg Bott, University of Birmingham, and Dr. David T. Allen, UCLA, who contributed to reviewing and evaluating some of my critical research work. In addition, my gratitude goes to my brothers Suhail and Samer Hout for their respect, moral support, and dedication to our family traditions that emphasize care and compassion to others, and also to our sister Jaime for her unconditional love and undivided attention to our family. My mind and heart are filled with fond memories from our life-time friendship with Julius (Jay) and Judy Szalai who supported our families over the years. Brotherly communications, advice, and support from Robert and Susan Purkiewicz bridge our camaraderie relationship. My gratitude is expressed for the exemplary role model that my uncle Dr. Hisham A. Barakat played in influencing my motivation and upbringing.

Author

DR. SAM A. HOUT is a chartered chemical engineer and certified in business management by the American Production and Inventory Control Society (APICS). He also is a member of the International Society of Pharmaceutical Engineers (ISPE) and specializes in process engineering and business process improvement. He received his education and training in the UK (Ph.D. Chemical Engineering, University of Bath) and in the United States (MBA, University of California). Dr. Hout is currently VP, technical operations at KC Pharmaceuticals, California. For the past 20 years in drugs manufacturing, he held the position of Sr. Director of Engineering, project management, and technology process transfers at SIEGFRIED Pharmaceuticals. Previously, Dr. Hout has held the position of Sr. Manager of Engineering at TEVA Pharmaceuticals, parenteral medicines in Irvine, California. These specialty divisions produced specialty drugs for hospital institutional markets worldwide.

Prior to assuming the position at Teva, Dr. Hout was Director of Operations at PHENOMENEX, the leading global company in HPLC (High Performance Liquid Chromatography) for drug separation and analysis. During the past thirty years, Hout has helped to guide leading companies in technical positions such as J&J, Medtronic, and Fischer Scientific in developing and producing oral solid dosage, medical devices, and diagnostic standards that combat disease and promote healthier lives. In working with a global market, Hout travels frequently throughout the United States, Europe, and the Far East. Major accomplishments include design and construction of various plants; introduction of many new products; optimization of several bio and pharmaceutical operations. Hout consulted in many chemical engineering affiliated industries.

Introduction
Heat, Mass, and Momentum Transfer

The basis of heat, mass, and momentum transfer lies in the laws of chemical combination, conservation of mass in chemical reactions, and laws of definite, multiple, and equivalent proportions. For example, C^{12} is the isotope accepted as a standard, and the Avogadro number as the exact value $N = 6.02214076 \times 10^{23}$ and redefined the mole as the amount of a substance under consideration that contains N constituent particles of the substance. The MOLE (mol) is a unit of measurement that is the amount of a pure substance containing the same number of chemical units (atoms, molecules, etc.) as there are atoms in exactly 12 grams of carbon-12 (i.e., 6.022×10^{23}), e.g., a mole of O_2 molecules = 32 g.

The intensity of electrons is dependent on the number of electrons and is directly proportional to the intensity of light. However, the speed of electrons (energy of electrons) is directly proportional to the frequency of light. Therefore, not any light can produce or emit an electron from a body of mass, the light must have a minimum amount of energy to emit electrons. De Broglie modeled the duality of lights based on photons and waves. Schrodinger based his equations of change on the De Broglie assumptions that electrons can behave as waves. When the wave equation is applied to electrons, an implied correlation between energy and the position of the system is concluded. The position of electrons will be in a sphere around the nucleus and is determined by the energy level.

This book will discuss and chronicle various types of manufacturing processes, for example, casting and molding, machining, joining, shearing, and forming. In addition, it will refer to repetitive manufacturing, discrete manufacturing, job shop manufacturing, process manufacturing (continuous), and process manufacturing (batch). It will cover detailed examples from the nuclear, electronic, plastics, adhesives, inks, packaging, chemical, and pharmaceutical industries.

The various disciplines in chemical engineering will be presented in chapters to lay the foundation of basic fundamental principles behind the design and operation of chemical plants. Operations in the chemical manufacturing sector include basic chemicals facilities producing chemicals by basic processes, such as thermal cracking and distillation. Products include petrochemicals, industrial gases, synthetic dyes and pigments, and many other organic and inorganic chemicals. The chemical manufacturing subsector is part of the manufacturing sector. The chemical manufacturing subsector is based on the transformation of organic and inorganic raw materials by a chemical process and the formulation of products. These categories are industrial inorganic chemicals; plastics, materials, and synthetics; drugs; soap, cleaners, and

DOI: 10.1201/9781003384199-1

toilet goods; paints and allied products; industrial organic chemicals; agricultural chemicals; and miscellaneous chemical products.

The boundaries of the chemical industry, then, are somewhat overlapping. Its main raw materials are fossil fuels (coal, natural gas, and petroleum), air, water, salt, limestone, sulfur or an equivalent, and some specialized raw materials for special products, such as phosphates and the mineral fluorspar.

The industry includes manufacturers of inorganic- and organic-industrial chemicals, ceramic products, petrochemicals, agrochemicals, polymers, rubber (elastomers), oleochemicals (oils, fats, and waxes), explosives, fragrances, and flavors. The key segments of the chemical industry are commodity chemicals, specialty chemicals, pharmaceuticals, agrochemicals, and consumer products. Among the chemicals that have seen major price increases recently are butadiene, ammonia, and benzene. Some chemicals such as acetonitrile, aniline, and acetone have also seen normalizations in prices.

The pharmaceutical industry discovers, develops, produces, and markets drugs or pharmaceutical drugs for use as medications to be administered to patients, with the aim to cure them, vaccinate them, or alleviate symptoms. Pharmaceutical companies may deal in generic or brand medications and medical devices. They are subject to a variety of laws and regulations that govern the patenting, testing, safety, efficacy using drug testing, and marketing of drugs. The global pharmaceuticals market produces treatments worth thousands of billions of dollars annually. Pharmaceutical manufacturing is the process of industrial-scale synthesis of pharmaceutical drugs as part of the pharmaceutical industry. The process of drug manufacturing can be broken down into a series of unit operations, such as milling, granulation, coating, tablet pressing, sterile liquid filing, lyophilized products, and others.

Organic synthesis is a special branch of chemical synthesis and is concerned with the intentional construction of organic compounds, which is a major activity in the development of new drugs. Organic molecules are often more complex than inorganic compounds, and their synthesis has developed into one of the most important branches of organic chemistry, especially in therapeutic drug medication developments. A medication is a drug used to diagnose, cure, treat, or prevent disease. Drug therapy (pharmacotherapy) is an important part of the medical field and relies on the science of pharmacology for continual advancement and on pharmacy for appropriate management.

In chemical engineering and related fields, a unit operation is a basic step in a process. A raw material going into a chemical process or plant as input to be converted into a product is commonly called a feedstock, or simply feed. Unit operations involve a physical change or chemical transformation such as separation, crystallization, evaporation, filtration, polymerization, isomerization, and other reactions. For example, in milk processing, the following unit operations are involved: Homogenization, pasteurization, and packaging. These unit operations are connected to create the overall process. A process may require many unit operations to obtain the desired product from the starting materials or feedstocks.

1 Instrumentation

Pressure measurements are critical in process control. Absolute pressure correlation to identify vacuum requirements against gauge pressure above atmospheric pressure define process conditions to establish fluid flow effectiveness as required under specific temperatures in batch or continuous operations. Pressure selection factors are as follows:

1. Measure pressure to decide what needs to be controlled and where
2. Operating pressure range to be measured
3. Are pressure measurements recorded or indicated for operating suitability?
4. Accuracy, precision, and sensitivity requirements for the instrument to do the job
5. Response
6. Process hazards, e.g., corrosion, explosion, and vibration
7. Shape and size limitations
8. Reliability, ruggedness, availability of parts, and maintenance
9. Cost: Capital, operating, and delivery lead-time
10. Permissible time-lag

The main types of measurement are for liquid for level and flow rate, temperature in various stages of unit operations, and pressure in tubular flow or processing chambers and reactors. Liquid level measurements include direct reading, float, pressure, electronic capacitance, and radiation. Temperature is measured by expansion, change of state, electrical generation or resistance, and radiation frequencies.

Thermistors are semiconductors with a negative temperature coefficient of resistance, i.e., temperature rises as resistance falls. The lag or time delay in reading the measurement is based on the sensor level of response. The position and location of the sensing element are important to minimize lag.

CONNECTION OF PRESSURE GAUGES IN CHEMICAL PLANTS

Any pressure measurement depends on the proportionality of force per unit area, which will accordingly define the pressure measurement:

1. Differential (difference between two pressures in the system)
2. Gauge (pressure above local atmospheric)
3. Vacuum (pressure less than local atmospheric)
4. Absolute (total pressure)

Pressure calculation would determine whether a single pressure gauge or a connected system of pressure gauges is needed for the process. The specific type of pressure

DOI: 10.1201/9781003384199-2

gauges is chosen according to the suitability of the instrument to provide for plant controls. For example, the bourdon tube and pressure transducers are relevant choices for differential pressure gauges that require no measurement element.

The static head of liquid should be accounted for between the pressure sensor and the measurement point. If static head corrections are needed, pressure drop calculations should be conclusive to define the kind of signal expected: pneumatic or electrical. As a process control factor, pressure may have a direct or indirect effect on other process characteristics, e.g., composition in distillation. In such cases, accurate control of pressure is essential, and care must be put into the design of installation and isolation points.

In the case of compressible fluids, considerations for vapor pressure measurements and compatibility of the gauge with the process are of utmost importance. Fluid properties' impact on the gauge such as volatility or toxicity of the fluid is a key factor to determine suitability. The gauge needs to be calibrated before usage in process controls. In addition, gauge materials should be free from creep and fatigue tendencies. The gauge should show no stress hysteresis (deformation) and made of materials that are corrosion resistant. In connecting the gauge to the process, a direct connection is often used. Nonetheless, seals for corrosive, toxic, hazardous, or elevated temperature systems should be utilized in diaphragms, bellows, or others to combat innocuous fluids.

By introducing purges, the use of inert gas or liquid to keep the pressure gauge free from undesirable fluids, a purge must be compatible with the process. Pulsation dampers are used to smooth pressure fluctuations in difficult-to-read instruments, or when the instrument failure mode can lead to fatigue failures.

Differential pressure flow meters work on the principle that fluid velocity generates a difference, which becomes the analog of the flow and can be measured on a gauge. Fluid flow from a larger to a smaller cross section of the pipe generates an increase in velocity.

This increase in kinetic energy is provided by a reduction in pressure energy resulting in a pressure drop, which can be measured. This is correlated with the following equation:

$$V_2^2 - V_1^2 = (2g/\rho).\ \Delta P$$

V—fluid velocity,   g—gravity acceleration constant, ρ—fluid density,
ΔP—pressure drop

Examples of differential pressure flow meters are orifice plate, venturi tube, venturi nozzle, and pitot tube. The general rules governing differential pressure flowmeters are that they should not distort, should have a vent hole for air escape, and must be concentric with a pipe orifice. The pipe upstream is straight with vanes as needed to ensure uniform flow as the fluid approaches the plate and control differential pressure yaw (head).

Temperature measurements are accomplished by using thermocouples or RTDs. Thermocouple (T.C.) electric effect is based on two wires of different metals that are joined at their ends (two junctions). The wires are separated except at the joints.

Upon heating one junction to a higher temperature, an e.m.f. is produced causing a current to flow in the circuit loop. The current varies with the temperature difference between the two junctions' (T.C.) output resistance. A capacitance transducer, the e.m.f. values are additive at different temperatures. T.C. is affected by the atmosphere and may suffer oxidation. Therefore, T.C. suitability must be checked for application. The resistance of the wires is a factor to be considered in a T.C. Using thin wires reduces heat conduction between the wires. The pocket wall dimension and air space are a factor in producing a response time lag of the thermocouple to a temperature change. T.C. are placed in protective sheathing that can withstand hot temperatures, and sheathing electrical properties are temperature independent. The response and sensitivity characteristics of T.C. are employed when a quick response for a small temperature change is required to be detected.

Electronic transducer devices change signals from one form to another, usually non-electrical to electrical signals, e.g., convertor. An inductive transducer (transformer) whereby the position of the core between the coils will convert the displacement into voltage signals. The resistance transducer converts an input variable position into an output variable resistance. A capacitance transducer utilizes an input variable pressure against a diaphragm to generate an output variable capacitance.

Pneumatic control valves are the correcting units of process control. If the flow rate situation is increased due to some fault condition, the valve will open due to increased pressure. These valves are usually consisting of a motor, a positioner, and a correcting element. Flow through the valve is related to air pressure (0.2–1.0 bar). Valve characteristics are quick opening, linear, equal percentage of flow, and valve lift. These control valves are usually installed with isolation valves and a bypass. The control valve may be connected to an alarm system by giving an electric or a pneumatic signal. The capacity of the valve should be designed to allow for higher output than input. The expelled materials when the valve shoots should be directed. Instrumentation is concerned with measurements of process variables (Table 1.1).

TABLE 1.1
Significance of Process Parameters

Process Parameter Measured	Occurrence (%)
Temperature	34.7
Flow	7.4
Level	11.8
Pressure	11.7
Composition	5.6
Electrical	4.6
Humidity	3.5
Speed	2.1
Density	1.8
Other	6.7

Process in static behavior is when instruments are measured in a steady state. The accuracy of the measurement is its closeness to the true value. The precision of measurement is more about the reproducibility of repeated measurements. The sensitivity of an instrument is the ratio between a change in the output signal to the change in a measured variable, which implies the smallest change in measured value to which an instrument responds. For example, thermocouples are more sensitive than ordinary mercury thermometers for measuring temperature because a slight change in temperature can be measured with a thermocouple, but not a thermometer.

Pressure liquid column-type measurements are accomplished using manometers. These U-tubes of the uniform bore can be glass or metal and contain a liquid of known density. Manometry reads differential pressure balancing weights of liquid against applied pressure. The applied pressure may be atmospheric or a vacuum. The length of the tube is usually limited to 1 m. The liquid density used is usually 0.8 s.g. for oil or 13.6 s.g for mercury. Pressure is expressed as "length unit of compound," e.g., mmHg gauge, mmHg absolute, or mm H_2O gauge. The necessary properties of manometer liquids are as follows:

- Mainly used on gases (Hg for liquids)
- Compatible with the process
- Involatile (loss of liquid)
- Non-toxic (breakage)

Inclined tube manometers increase the scale length and provide more accurate measurements of small pressure drops. Most pressure gauges are calibrated using a dead-weight piston gauge.

A Bourdon tube gauge is a flattened tube that assumes a circular cross section when internal pressure is applied. Under low pressures, the tube will deflect into a spiral shape. For high pressures, it will take the shape of a small-diameter helix. For the same temperature dependence, this type of gauge is accurate to 1.5%.

A more sensitive gauge type is a diaphragm gauge. Pressure is applied to the elastic diaphragm, which bends and displays pressure readings as calibrated depending on the applied pressure correlated to the diaphragm bend deflection. This is useful for low-pressure measurement in the range of 0.1–1 bar.

British Standard 1042 reference specifications are used for the flow of gases and liquids and related measurements. These measurements are based on the rate such as kg/s or m^3/s, which is important in a continuous process, and total flow such as kg or m^3, which is important in the batch process. For flow rate, the measurement is concerned with fluid velocity. Total flow measurement is based on volume or displacement. Velocity meters can be direct velocity, variable area, or differential pressure measurements. Differential pressure flow meters are made such that the fluid velocity is made to generate a differential pressure, which becomes the analog of flow and can be measured on a gauge. The Bernoulli theorem correlates noncompressible fluid flow from a larger area to a smaller cross section of the pipe, and thus the velocity increases. This increase in kinetic energy is provided by a reduction in pressure energy resulting in a pressure drop, which can be measured. For example, a pitot tube measures the velocity of a stream by converting the kinetic energy of a small cross section entirely into a pressure head. This is called impact pressure and is a function of velocity.

Variable area meters such as rotameters are made of a standard float inside a graduated cylinder that would be calibrated for direct flow reading. Tapered plug tubes such as compressed can also be used. Flow generates a constant differential through an orifice where size varies with flow rate. Either tube or float moves up and down, the flow area changes, and the weight of the float balances the force due to pressure drop developed across the annular orifice by the float. For larger flows, e.g., 4" pipe or bigger, rotameters are installed in a bypass line across an orifice plate, thus measuring a proportion of the flow.

The chemical analysis utilizes techniques for measurements such as high-performance liquid chromatography (HPLC), gas chromatography (GC), ion chromatography, titration, and other methods. For example, high-frequency titration meters utilize a coil that emits an oscillating radio signal of a particular frequency. The power required to maintain a radio signal depends upon the solution composition. Oxygen meters operate on the paramagnetic properties of O_2. Part of the gas stream is diverted through a central tube under the influence of a permanent magnet. The resultant flow causes a current in the solenoid that is dependent on oxygen concentration.

Other techniques such as the measurement of sonic velocity vary with temperature and molecular weight. Therefore, knowing the temperature, sonic velocity can be calibrated against chemical composition. Similarly, refractive index fringe shifts are related to chemical composition.

GC depends upon differential absorption/desorption rates of chemicals on solid packings. Path length in absorption medium and temperature determine the separation of components. The resulting separated slugs are continued a carrier gas stream to thermal conductivity or other devices for electrical output.

Material radioactivity can be used in several applications. The Geiger-Muller tube (1,000–1,200 V) charged electrode in an inert gas atmosphere emits a particle, which upon striking the electrode will generate an electrical pulse that can be amplified to pulse counter measuring α, β, and γ rays. Ionization gauge similarly lower voltage proportional to total radiation received. A scintillation counter is whereby a radiation strikes zinc sulfide, which produces a light pulse that activates a photomultiplier which in turn detects radiation. Neutron flux detectors involve tungsten electrodes in a boron fluoride atmosphere. The neutron causes the fission of the Boron atom pulse to the electrode. The electrode is coated with enriched uranium oxide fission of nucleus ionization of the surrounding gas atmosphere, thus generating an electrical pulse.

Flow measurement devices can be based on variable area and variable head measure nets, mechanical deflection of vanes, fans, or propeller inferential velocity readings. By notation, ultra-sonic measurements depend on the velocity of sound as it travels through a medium.

Level measurements can be direct, using a float, utilizing pressure differentials, capacitance, or radiation. The buoyancy force is measured by torque as the force upthrust depends on the specific gravity of the liquid medium. Capacitance level probes are based on the property that the dielectric constant of liquids is greater than

TABLE 1.2
Common Thermocouple Materials

Positive Wire	Negative Wire	Maximum Temperature, °C	Mid-Range Accuracy, °C	Output/100°C mV
90% Pt 10% Ra	Pt	1,400	±1	1
Cu	60% Cu 40% Ni	400	1.5	5
Fe	Constantan	850	2.5	5
Chromel 90% Ni, 10% Cr	Alumel 94% Ni, 6% Al	1,100	4	4

that of gases. By contrast, temperature measurements (B.S. 1041) are based on principles of expansion, change of state, electrical generation, and electrical resistance. For example, solid bimetallic strips or helices are based on different coefficients of expansion for two dissimilar materials (Table 1.2).

Thermistors are metal oxide semiconductors with a negative temperature coefficient of resistance, i.e., resistance falls as temperature rises. Typically, there is truly little temperature lag and high sensitivity; $\Delta T = 0.001°C$.

Radiation pyrometers are based on measuring radiation energy emitted from a body in the ultraviolet (UV), infrared (IR), and visible spectrum.

Recorders are important tools that document process conditions for monitoring and regulatory reasons. They provide a printed record of conditions and show measured variables against time. They are circular or linear with a single channel or multipoint recording. Alarms are fitted where dangerous or undesirable conditions can occur. Usually, alarms are pointed to recover process conditions in a fail-safe mode. Alarms may sound or flash warning lights. Operators can accept alarms by resetting a button, but only after conditions are recovered and corrected. Auto-alarms might be for alert or action level and might shut down unsafe operations. In connection to alarms, control valves are correcting units of process control. They tend to alter the flow rate of heating and cooling of process fluids. They consist of a motor mechanism, positioner, and correcting element. Flow through the valve is related to pressure or controlled through a solenoid.

INSTRUMENTATION AND CONTROL OF DISTILLATION COLUMNS

Effective control of distillation columns requires a detailed process design. The difficulty arises in continuous measurements of streams including the use of IR analyzers, refractometers, and chromatographs, which have the disadvantage of larger time lag and are less dependable than temperature measurements. Stream measurements are affected by composition and pressure, which influence boiling point such that minor changes in minor components have a pronounced effect on B.P. The large hold-up of liquid and vapor contributes to larger capacity lags in the order of a few hours that may be needed for a change in feed composition. Controls are usually applied by measurement of composition, heat input and output, pressure, and reflux start. Flow control applies when feed rate and composition are constant. The most direct control is achieved by using a signal from the composition transmitter to adjust the flow rate. The averaging level controller increases or decreases an equal percentage in the required flow stream. The reflux drum serves as a liquid seal and disengaging space for non-condensable. It also helps in damping fluctuations in flow and functions as a reservoir for reflux materials. Steam flow is controlled by an upstream pressure regulator, thus maintaining constant vapor flow. In addition, vapor flow could be controlled by maintaining a pressure drop measurement across the column. Pressure drop depends on the velocity of the liquid, which is constant. The location of sensing elements is important since often lags may be minimized by careful consideration of location. Temperature bulbs should be placed accordingly to reduce lags in the system. Temperature bulbs are placed on a few plates from the top where the composition boiling point slope is steepest. With multicomponent distillation, the problem becomes greater due to the influence of minor components. Thus, the sensing element should be placed as sensitivity permits with considerations due to rapid changes in the concentration of key components in the column.

Chromatography is used as a means of analyzing process streams. Distillation samples from trays can be analyzed quickly and accurately. The signal from these units can be suited for computer control or electronic controllers. Response time is fast in the order of a minute. Chromatographs, refractometers, IR, and UV analyzers can be easily installed for batch or continuous stream measurements with fast and accurate response time. Other process units play a major controlling factor in the feed rate. Therefore, it is recommended to use a feed surge tank and flow controller to adjust for a more balanced process. A proportional controller might be used to control the stream rate to the reboiler in proportion to the federate. Ratio control is achieved through corrective action as soon as feed disturbance is noticed.

In many cases, the variation in the boiling point of a liquid with changes in pressure is much more significant than that due to composition. The steady-state performance is not strongly pressure dependent. Therefore, the control of BP uses a differential transmitter, where pressure is measured at the top of the column, rather than the middle or in the reboiler.

In vacuum distillation, a vacuum is usually generated by means of a steam ejector and pressure is controlled by a bleed of inert gas or air into the line from the reflux drum to the ejector. These techniques are limited to work effectively in a narrow range of steam pressures and are not effective as a means of process control. The effect of the non-condensable present in the overhead product affects the method used to maintain the vacuum. This arises because even the presence of 1% gases in a vapor stream can reduce heat transfer by almost 50%. The return of reflux at constant temperature is important. However, automatic control of feed temperature is easily achieved by feedback control loop and steam header, but changes in feed temperature may not be as important as changes in flow.

INSTRUMENTATION OF HEAT EXCHANGERS

The heat transferred in a heat exchanger is described by

$$Q = U.A.\Delta T$$

where Q is the quantity of heat correlated to the overall heat transfer coefficient U, surface area A, and differential temperature gradient that develops the kinetics behind the transfer.

In a shell and tube heat exchanger, the most common method is to regulate the flow of steam into the hell side. Regulating the flow of condensate alters the level in the shell, which affects the rate of heat transfer. In liquid–liquid systems, fixed flow rates and maximum allowable temperature are system constraints that need to be considered. Usually, 10–30% of the process fluid is bypassed for operational considerations and rapid control.

Liquid level control is significant in each vessel, which can be critical to the process. If the level is too high, it might create disturbances of reaction equilibria. It might also lead to the spillage of valuable or dangerous materials. If the level is too low, reaction equilibrium might be disturbed, and loss of heat transfer surface might lead to burnout in heated vessels.

Continuous-level measuring devices provide continuity of measurement over a range, which are mostly adaptive to pneumatic or electronic controllers for automatic control. Fixed point level measurements provide response at one or several specific positions. These are used for alarm systems and might be rigged in multiple banks for semi-continuous applications. In continuous system processes, accumulators or storage vessels between stages function as surge tanks to allow process disturbances not transmitted through the whole stream. The controller will automatically reset on a level error such that flow to the subsequent stage is not affected.

In maintaining pressure below the maximum safe working pressure, reliability is more important than accuracy. As a process control factor, pressure may have a direct or indirect effect on other process characteristics, e.g., composition in distillation. In such cases, accurate control of pressure is essential.

Certain precautions in the installation of instrumentation systems are necessary in which contact between the sensing device and controlled material is made (Figure 1.1). A meaningful reading is obtained as the required contact is maintained following installation. The sensor element must have a low capacity for the controlled property, i.e., insertion of probe, temperature lag, or heat capacity of probe or

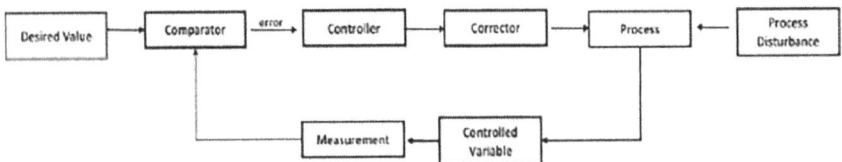

FIGURE 1.1   Closed-loop control: Block diagram of the feedback control system.

pressure drop across element may affect performance. Incompatibility of materials of construction of probe with process fluids may result in a change in temperature owing to exothermic corrosion reactions, evolution of gas or impurity, and destruction of the instrument. Specific consideration in gas systems where condensation of a liquid on the element might lead to incorrect readings specifically temperature elements. Systems' operation in remote non-contact mode is of particular importance in temperature instrumentation. For example, a radiation-type element will only give a true reading if the body is black, and the intervening space is transparent to the radiation in question. Accurate reading depends on the body enclosure and the housing temperature. The surface on which the instrument is focused has a known emissivity and the temperature of the surroundings is known, and correction factors may be calculated.

2 Chemical Plant — Engineering Materials

It is important to realize that the dividing lines between solids, liquids, and gases are not absolute, i.e., materials change phases (Figure 2.1). The transition between phases will occur for every element or compound when the prevailing conditions of temperature, pressure, and environment undergo changes. For engineering design purposes, it is convenient to categorize materials into types.

1. Metals, e.g., mild steel, stainless steel, and cast iron for ferrous. For non-ferrous, e.g., aluminum, copper, nickel, zinc, lead, titanium, zirconium, and so on.
2. Ceramic materials, e.g., glass, cement, silica, and so on.
3. Organic, e.g., PVC, PE, PTFE, rubbers, and so on.

Thermoplastic polymeric materials can be repeatedly softened and shaped by heating. They are usually produced by addition polymerization. Monomers combine by addition to themselves with no byproduct to form a polymer with only carbon in the backbone of the chain in cases of organic materials. Compounds with stabilizers, plasticizers, and pigments are resistant to attack by chemicals, but unstable to heat and UV light. They are used in certain construction materials applications and for electrical and other insulation needs.

Important physical properties of materials used in chemical plant construction include mechanical properties. These properties focus on strength, e.g., tensile strength in pressure vessels, compressive strength for bases and foundations, shear in shafts, and impact in mills. Stiffness and hardness (elastic moduli and Poisson's ratio) are important in pipework and reduction in cross section when it passes its elastic limit.

Working considerations with respect to ductility, malleability, toughness, and hardenability to cover rolling plates, and the ability of materials to be heat treated to alter mechanical properties; time stability to withstand longer periods of time with respect to dimensional creep under load; fatigue, which is indicative of a sudden failure of moving parts including damping effects related to noise pulsation; and other factors such as density and wear resistance against hardness and friction, which indicates the ability of materials to resist erosion. For example, high coefficient of friction renders materials unsuitable for bearings.

Thermal properties such as melting point or softening, thermal expansion coefficient, conductivity, thermal stability, specific heat, emissivity, and absorptivity are all major considerations in materials selection. When designing fluid flow systems, interfacial properties, wetting as defined by the contact angle, are to be noted:

DOI: 10.1201/9781003384199-3

FIGURE 2.1 Materials of construction.

Electrical

1. Conductivity
2. Dielectric constant
3. Magnetic behavior

Optical

1. Transparency
2. Refractive index
3. Color

Radiation

1. X-rays
2. γ-rays
3. α, β-radiation
4. Electromagnetic radiation

Chemical

1. Solvency
2. Corrosion resistance
3. Process compatibility

Economic

1. Raw materials
2. Fabrication
3. Maintenance
4. Recycle

Comparison of materials for strength properties reflects those brittle materials such as concrete or cast iron have different properties in tension against compression. The intermolecular attraction between molecules is weaker than repulsion; therefore, it is practical to use these types of materials in compression. Notable change may be brought about by alloying metals, e.g., with Cr, Mn, Si, or Cr, Mo, Ni to form high-tensile steel. In addition, low strength of plastics gains enormous improvement in strength when glass or carbon fibers are added (Tables 2.1 to 2.3).

The microscopic level scale shows that molecules have structural regularity caused by atomic bonding and hence packing, whereby when repetition occurs, molecules tend to arrange themselves in an orderly three-dimensional pattern, which constitute a crystalline structure. If there is no repeating pattern of molecules, then the solid composition is amorphous. The amorphous state is metastable due to less packing of

TABLE 2.1
Strength of Materials

Material	Strength per Unit Weight, MN/M2/g
Concrete tension	1.2
PTFE	7
Concrete compression	12
Polyethylene	13
Cast iron tension	21
Mild steel	47
Polystyrene	57
Oak	60
Bronze	63
Brass	70
Nylon	79
Cast iron compression	100
Pine	125
Titanium alloy	153
Aluminum alloy	154
Magnesium alloy	156
High-tensile steel	158
Glass fiber resin	800
Carbon fiber resin	875

TABLE 2.2
Stiffness of Materials

Material	Stiffness per Unit Weight, GN/M2/g
PTFE	0.2
Polyethylene	0.2
Nylon	1.8
Polystyrene	3.3
Concrete	6
Bronze	11
Brass	11
Oak	14
Cast iron	20
Mild steel	25
Magnesium alloy	25
Aluminum alloy	25
High-tensile steel	25
Titanium alloy	27
Pine	28
Glass fiber resin	30
Carbon fiber resin	105

TABLE 2.3
Strength Properties of Engineering Materials

Material	Proportional Limit MN/M2	Ultimate Strength MN/M2	Elongation Tensile Fracture	Young's Modulus GN/M2	Specific Gravity	Coefficient of Linear Expansion/°C × 10–5
Mild steel	280	370	0.30	200	7.8	1.2
High-tensile steel	770	1550	0.10	200	7.8	1.3
Aluminum alloy	230	430	1.10	70	2.8	2.3
Titanium alloy	385	690	0.15	120	4.5	0.9
Magnesium alloy	155	280	0.08	45	1.8	2.7
Brass (80% Cu)	450	600	0.04	100	8.6	1.9
Bronze (90% Cu)	170	560	0.10	100	8.9	1.8
Cast iron—tension		155		140		
-Compression		700		140	7.2	1.1
Concrete—tension		3		14		
-Compression		30		14	2.4	1.2
Oak—grain direction	30	49		11	0.8	
Pine—grain direction	33	50		11	0.4	
Nylon	77	90	1.0	2	1.1	10
Polystyrene	46	60	0.03	3.5	1.1	10
Polyethylene	6	12	5.0	0.2	0.9	28
PTFE	8	15	2.0	0.4	2.2	11
Glass fiber resin		1600			2.0	
Carbon fiber resin		1400		170	1.6	

molecular units and higher density, e.g., conversion of graphite (amorphous carbon) to its crystalline form (diamond) by the application of heat and high pressure.

Only a few elements are used in their pure state, e.g., copper as an electrical conductor. The majority of metals are used as alloys whose function is to improve corrosion resistance, tensile strength, hardness, and so on. When a second element is added to a pure material, atoms of the new element form a solid solution within the parent element, in single phase. In addition, atoms of the second element form a new second phase usually containing some atoms of the parent element. Replacing parent element atoms with atoms of the allying element is a substitutional solid solution. Common alloys are brass, steel alloys, Monel, and amalgams. Solid solution can be through random substitution or orderly one that would for a superlattice upon heat treatment, which reduces residual stress caused by distortion.

Crystal imperfections are a characteristic of all crystals. Several engineering properties are affected by these defects, which may be point, line, or plane defects. Point defects exhibit vacancies whereby one or more atoms are missing from the crystal lattice. This can arise when packing during crystallization was not perfect or because an atom has been displaced due to thermal vibrations at elevated temperatures. In ionic solids, this type of vacancy usually involves pairs of multiples of ions to maintain the charge balance. Interstitialties occur when extra atoms may be lodged in the crystal structure particularly if the packing factor is low. Such an imperfection produces distortion of the lattice unless atoms are smaller than the remaining atoms in the crystal. Another type of interstitial defect may be caused by an ion being displaced from the normal position to an adjacent one, which will result in distortion. Close-packed structures tend to have fewer interstitial types of defects than vacancies because additional energy must be provided to force atoms into new positions. Foreign atoms (impurities) may occupy positions between atoms in a lattice to become interstitial or vacancies.

Line defects are caused by any succession of point defects, which provides a straight or curved line. In practice, these distortions are called dislocations. Edge dislocation occurs when a linear defect is caused by an extra line of atoms in the structure. The dislocation is the edge of the extra plane of atoms within the structure. It is accompanied by zones of compression and tension within the lattice, thus producing a line of weakness. Screw dislocations are often associated with shear stresses where one plane of atoms tends to slide against another plane. Both edge and screw dislocations are associated with the crystal growth process. Edge defects arise when there is a slight mismatch in adjacent parts of the growing crystal and an extra row of atoms is introduced or eliminated.

Plane crystal defects may extend in two dimensions as a boundary surface of the crystal. The atoms at the surface of a crystal open to the atmosphere or liquid environment have neighbors on one side and have higher energies than internal atoms. The surface energy is exemplified by the adsorption of gases and liquids onto solid surfaces. The grain boundaries of solid crystals have growth in various orientations. The shape of each grain is determined by the presence of surrounding grains. Packing random disorder is characterized by a void fraction at the boundary between grains. Small angle tilt boundaries occur when the crystal tends to grow with the opposite angle between faces. This produces a series of dislocations, e.g., when metals are worked, or stressed failure tends to occur at imperfections.

TABLE 2.4
Comparison of Materials' Cost

Pipe Materials (100 mm Ø, 30 m Length)	Cost, $
Mild steel	100
PVC	160
Al/Mn alloy	280
Glass	400
Rubber lined steel	400
Epoxy/glass fiber	400
Stainless steel	600
Titanium	1200
PTFE-lined steel	2000
Glass-lined steel	2400
Hastelloy	2800

Non-metallic crystals are less ductile than metallic crystals because of the relative location of atoms when ionic or covalent bonds are involved. For deformation to occur, bonds must be permanently disrupted, and the material tends to crack catastrophically. In metals, however, atoms are not so intrinsically bonded, and atoms may move relative to one another, hence slip, which implies ductility. Similarly, electrical and thermal conductivity is higher in metals due to the greater mobility of electrons in the metallic structure.

Crystal size in solid systems depends on several factors. The type of system, alloy or compound, initiation of crystallization (seeding), rate of cooling, and mechanical and heat treatment after solidification all play a role in crystal size formation. In general, smaller size crystals are associated with greater strength. Glass formation can be regarded as a supercooled liquid. During cooling, the viscosity of the melt increases due to decreased mobility of the bulky, large, and complex shapes of the glass molecules. Eventually, the mobility is reduced such that rigidity occurs before the onset of crystallization (fused silica). Classification of glass is by composition. For example, borosilicate laboratory glassware has a low coefficient of thermal expansion. This increases thermal shock resistance from 100 C (lime soda-silica) to 350 C.

Pipe material costs may be divided into purchasing of materials, fabrication and assembly, installation, and in-service costs.

RHEOLOGICAL BEHAVIOR

The range of variety of responses to stress in the various states of matter is best summarized in correlations that relate strain against time and stress against strain. Characterization of the rheological flow of materials follows Hooke's law, which applies to ideal elastic solids, and Newton's equation of flow for viscous liquids and fluids in general. Hookean solids display a linear relationship between stress and strain. Strain is independent of time and disappears immediately as the stress is removed. This simple model describes the behavior of metals below the recrystallization temperature. Non-Hookean solids show a non-linear relationship between structural components as stress is increased. With such materials, strain is independent of time and disappears immediately as the stress is removed. With recoverable solids such as rubber, strain decays over a finite period, which is measurable. Plasto-elastic materials yield stress and exhibit deformation, which is partly recoverable upon the removal of stress. However, plasto-inelastic materials yield stress deformation that is not recovered after the removal of stress. Bingham materials' strain increases linearly with time, but with no recovery yield stress. Visco-elastic liquids flow with stress, which exhibits viscous behavior, e.g., paints. Visco-inelastic liquids flow with stress, but with no recovery, e.g., emulsions or dispersions. Newtonian liquids such as water flow with stress in a linear relationship with time and strain. Although practical concerns related to materials of construction are mainly solids, the liquid-like behavior of solids during fabrication might be utilized in the manufacturing of paints.

Materials may fail in service in a ductile or brittle fashion. Tough materials such as metals, fiberglass, and so on usually constitute crystalline structures and amorphous structures above the glass transition temperature, T_g, and fail in a ductile manner. Weaker materials such as glass and cement are brittle and fail by the brittle manner of fracture. Therefore, toughness may be defined as the resistance of materials to crack propagation. When brittle materials are struck, they appear to shatter instantaneously. Cracks are formed and propagate quickly upon application of stress. By contrast, ductile materials tend to be more time-dependent in response to stress due to plastic deformation. Although toughness and ductility may be improved by alloying, the ductility of a fabricated component is limited to processing conditions, which might be enhanced by annealing for in-service use. Static strength is always greater than cyclic strength whereby materials will eventually fail due to fatigue, e.g., crankshafts. Materials fail by fatigue due to repeated cyclic stresses, which cause an incremental slip. This gradual reduction in ductility results in the formation of sub-micro cracks, which ultimately propagate to fracture. Fatigue is alienated by polishing the surface of materials to reduce surface roughness and hence reduce the notches and other irregularities in the surface.

For non-cyclic loading, the temperature effects on the stress/strain behavior of materials can be illustrated by showing the variations of Young's modulus for a polymer with temperature. Thermal expansion of steel is exhibited at higher temperatures, but change is less gradual. Thermal expansion may not be in each direction of the material is anisotropic. Nonetheless, metals expand uniformly since their crystal structure consists of randomly arrange crystals.

Due to differential rates of expansion, materials will be subject to internal stresses (heating—compressive; cooling—tensile), and failure may result if the materials are not capable of relieving them. Steady-state distribution exhibits thermal stress, while unsteady-state transient temperatures allow for thermal shock.

3 Pressure Vessels

By increasing pressure on a single system, you can shift the equilibrium reaction to the right of the chemical reaction balance. Pressure vessels are designed to withstand pressure. Typically, we deal with cylindrical pressures in a design because of the following reasons:

- Cylinders are easier to make.
- Cylindrical shapes are strong and symmetrical.
- Pressure is equal in all directions inside the cylinder.
- Radial stress is exerted against the cylinder wall.
- Circumferential stress is evenly spread along the cylinder circumference.

STRESS = FORCE/AREA OVER WHICH IT ACTS

The axial stress is generated along the length of the cylinder. The axial stress is half of the circumferential stress. A radial stress is compressive as it tends to compress the wall. If the conditions in the pressure vessel are based on vacuum, then the stress effect is tensile. In order to make the vessel safe, the circumferential stress should be calculated to produce a safe design.

Spherical vessels are used in the storage of high-pressure gases. In this case, there are no axial stresses. The force is acting on diameter over the projected area:

$$F = P. \pi r^2$$

Spheres are stronger shapes than cylinders but are exceedingly difficult to make out of a sheet of materials. Domed end hemisphere (B.S. 1500, 1515) is difficult and expensive to make. It is usually designed for high pressure with the radius of the domed end equal to the radius of the cylinder. Similarly, it has applications with the semi-ellipsoidal end. To reduce the cost, dished and flanged ends are used. The wall thickness, t, of a pressure vessel can be calculated:

$t = (p.D.k1.k2)/(2\sigma.J.k3) + C1$ σ = Design stress
P = Pressure D = Diameter $k1$ = Flatness of end, if $h/D > 0.25$ $k1 = 1$
$k2$ allowance for opening, nozzles, (small pipes, $k2 = 1$)
$k3$ allowances for manufacturing method (if one piece, $k3 = 1$)
J = Weld factor (max. = 0.95) $C1$ = Corrosion allowance
$\sigma = (P_i r_i^2 - P_o r_o^2)/(r_o^2 - r_i^2) = (P_i - P_o) r_i^2.r_o^2 / r^2 (r_o^2 - r_i^2)$
P_i = Internal pressure P_o = External pressure r_i = Internal radius
r_o = External radius r = Radius under consideration

DOI: 10.1201/9781003384199-4

Joints and connections can be bolted, riveted, or welded. In conjunction, joint failure might occur due to rivet shear failure, bearing pressure failure, plate tensile failure, plate shearing failure, and plate edge tensile failure. These failures apply to bolts as well:

Joint efficiency = Strength with joint/Strength of homogenous sheet × 100%

Storage vessel considerations are based on materials stored in these vessels, resulting from physical properties, e.g., storage of solids, liquids, or gases. Whether the material is flammable, toxic, unstable, viscous, volatile, or corrosive in nature. Design criteria include capacity, measuring contents, level, cost, position, maintenance, and inspection requirements, ensuring integrity for loss prevention as a result of over pressure, under pressure, over filling, explosion, fire, corrosion, leak detection, and human error, e.g., leaving drain open.

For solids storage, the angle of repose is a function of angularity and particle size distribution in creating a pile. The angle of repose for dry solids,

$$35° < \alpha < 45°, \quad \text{and} \quad 45° < \alpha < 90° \text{ for wet solids. For very wet solids,}$$
$$\alpha \text{ approaches zero.}$$

Solids are stored in silos made out of concrete, wood, or plastic; they can also be handled by using bins and hoppers and are transferred by vacuum, or through augured hoppers. The main concern is dust explosion, which requires finely divided flammable solid in air with a source of ignition.

Liquid storage is in tanks, which can have a sloped bottom, cone, or dished bottom. Indoor open tanks should contain non-flammable, non-toxic, or non-volatile materials. When handling volatile or flammable materials, it is important to bond and ground the tank to prevent ignition from static charges. When handling oil, tanks must be fitted with vents, sumps, and permanent fire protection.

Aspect ratio = Height/Width (Diameter)

Silos are usually 3:1 or 5:1. Nonetheless, solid compaction on the base of silo should be calculated based on distributed load hydrostatic. In many regards, the use of plant area for tanks and pipework and optimized calculations are needed to minimize the cost.

For liquified petroleum gas (LPG) in spherical storage, a pressure relief valve on top of the sphere is necessary. In addition, the sphere must be installed in a bundle with a double block and bleed valving set up. For the gas holder tank, a water seal floating head is recommended and the in/out of the tank is connected to the bottom.

For cylindrical compressed gas tanks, a permanent connection to high-pressure equipment should have a PI for both line pressure and tank pressure plus a check valve and a double block and bleed valves in line.

Pressure vessels require necessary protection for safe operation and pressure relief if needed:

- Set pressure < hydraulic test pressure of the vessel
- 1.1 × maximum working pressure < hydraulic test pressure

TABLE 3.1
Valve Type Selection

Valve Type	Tight Shutoff	Flow Control	Position Indication	ΔP Open	Material	Cost	Comments
Gate	Yes	Fair	Good or poor	Low	Metal	High	Depends on construction
Globe	Yes	Good	Good or poor	High	Metal, glass	Medium	Depends on construction
Needle	No	Good	Poor	High	Metal	High	
Angle	Yes	Good	Poor	Medium	Metal, glass	Medium	
Plug cock	Yes	Poor	Good	Low	Metal, glass, plastic	Medium	Temperature limit 250C differential expansion
Ball	Yes	Poor	Good	Low	Metal, glass, plastic, carbon	Low	Temperature limit 250C flexible seats
Diaphragm	Yes	Fair	Poor	Medium	Metal, glass, plastic, carbon	Low	Pressure limit 4 bar Temperature limit 22°C
Butterfly	No	Fair	Good	Medium	Metal, plastic	low	

- Hydraulic pressure is 1.5–2 × working pressure
- Volumetric capacity of valve > maximum possible in flow
- Flow through the valve reaches a maximum at 10% over set pressure
- Valve cannot be permanently isolated from the vessel
- Define where would the vented vapors go
- Verify and check the valve to be operational

In addition, dump valves remain open when actuated by pneumatic, electrical, hydraulic signals. They need to be fast acting to accommodate large through put while all the contents are dumped. Additional safety features such as blow-out panels for explosion relief for ovens, furnaces, and cyclones (dust explosion) can be employed. On some pressurized equipment, a bursting thin metal disc clamped between flanges on vent branch.

Operability studies for pressure vessels include the following:

Functional analysis, deviations, root cause analysis, and hazard analysis. The examination sequences:

1. START—select vessel
2. Explain its function and lines
3. Select a line
4. Explain intention of line
5. Apply guide word (Flow, Temperature, and Pressure)

6. Develop meaningful deviations
7. Examine possible causes
8. Examine consequences
9. Detect hazards
10. Make record
11. Repeat for all deviations
12. Repeat for all guide words
13. Mark line as having been examined
14. Repeat for each line
15. Select an auxiliary, e.g., heating and mixing
16. Explain intention of auxiliary
17. Carry out items 5–12 for each auxiliary
18. Mark off auxiliary
19. Carry out 15–18 for each auxiliary
20. Explain function of vessel
21. Repeat 5–12 for vessel
22. Mark vessel as completed
23. Carry out 1–22 for all vessels on flow sheet
24. Carry out 1–23 for all flow sheets—END

4 Thermodynamics — Mass, Enthalpy, Entropy, Free Energy, and Internal Energy in Closed and Open Systems

As we define an internal system in equilibrium, i.e., it consists of a single phase, then that phase has uniform properties throughout the system. We can state that in the absence of heat exchange, or any external work done on the system, the properties of the matter in the system are unchanged. The state of matter in a system is defined by the property of that matter, i.e., when we write PV = nRT, then we can say that the property of that system is an ideal gas.

There are two classes of system property. Intensive (point) property is independent of the mass of the system. For example, the pressure in the room will not vary if divided into two parts. Pressure, temperature, refractive index, thermal conductivity, and viscosity are all examples of intensive property. Extensive (bulk) property value depends on the mass of the system, for example, volume, enthalpy, entropy, free energy, and internal energy.

Conceptually, the height of a three-dimensional system is independent of the path determined by the length and width of the system. Therefore, the amount of work done, and heat exchanged depends on the path. The parameters w and q are not path independent; therefore, they are not thermodynamic properties.

Adiabatic enclosure exists when there is no exchange of energy (heat) between the system and its surroundings. Heat is generated because of a temperature gradient (differential temperature). If the system and surroundings have the same temperature, then they are in thermal equilibrium with no heat exchanged. The change in Energy ΔE is described as heat exchange due to temperature difference, or work where force is implied through displacement.

In a closed system, there is no mass flow across the system. The cycle is stationary as the system passes through a series of thermos states. The matter in the system returns to its original state. For example, a cycle process of a working fluid returns to its original state retaining all its properties and having the same values at the end as at the beginning. An open system is defined by a volume in space rather than a fixed mass, whereby materials crossing the boundaries energy is converted in and out of the system, and work must be done to get mass into or out of the system. All these concepts are mathematically correlated and modeled by the various laws of

DOI: 10.1201/9781003384199-5

thermodynamics. For a steady state, there is no accumulation in the system. For example, a turbine operating under steady conditions will have a steady flow. For a chemical reactor, there will be a rate for heat transfer and there may be energy release due to chemical reactions.

Work can be converted into heat (the mechanical equivalent of heat) including energy loss. For example, the most efficient power station can only convert 30–40% of its heat input into work. The reason is it is impossible to operate a continuous process where all heat taken in is turned into its equivalent amount of work. In practice, a cyclic process is required, which involves taking heat at one temperature and rejecting it at another lower one and this imposes limits on the efficiency of such a cycle.

Any process in which the sole net result is equivalent to the transfer of heat from a lower temperature to a higher one is impossible. In any reversible change, the change in entropy and its surroundings is zero. An irreversible change is always accompanied by an increase in the entropy of the system together with its surroundings. For example, if a gas is confined in a piston-cylinder system when the piston moves and cannot be moved to return the system to its original state, an irreversible process has taken place. If a spring is added, then the energy transmitted is displaced and a balanced situation occurs. The stored energy in the spring is available to reverse the process without additional work to be supplied, which implies a balanced reversible process, i.e., a reversible change.

In summary, the first law of thermodynamics (TD) correlates the quantity of heat transferred, q to change in energy, ΔE plus the work, w that happens to create displacement such that $q = \Delta E + w$. The second law of TD involves a temperature gradient $T2 > T1$, which results in the transfer of heat. For reversible systems and surroundings, the entropy, $S = 0$. For irreversible systems, $S > 0$ for surroundings, and $S = 0$ for the system. The availability of energy in closed systems and steady flow is correlated in a functional relationship between energy, G, enthalpy, H, temperature, T, and change in entropy, ΔS such that

$$\Delta G = \Delta H - T. \Delta S$$

THERMODYNAMIC PROPERTIES OF FLUIDS

First law steady flow with relatively negligible kinetic and potential energy changes:

$$q = w + (h_e - h_i)$$

For example, a turbine compressor in ammonia refrigeration with $q = 0$ and the compressor is both adiabatic and reversible:

$$-w = (h_e - h_i),$$

knowing upstream and downstream conditions, the reversible work can be evaluated. The enthalpy change will predict temperature, pending the thermodynamic properties of the system. Two intensive properties, mass and phase composition, should be fixed. Using P, T, V, and composition is practical for measurements. As a function of these basic properties, thermodynamic properties can be tabulated. To carry out requires P-V-T data for phase over the composition range, or for pure substances, and similar heat capacity data evaluated at a convenient pressure. One direct model for the P-V-T equation of state can be used to produce thermodynamic properties. An indirect chemical potential model can also be utilized. This is useful in handling equilibrium situations for phase or reaction equilibria and for certain irreversible separation processes. These models can apply to both gases and liquids for pure substances and mixtures.

For a single ideal gas, a single-component phase system specifies a mole number, T, and P. V is fixed as a function of P and T:

$$PV = nRT \qquad R = \text{universal gas constant} = 8.31 \text{ J/mol.k}$$

THE IDEAL GAS MODEL ASSUMES

1. The molecules are point masses (no volume).
2. There are no attractive forces between molecules.
3. $Cp - Cv = R$, where Cp and Cv are determined experimentally at a specified pressure or estimated from statistical thermodynamics.

The indirect chemical potential model can be expressed for either an ideal gas mixture or the ideal liquid solution:

$$\mu = \mu^0 + RT. \ln(p/p^0) \qquad \mu = \text{Chemical potential} \qquad p^0 = 1 \text{ atm., } 760 \text{ mmHg @STP}$$

For real gases, the direct model is a modified version of the ideal gas equation, and the indirect model replaces the pressure of the gas with a dummy property, namely, fugacity, f. This thermodynamic property of a real gas which is substituted for the pressure or partial pressure in the equations for an ideal gas gives equations applicable to the real gas. The vapor pressure of a vapor is assumed to be an ideal gas obtained by correcting the determined vapor pressure and is useful as a measure of the escaping tendency of a substance from a heterogeneous system. Fugacity is a measure of the "real" partial pressure or pressure of a gas in comparison to an ideal gas. It is the effective partial pressure or pressure—a measure of thermodynamic activity. Fugacity is also a measure of chemical potential. Practically, fugacity is a measure of Gibbs molar internal energy. Therefore, fugacity, f, is a measure of the molar Gibbs energy of a real gas. It is evident from the definition that the fugacity has the units of pressure; in fact, the fugacity may be looked at as a vapor pressure modified to correctly represent the tendency of the molecules from one phase to escape into the other.

Hundreds of equations of state have been proposed since Van der Waals produced his modification of the ideal gas law based on variable conditions.

Equations of state serve three purposes:

1. Representation of PVT data for smoothing, interpolation, differentiation, and integration to calculate and derive properties
2. Prediction of gas phase properties of pure fluids and mixtures from a base minimum experimental data

PREDICTION OF VAPOR–LIQUID EQUILIBRIA
OF MIXTURES AT HIGHER PRESSURE

Chemical engineering is concerned with the processing of materials in gaseous, liquid, and solid states. Equilibrium methods are not particularly useful with the solid state as equilibrium is reached very slowly in solid systems.

In the case of ideal gases, we assumed that there was no interatomic, intermolecular, binding energy. For liquids, the atom's kinetic energy is comparable with the binding energy, which eliminates the construction of an ideal liquid model similar to the ideal gas law. If we construct theoretical simple models for equations of state or chemical potential for pure liquid, these will yield complex equations. If liquids are modeled as dense gases, then for engineering purposes, the thermodynamic properties of liquids can be established in a similar way as for gases. For example, using P-V-T data either directly or through equations of state and minimum specific heat data.

The analysis of industrial processes and the solution of phase and reaction equilibria problems require that we have values of the thermodynamic properties of interest as a function of pressure, temperature, and composition of the phase in question. For this, we require P-V-T data and specific heat data. Calculations of enthalpy, entropy, and fugacity for both single substances and the mixture are modeled.

Variations of enthalpy, entropy, and fugacity are determined as a function of variation in pressure and temperature:

$$H = U + PV \text{ by definition. Differentiating gives}$$
$$dH = dU + Pdv + Vdp$$

Also,

$$dU = TdS - PdV$$

Following mathematical calculations, at constant temperature and composition,

$$RT. \, Dln.fi = Vi.dp$$

Approximation for a pure component which may be a gas or liquid can be applied to a mixture of unvarying There are several steps in the process of constructing a table or property chart for thermodynamic properties. We start by calculating enthalpy and entropy values for the substance, and then this data can be used for constructing desired tables. Datum pressure and temperature are selected, P_0 and T_0.

Enthalpy and entropy are computed at these parameter values. For each desired temperature, the temperature range is compiled. The heat capacity is experimentally determined. For each value of T, the values of H and S at desired pressure are computed through integration. PVT data can be obtained graphically or numerically.

Thermodynamic properties of a system consisting of two phases in equilibrium require that P and T are uniform through the two phases (either work or heat flow would occur to make them in equilibrium). The values of the extensive properties will be fixed only by the properties of each phase present. Thus, if X is the weight (or mass) fraction of phase α, with β as a second phase, using specific values on a mass basis of V, H, and S, then

$$\text{Specific volume of mixture, } V = V^{\alpha} X + V^{\beta} (1 - X), \text{ and}$$
$$\text{Enthalpy of mixture, } h = h^{\alpha} X + h^{\beta} (1 - X)$$

Engineering thermodynamic cycles can be analyzed for actual processes by assumptions considered for an ideal version of the process. These idealized processes are called pattern processes. The two processes commonly used are isothermal compression or expansion and adiabatic compression or expansion. In addition, an adiabatic reversible process is isentropic, $\Delta S = 0$.

These pattern processes are used as comparisons for the following:

1. Restrained adiabatic compression or expansion of a gas in a piston-cylinder system
2. Compression or expansion of a fluid in a steady flow, adiabatic compressor, turbine, or pump
3. Frictionless adiabatic flow of a substance through a duct
4. Adiabatic magnetization or demagnetization
5. Adiabatic polarization or depolarization

The actual process may not be as close, and a performance parameter is introduced to account for the departure from the pattern process. Thus, for example, for a turbine, we define an isentropic efficiency, n_s.

n_s = Work from actual adiabatic turbine/Work from isentropic process between inlet and discharge pressure. This must not be confused with an energy conversion efficiency of a heat engine.

For compressors, there are two classes:

1. Polytropic is a reversible but non-adiabatic compression ($q \neq 0$), and heat is supplied or removed. This case is approached by a reciprocating compressor.
2. Polytropic is irreversible ($ds > 0$) but adiabatic ($q = 0$), and there is a measurable thermal effect due to friction and turbulence but not an actual external transfer of heat. Turbo-compressors follow this case.

Heat engines fall into two classes:

1. External combustion engine

This includes engines using both single-phase working fluids and two-phase (vapor–liquid) working fluids. The working fluid is heated externally, i.e., the fuel is burnt and the heat is transferred to the working fluid, acting as a heat carrier.

2. Internal combustion engine

In these types of engines, the fuel is burnt inside the working cylinders, with no heat transfer. The fuel energy is liberated in place, and no heat transfer across metal surfaces is involved.

5 Fluid Dynamics — Crystallization, Sedimentation, and Particle Dynamics

Particle fluid dynamics is concerned with the relative motion of particles in fluids. The Navier–Stokes equation correlates the sum of all forces involved:

(Force to accelerate in unsteady flow, unit mass of fluid) + (Momentum, inertia of transport of fluid) = (Gravitational force) + (Static pressure gradient) + (Viscous resistance to shear)

If the properties and flow characteristics of each position in space remain invariant with time, the flow will be in a steady state. A time-dependent flow is designated as an unsteady state flow. For Newtonian fluids, the ratio of shear stress to the rate of shear is constant and is equal to the viscosity of the fluid. Incompressible fluids are in a liquid state with constant density, and their volume is independent of pressure.

An engineering approach to correlate fluid flow in pipes is using dimensionless correlation Reynolds number (Re):

$$Re = \rho.d.u/\mu \qquad \rho = \text{Fluid density; } d = \text{Pipe diameter; } u = \text{Fluid velocity;}$$
$$\mu = \text{Fluid viscosity}$$

Re is influenced by the surface roughness inside the pipe and is a correlation that measures fluids in laminar, smooth, or turbulent conditions.

Assuming spherical particles in fluids the total drag force (non-creeping motion), F is directly proportional to the area covered by the fluid and the kinetic energy of the fluid such as

$$F = C. A. \rho. U^2/2 \qquad C = \text{coefficient of drag}$$

Drag is the force exerted on a body from a moving fluid in the direction of fluid stream flow. It is usually indirectly measured by calculating the terminal velocity, Ut. The drag coefficient depends on Re and on velocity.

$$Ut = (Re) \, \mu/d \, \rho$$

DOI: 10.1201/9781003384199-6

TABLE 5.1
Propulsion of Fuels

Motor	Propellant	SI
Black powder	C/KNO_3	140
Solid	Polyurethane + Al	250
Liquid fuel	H_2O_2/kerosene	265
	Liq. O_2 (LOX)/kerosene	275
	Liq. O_2/H_2	385
	Liq. F_2/H_2	425
Nuclear	H_2	>1,000

Design considerations for the hydraulic classifier to model variations in fluid density, cyclone design to calculate centrifugal acceleration and pressure drop across vortex, and velocity head based on classifier geometry to complete clarification of materials are all examples of particle movement and fluid flow dynamic characteristics. In a circular tank, liquid radial spread based on central feed residence time and radial velocity are critical process parameters in the mathematical modeling of such systems. Mixing and middling increase as a function of fluid velocity and density. When particle acceleration changes, so does the kinetic energy driven to the surrounding fluid. This impulse energy can be shown, called fluid impulse that is equivalent to ½ mass of fluid displaced. Fuels are compared by their "specific Impulse," SI = Thrust produced/Weight of fuel burnet per second (Table 5.1).

6 Fluid Mechanics

FLUID FLOW

Transportation of fluid within unit operations following a flow diagram is accomplished through a piping system. Tubular pipes are most common with circular a cross section for achieving structural strength and greater cross-sectional area. Most problems are solved using experimentally determined coefficients. Many empirical formulas are used in pipe fluid calculation with limitations and can be applied only when the conditions of the problem closely approach the conditions of experimental derivation. For example, the Darcy formula can be derived by means of dimensional analysis; however, the friction factor must be derived experimentally. Darcy's correlation is widely used as it has wide application in fluid mechanics.

PHYSICAL PROPERTIES OF FLUIDS

Knowing the physical properties of the fluid in the calculation is a requirement. Accurate values of these properties affecting the fluid flow are important in deriving correlations that are significant for the design of unit operations and systems, namely, weight density, viscosity, and so on. Viscosity expresses the readiness with which a fluid flows when it is acted upon with external force. The absolute viscosity of a fluid is a measure of resistance to internal deformation (shear). Water viscosity is much less than molasses. Gases viscosity is quite small compared to water. For example, thixotropic properties of viscosity depend upon working fluids during processing such as printing inks, pulp slurries including cellulose binders, and catsup. It is important to use the CGS (metric) or British units as they are employed in consistent calculations. The utilization of simplified measurement through derivations is significant for the practice of design in chemical and processing plants. For example, the measurement of the absolute viscosity of fluids requires elaborate equipment and experimental skill. On the other hand, a rather simple instrument can be used for measuring kinematic viscosity (Saybolt Viscosimeter).

The weight density of a substance is its weight per unit volume. In steam flow computations, the reciprocal of the weight density is the specific volume is a commonly used unit. Specific gravity is a relative measure of weight density. Since pressure has an insignificant effect on the weight density of liquids, temperature is a condition that must be considered in designating the basis of specific gravity. Thus, the specific gravity of a liquid is the ratio of its weight density at a specified temperature to that of water at a standard temperature, 60 F. A hydrometer is used to measure the specific gravity of liquids directly. API is used for oil, and Baume scales are used for lighter and heavier than water liquids. The specific gravity of gases is defined as the ratio of molecular weight (or gas constant) of gas to that of air.

DOI: 10.1201/9781003384199-7

The Bernoulli theorem is a means of expressing the application of conservation of energy to the flow of fluid in a conduit. The total energy at any point above the datum plane is equal to the sum of the elevation head, the pressure head, and the velocity head. In practice, losses or energy increases must be included in the Bernoulli equation. Thus, an energy balance may be written for two points in a fluid profile. The pipe friction loss from point 1 to point 2 in the pipe is referred to as the head loss. Therefore, all practical formulas for the flow of fluids are derived from the Bernoulli theorem with modifications to account for losses due to friction.

Measurement of pressure in manufacturing transport systems, fluid flow piping, or HVAC room-to-room air flows is important to capture. Perfect vacuum or absolute zero pressure is a convenient datum for pressure measurement. Therefore, absolute pressure is a measurement of vacuum below atmospheric pressure. Hence gauge pressure is any pressure above atmospheric pressure. Barometric pressure is the level of atmospheric pressure above a perfect vacuum. The standard atmospheric pressure is 14.7 psi, or 760 mm of mercury. Vacuum is usually expressed in inches (mm or μm) of mercury, which is the depression of pressure below the atmospheric level. In practice, pressure drop calculations are critical to pipe fluid measurements and design calculations in plant design and construction. Darcy's formula can be expressed in feet of fluid, but the more common equation is written to express pressure drop in psi:

$$\Delta P = \rho f L v^2 / 144 D\, 2g$$
$\rho = Density; f = Friction\ factor; L = Length\ of\ pipe; v = Velocity\ of\ fluid;$
$D = Pipe\ ID\ diameter; g = Acceleration.$

Friction loss in the pipe is sensitive to changes in internal diameter and roughness of the pipe. For a given flow rate and fixed friction factor, pressure drop per foot of pipe varies inversely with the fifth power of the diameter. For example, a 2% change in pipe ID causes an 11% increase in pressure drop. A 5% change in pipe ID increases pressure drop by 29%, respectively. This is exacerbated by the fouling of pipes on the inside wall due to chemical reactions heat transfer effects and diffusion of precursors toward the surface.

COMPRESSIBLE FLUID FLOW IN PIPE

An accurate pressure drop determination of a compressible fluid flow in a pipe requires knowledge of the relationship between pressure and specific volume Adiabatic and isothermal conditions are considered. The density of gases changes with pressure. In addition, velocity will change with a pressure drop across a pipe distance. For compressible fluids such as natural gas, air, and steam, the following limitations apply:

1. If the pressure drop is less than 10% of the inlet pressure, the specific volume should be based on upstream or downstream conditions.
2. If $10\% < \Delta P < 40\%$ of inlet pressure, use average specific volume for upstream and downstream conditions.
3. If $\Delta P > 40\%$ (long pipe), use critical velocity or flow rate based on expansion factor compressible flow, and velocity head loss.

STEAM

The amount of heat necessary to cause the temperature of water to rise by one degree is equal to one Btu. This is applicable if the pressure does not exceed 50 psia. Similarly, if zero heat content at 32 F, one lb. of water at 212 F contains 180.2 Btu, this is called sensible heat (single-phase flow). To change to vapor at atmospheric pressure (14.7 psia), 970 Btu must be added at a temperature of 212 F. This is called the latent heat of evaporation. So, the total heat to boil (two-phase flow) water is 1150.5 Btu (180.2 + 970.3).

Saturated steam is in contact with liquid water at boiling temperature and condensing point of steam equilibrium. Dry or wet steam might be formed depending on generation conditions. Dry saturated steam is free and water particles. Wet saturated steam contains water particles in suspension. Saturated steam at any pressure has a definite temperature. Superheated steam is heated to a higher temperature than saturated steam at a given pressure.

Since complex chemical plant installations contain a considerable number of valves and fittings, fluid mechanics calculations require estimating their resistance to the flow of fluids, which is necessary to determine the characteristics and practicality of the transfer of fluids within a complete piping system.

Valves such as gate, ball, plug, and butterfly tend to have lower resistance to flow, while globe and angle valves would have a relatively high resistance to flow. Fittings could be branching, reducing, expanding, or deflecting flow, for example, tees, crosses, side outlet elbows, 90-degree elbow, and 45-degree elbow. Reducing or expanding fittings change the area of the fluid pathway, e.g., reducers and bushings. Deflecting fittings change the direction of flow, e.g., bends, elbows, returns, and so on.

The resistance coefficient K is directly proportional to the equivalent length, L/D, and friction factor f

$$K = f\,(L/D) \qquad D = \text{Pipe diameter}$$

The flow coefficient C_v of a valve is defined as the flow of water (gpm) at 60°F with $\Delta P = 1$ psi across the valve.

The head loss through a pipe with valve and fittings is generally given in terms of resistance coefficient K which indicates static head loss through a valve in terms of velocity head, or equivalent length in pipe diameter L/D that will cause the same head loss as the valve

$$h_L = K.\ v^2/2g$$

7 Heat Transmission— Natural Convection, Forced Convection, and Single- and Two-Phase (Boiling) Flow Systems

Heat Transfer, $Q = h. A. \Delta T$

h = Heat transfer coefficient, A = Surface area, and ΔT = Differential temperature gradient.

h depends on the following:

1. The fluid involved in the heat conversion process
2. The geometry of the system
3. The dynamics of flow (forced or natural, laminar, or turbulent flow)
4. Thermal conditions

Heat transfer by conduction is a function of heat transfer surface thermal conductivity described as follows:

$dQ/dt = k/x. A. \Delta T$ where k = Thermal conductivity; x = Length.

The following dimensionless groups describe correlations in heat transfer:

$Nu = h. d/k$ d = Diameter of tube or pipe (Nusselt number);

$Re = \rho. d. u/\mu$ (Reynolds number);

P = Density of the fluid;   u = Fluid velocity in pipe; μ = Viscosity of the fluid;

$Pr = \mu. Cp/k$ (Prandtl number);   Cp = Heat capacity;

$Nu = 0.023. (Re)^{0.8}. (Pr)^{0.4}$.

It is experimentally proven that counter-current flow gives better heat recovery than co-current flow. Over time, there is a buildup of deposits at the heat transfer surfaces due to precursors in the fluid that have a chemical reaction at the heated surface coupled with diffusion toward the surface. This phenomenon has been studied and classified as fouling and described as a fouling factor due to scaling or other mechanisms that will add to the resistance to heat transfer and therefore considering its

DOI: 10.1201/9781003384199-8

value in the design of heat exchangers and reactors to provide for the protection of the processing units over time.

In the selection and design assumptions of heat transfer systems, the lowest pressure drop is put through the shell side of a heat exchanger for condensing vapors or steam. Higher pressure is applied where the smallest diameters are applied in heat exchanger tubes. In addition, the installation and layout of a stack of heat exchangers are of particular interest to ensure access to the tube bundle at the withdrawing end for service due to fouling cleaning or tube leakers. For multi-pass, the calculation for heat transfer including the incorporation of fouling includes the mean differential temperature across the length or area of heat transfer. The introduction of an agitator to improve heat transfer dynamics involves a baffled external jacket or an internal helical coil. The use of turbine, propeller, or anchor-type mixers for higher viscous fluids might be employed.

Condensers' layout and inclination play a role in their installation for effective performance. Routinely, horizontal, or inclined condenser with cooling medium bottom feed is the preferred process to counteract air pockets and splashing effects.

A pre-cool condenser for vapor, followed by a main condenser to transfer vapors into liquids, followed by a sub-cool condenser to bring liquid into operating conditions could be the setup of a condenser train in series. Design considerations for transition to turbulent flow should be calculated for vertical condensers. Similarly, the calculation for stratified or annulus flow should be evaluated in horizontal condensers. Pressure drop calculations for a condenser placed at the top of the distillation column are proportional to the height of the top plate between the distillation column and condenser and the fluid density. Pressure drop in condensers increases due to submergence. To minimize pressure drop, assuming isothermal condensation, tube diameter would be increased, and baffling decreased as part of the design considerations.

Submergence = Sensible heat to remove/Condensing load total heat removed

The condensing range can be described in terms of dew point, which is the first drop to condense, and bubble point, which is the last drop to condense. The relationship of phase change from vapor to liquid can be correlated as

$$P + F = C + 2 \qquad P = \text{Phase} \qquad F = \text{Degree of freedom;} \qquad C = \text{Components}$$

1. For single vapor, $2 + F = 1 + 2$, which implies $F = 1$ and pressure needs adjustment to achieve optimal temperature.
2. For binary mixture, $2 + F = 2 + 2$, which implies $F = 2$ and pressure and composition need adjustment to achieve the optimal temperature.
3. For binary immiscible mixture, $3 + F = 2 + 2$, which implies $F = 1$ and isothermal condition.

Note that if the composition changes and temperature changes, then the mean heat transfer coefficient, h_m, changes. Sensible heat, Q, is correlated as a function of the temperature gradient

$$Q = w \cdot C_p \cdot (T_{end} - T_{start}) \qquad w = \text{Mass;} \qquad C_p = \text{Heat capacity}$$

Reboilers (vaporizers) are conceived as heat transfer units that utilize submerged heat transfer surfaces with forced convective boiling. The boiling tube's surface conditions affect wetting conditions. Partial wetting is inefficient and creates larger bubbles. Polished surfaces improve wetting and produce relatively smaller bubbles with higher boiling efficiencies. Additional agitation improves nucleate boiling and increases heat transfer coefficient, h. Rapid formation of bubbles might lead to vapor locks, or burn-out effects, which is the critical point [Do Not Burn (DNB)] in going from nucleate region to transition to film region. Need to operate below the DNB point.

The arrangement of reboilers targets effective boiling with calculations to quantify bubble formation from single-phase flow to two-phase flow. Annular flow is optimized to develop vapor in the middle with liquid at the wall. Forced convective boiling inside vertical tubes or natural circulation unless fluid viscosity is high.

Circulation ratio = Vapor leaving/Liquid leaving
Circulation rate = Actual throughput/Required throughput

The advantages of vertical boilers are that they have longer windows for cleaning as there is less scaling, they are easily cleaned, and they have a high overall heat transfer coefficient. The disadvantages are the requirement for higher elevation and corrosion might develop in the shell due to the steam/air mixture.

Design considerations should include the use of empirical "pool boil up" data, estimated overall heat transfer coefficient with appropriate account for differential temperature, and scale-up allowances. To prevent a surge point, where extreme oscillation due to the suction of vapor into the bottom of the heater might occur, avoid a large vapor outlet, and place a control valve on the liquid inlet (Table 7.1).

Design assumptions for heat exchangers should include the following:

1. Maintenance, cleaning, and corrosion resistance factor (allowance)
2. Pressure rating, pressure drop, low-pressure operation, condensation, and high pressure in tubes
3. Type (1:1, 1:2 faster, . . . , 2:4 better heat recovery)
4. Shell length and diameter, tube length and diameter
5. h, R, $h = f(v^{0.8})$
6. μ, Nu, Re, Pr
7. Pitch (triangular, square)
8. Thermal expansion, thermal duty
9. Shell fluid temperature, constant overpass, constant flow rate, and heat capacity
10. Minimize heat loss and maximize heat recovery
11. No phase change
12. Equal area in each heat transfer pass

It is important to realize and recognize that allowances in heat transfer calculations are based on empirical knowledge. Calculations accuracy can be ±20%. In addition, corrections for viscosity changes should be in consideration. If there is a significant

TABLE 7.1
Boiling Heat Transfer

Sub-Cooled Liquid	Froth, Slug	Annular, Fog, Dry	Gas
Convection to liquid	Saturated nucleate boiling	Convective boiling to	Convection to
Sub-cooled nucleate boiling	to bulk	vapor plus droplets	superheated
(Bubble at wall)	H= hconv. + hnucl.	(Burn-out point)	vapor

change between T-wall and T-bulk, then a viscosity correction is required for viscous fluids:

$$\text{Viscosity ratio} = (\mu b/\mu w)\ 0.14$$

The differential temperature is expressed as log mean temperature difference (LMTD)

$$\Delta T_m = (\Delta T1 - \Delta T2)/Ln\ (\Delta T1/\Delta T2)$$

The overall heat transfer coefficient is correlated to allow for fouling factor due to deposition that adds to the resistance to heat transfer across the heat transfer surface area. R, fouling factor, is included in the calculations to protect the design from deterioration in service. This allowance provides for scale build-up between cleaning operations. This is empirical data, which is dependent of fluids and precursors, monomer that can polymerize due to heating, or polymers that can deposit at the heating surfaces through chemical reactions or diffusion toward the wall:

$$1/U_D = 1/U_C + R \qquad U_D = \text{Overall heat transfer coefficient for dirty conditions}$$
$$U_C = \text{Overall heat transfer coefficient for clean conditions}$$

The geometry and pitch of tubes in a heat exchanger are important to develop enhanced heat transfer conditions. In addition, the direction of flow including mixing enhancement such as baffles will increase turbulence and improve heat transfer rates. The tube design can be fixed inside the shell or with a removable bundle to assist in cleaning operations. This is helpful with heavy shell side fouling. Nonetheless, cleaning inside the tubes requires a special operation that might use other types of aggressive cleaning agents that are process specific to remove internal deposits.

General design considerations target the selection of process conditions and fluid properties. For example, condensing vapors or high viscous fluids are placed on the shell side to minimize pressure drop. For high-pressure fluids, the pass should be through the smallest diameter, i.e., tubes. For corrosive fluids inside tubes, use expansive specialized materials of construction that are resistant to corrosion. Cleaning is of major importance for dirty fluids such as polymers, impure water, and suspensions.

Installation and layout of heat transfer stacks should allow for tube bundle access and removal. Multi stacks should take into consideration piing interference and provisions for back flushing. Lifting gear should be included in design considerations to ensure endurance to handle thin destructible tube bundles with utmost care.

Design specifications codes for pressure vessels, tube bundles, and end plates heat exchangers should cover fabrication tolerances, corrosion allowances, manufacture, testing, and inspection. Heat exchangers follow T.E.M.A, BS 3274, and BS 2041. Pressure vessels follow A.S.M.E sect. VIII. Specifications should include materials of construction and physical properties of both streams including viscosity, thermal conductivity, and fluid density. Fouling factors required should allow for the maximum possible pressure drop. In addition, all limitations on shell diameter or tube length should be spelled out. Similarly, exchanger type, i.e., 1:2, 2:4, floating head, end details, thermal duty limits, pressure rating, and any special requirements must be stated (Tables 7.2 and 7.3).

Heat transfer in process vessels may require the addition of jackets when the amount of heat that needs to be taken out is not constant. Hence, limitations on reactor size, fermenters, and so on are process constraints unless a jacket or coil is added. A jacketed pan or kettle frequently used in process industries may require the addition or removal of heat by incorporating steam or a coolant. These applications involve an external jacket, which might be helically baffled for positive liquid circulation, or might be fitted with half-coil jackets. In addition, internal helical coils might be added to the process vessel. Agitation is generally employed for the process to obtain even temperature distribution in the vessel and to improve heat transfer, e.g., polymerization, reactors, and fermenters. Unless fluid flows continuously through the vessel, heat transfer to/from the jacket or coil is in an unsteady

TABLE 7.2
Viscosity Data of Petroleum Fractions

Fraction	Temperature, F	Viscosity, Cps
Crude oil	300	1.3
Crude oil	200	3.4
Petrol	120	0.4
Petrol	80	0.5

TABLE 7.3
Nominal Fluid Velocity Ranges in Heat Exchangers

Fluid	Tube-Side Velocity, ft./S	Shell-Side Velocity, ft./S
Water	3–6	2–4
Non-viscous liquid	2–5	2–3
Viscous liquid	2–4	1–2
Low-density gas	50–150	30–60
High-density gas	20–80	20–40

Or internal coil

TABLE 7.4

U, Overall Heat Transfer Coefficients for Range of Jackets (B.T.U./h. ft². °F)

Jacket	Vessel	Uc, Clean Overall Coefficient
Steam	Boiling water or dilute aqueous solution	175–250
Water	Water	100
Water	Aqueous solutions	75–80
Water	Hydrocarbons	50
Water	Organics (0.5–1 Cps)	10–20

Note: In all cases, neither side is refrigerated.

state. Normally, jacketed vessels are batch operated, which employ unsteady state equations allowing for variation of ΔT during the heating/cooling process. There is no suitable correlation of film coefficients for vessels without agitation. Therefore, the design is based on overall coefficients. These are best used only as a preliminary guide since considerable variations may arise due to variations in the viscosity of fluids, degree of agitation and agitator design, cleanliness of heat transfer surface, amount of non-condensibles, and so on (Table 7.4).

Agitators can be propellers, anchors, or turbines. Many turbine designs include straight blade, curved blade, angled blade, six-blade, or three-blade types. Heat transfer equations should be selected for each specific design (Ackley, 1960, Chem. Eng.; Chapman and Holland, 1965, Chem. Eng. Jan–Feb).

To derive the heat transfer equation for agitated systems, assume liquid flow in the jacket or coil to be constant. Heating or cooling media has a constant inlet temperature. Using average specific heat, although there might be hot spots, agitation for isothermal heating has uniform temperature throughout the batch. Average physical properties are used in heat transfer coefficient, hi, and ho calculations. The overall heat transfer coefficient, U, is assumed constant over surface, and there is no phase change with minimum heat loss. Therefore, to predict U

$$Q = M.\ Cp.\ \Delta T = U.\ A.\ \Delta T$$

Batch heating and cooling would be accomplished under isothermal conditions by the use of jacket tanks or coils. This would include boiling, vaporizing, or refrigeration. Batch non-isothermal heat transfer can be accomplished in a single phase such as hot liquid flowing through a coil or tube with no phase change (sensible heating). Non-isothermal cooling can be accomplished through cooling water, glycol, or brine from a reservoir.

In design considerations for condensers, the heat transfer surface might have a wetted film (film-wise) and the resistance is conduction across the laminar film or a non-wetted surface (drop-wise), which is more effective as the drops fall off revealing a fresh heat transfer surface. A stagnant film might be present for non-condensable fluids. The following are applicable factors in condensers:

- Steam only, pure vapor normally
- Mixtures of steam and hydrocarbons
- Highly polished surfaces
- Use film-wise in design
- Pre-cool vapor/condenser (vapor to liquid)/cub-cool liquid

The mean value of the heat transfer coefficient for a condenser inclined surface might be estimated from the following:

$$h_m = 0.942 \ (k^3 . \ \rho^2 . g . \ \lambda \sin \alpha \ / \ \mu. \ L. \ \Delta T)^{1/4} \quad \text{Laminar flow}$$
$$\rho, \mu - \text{Liquid properties}; \quad \lambda = \text{Latent heat}; \quad \alpha = \text{Angle of inclination}$$

EXTEND TO HORIZONTAL SURFACE

$$h_m = 0.725 \ (k^3 . \ \rho^2 . g . \ \lambda \ / \ \mu. \ D. \ \Delta T)^{1/4} \quad D = \text{outside diameter}$$
$$D < L \quad h_m, \text{Btu/hour.ft}^3.°F \ (\text{Tables 7.5–7.7})$$

Flow inside condenser tubes might be stratified or annulus, which is dependent on interfacial tension, fluid viscosity, and vapor phase specific gravity among other physical properties and system geometry. Conditions could follow isothermal condensation, i.e., single phase, whereby the condensation range from dew point to bubble point could be binary or multicomponent. This could be for pure substances or mixtures.

Boiling heat transfer involves two-phase flow conditions. Submerged surfaces in the kettle reboiler or vaporizer include heat transfer from tubes to vaporize the liquid in the reboiler shell. Alternatively, a vertical thermosyphon reboiler has boiling inside tubes through forced convective boiling:

Submergence = Sensible heat/Condensing load

In both cases, there is a phase change from liquid to vapor with fluid physical properties in consideration. To form a bubble,

$$Pr = Ps - 2\sigma/r \quad Pr = \text{Pressure inside curved surface};$$
$$Ps = \text{Saturated vapor pressure}; \quad \Sigma = \text{Surface tension}.$$

Hence, these conditions need superheat above liquid saturation temperature:

TABLE 7.5
Condensation Values

Vapor	Design	h_m, Btu/hour.ft3.°F	ΔT
Steam	Horizontal, d = 1–3 inch	2000–4000	5–40
Steam	Vertical, L = 10 ft	1000–2000	5–40
Organics	Horizontal	200–300	20–100
Organics	Vertical, d = 2 inch	300–450	10–40

TABLE 7.6
Heat Transfer Units Conversion Factors

Use: Multiply numerical value in British Units by Conversion Factor to give equivalent in SI Units.

or

Multiply numerical value in SI Units by Reciprocal Conversion Factor to give equivalent in British Units.

QUANTITY	BRITISH UNITS	SI UNITS	CONVERSION FACTOR	RECIPROCAL CONVERSION FACTOR
Mass	lb	kg	0.4536	2.2045
	slug	kg	14.594	0.06852
Length	ft	m	0.3048	3.2808
Force	lbf	N	4.4482	0.2248
Energy	Btu	J	1055.06	9.4781×10^{-4}
	ft lbf	J	1.3558	0.7375
Power	550ft lbf/s = 1 h.p.	W	745.69	1.3410×10^{-3}
Q	Btu/h	J/s(or W)	0.2931	3.4118
q	Btu/ft^2h	J/m^2s	3.155	0.3169
h,U	Btu/ft^2hdegF	J/m^2sdegC	5.678	0.1761
k	Btu/fthdegF	J/msdegC	1.731	0.5777
c_p	Btu/lbdegF	J/kgdegC	4186.8	2.388×10^{-4}
ρ	lb/ft^3	kg/m^3	16.0185	0.06243
μ	lb/ft h	$kg/ms(or\ Ns/m^2)$	4.134×10^{-4}	2.4189×10^3
ν,χ,ϵ,D	ft^2/h	m^2/s	2.581×10^{-5}	3.8744×10^4
τ,P,p	lbf/ft^2	N/m^2	47.880	0.02089
τ,P,p	lbf/in^2	N/m^2	6.8948×10^3	1.4503×10^{-4}

TABLE 7.7

Phase and Composition Effects on Process Parameters

Application Mixture	Correlation Phase + F = Component + 2	F	Result
Single vapor	2 + F = 1 + 2	F = 1	Fix pressure, fix temperature
Binary mixture	2 + F = 1 + 2	F = 2	Must fix P and composition to fix T
Binary immiscible	3 + F = 2 + 2	F = 1	Isothermal condensation

TABLE 7.8

Limited Heat Flux Practice

Materials	Process	Max. Allowable Flux, Btu/hour.ft²°F	Max. h, Btu/ hour.ft²°F
Organics	Forced circulation	20,000	300
Organics	Natural circulation	12,000	
Aqueous	Forced/natural	30,000	1,000

$$T_{wall} > T \text{ (saturated vapor)}$$

Surface evaporation happens through natural convection and is described by Nusselt dimensionless number:

$$Nu = 0.61 \, (Pr. \, Gr)^{1/4} \quad Pr = \text{Prandtl number} \quad Gr = \text{Grashof number}$$

During nucleate boiling, good agitation will enhance heat transfer and increase the heat transfer coefficient, h. Transition to film whereby bubbles form rapidly and many sites have increased temperature, which will lead to a stable film. Surface conditions play a major role in phase transition during boiling. A clean surface will become fouled through deposition and will develop large bubbles which are not conducive to effective heat transfer. Dirty surfaces allow vapor formed to escape from the surface because vapor disengagement is not easy, which implies that the dirt part of the heat transfer surface in a heat exchanger is disengaged and not efficient (Table 7.8).

The arrangement of reboilers is based on empirical correlations of pool boiling data. The circulation ratio depends on the application of natural convection or pumped fluid:

$$\text{Circulation ratio} = \text{Actual throughput/Required throughput}$$

Also

$$\text{Circulation ratio} = \text{Vapor leaving/Liquid leaving} \approx \tfrac{1}{4}$$

The effect of excess temperature for hot surface and boiling liquid (pool boiling) varies as temperature increases from interface evaporation (natural convective heat transfer) to nucleate boiling (bubbles condense in liquid to surface), to the transition phase of partial nucleate boiling plus unstable film at the surface, and to film boiling with stable film formation at the heating surface. Burn-out point is critical where the irreversible deposit will foul the surface. The temperature differential is described by

$$\Delta T^\circ F = T_{wall} - T_{saturated\ vapor}$$

8 Transport Phenomena

Fick's law of diffusion is fundamental in many mechanisms related to chemical engineering processes:

$$J = -D_{AB} \cdot d\rho/dy \qquad J = \text{Mass flux}$$

Assumptions

1. B is insoluble in A
2. The level of a volatile liquid is fixed
3. Steady state
4. No chemical reaction
5. Ideal behavior of gas
6. Constant temperature and pressure (constant physical properties)
7. $P_a = y_a \cdot P_T$
8. D_{AB} is constant (independent of concentration)

MASS BALANCE (STEADY STATE)

$$\text{In} = \text{Out} + \text{Reacted, Accumulated} = 0$$

The dimensionless analysis involves the correlation of physical properties, process parameters, and equipment geometry to reach a number that is dimensionless for further mathematical modeling and calculations. For example,

Reynolds #—$Re = \rho.d.v/\mu$
Nusselt #—$Nu = h.d/k$ $\qquad$ h = Film coefficient; $\quad$ k = Thermal conductivity
Grashof #—$Gr = d^3\rho^{\,2}. g. \beta. \Delta T/\mu^2$ $\quad$ β = Coefficient of cubic expansion
Prandtl #—$Pr = Cp.\mu/k$
Power # = $P/d^5N^3\rho$ $\qquad\qquad$ P = Power
Froude #—$Fr = d. N^2/g$ $\qquad\quad$ g = 32.2 ft/s^2
Euler #—$Eu = F.g.C/\rho.v^2.L^2$ $\qquad$ F = Exerted force
Schmidt #—$Sc = \mu/\rho. D_{AB}$ $\qquad$ D = Diffusivity $\qquad$ v (kinematic viscosity) = μ/ρ
Stanton #—$St = Nu/Pr.Re = h/\rho.Cp.v$

Rayleigh's method is based on a statement of dependent variables of given dimensions that depend on a relationship with a group of variables. Then, the individual variable must be related in such a way that the net dimensions of the relationship are

DOI: 10.1201/9781003384199-9

identical to those of the dependent variable. For example, the continuity of mass flow in a pipe:

$$G = \rho.v.A \qquad \text{G is a dependent variable}$$

Pressure drop is a dependent variable in flow in tubes:

$$\Delta P \text{ is dependent on d, L, v, } \rho, \mu, \text{ and g} \qquad \text{g} = \text{Gravitational constant}$$

In this case, pipe geometry (diameter) is applied. Similar setup for square pipe requires a different approximation.

Buckingham Pi (π) method is based on the postulate that the number of dimensionless groups is equal to the difference between the number of variables and the number of dimensions involved in these variables, and dimensional constants are included as variables. For example

$$8 \text{ variable, 5 dimension imply } 8{-}5 = 3 \text{ dimensionless groups}$$

Fick's law is involved in diffusion correlation.

Fourier's $\qquad qy = -K. \, dt/dy \qquad$ (Heat transfer)

Newton's $\qquad \tau_{yx} = -\mu. \, dvx/dy \qquad$ (Fluid flow viscosity)

Scalars are quantities having magnitude only, e.g., speed and time. Vectors have both magnitude and direction, e.g., velocity and force. A vector can therefore be resolved into three components in the rectangular coordinate system. Scalars are in zero order. Tensors and vectors are in first order. Second-order tensors such as shear stress and momentum flux can be resolved into components. Manipulations of transformations for the rectangular system can be equally applicable to cylindrical and spherical coordinates.

The vector v can be represented uniquely in the form:

$$V = iv_x + jv_y + kv_z \qquad \text{i, j, and k are unit vectors and x,y, and z directions}$$

Tensors can be expanded into their components:

$$\tau = ii \, \tau_{xx} + \cdots\cdots ij\tau_{xy} + \cdots\cdots ik\tau_{xz} + \cdots\cdots$$

This can be expanded into the form of an array, which is not a determinant. The laws of arithmetic apply to vectors, scalars, and tensors:

$$sv = vs \qquad \text{commutative}$$
$$r(sv) = (rs) \, v \qquad \text{associative}$$
$$(q + r + s) \, v = qv + rv + sv \qquad \text{distributive}$$

A vector by vector gives a scalar or dot product:

$$v. \, w = v.w. \cos \Theta \qquad \Theta = \text{angle between the two vectors}$$

and

$$v.v = v2$$

It is important to write down the terms in a systematic manner because the dyadic vw need not be the same as the dyadic wv.

Nabla or Del operator, ∇, is an operator used in mathematics (particularly in vector calculus) as a vector differential operator, usually represented by the nabla symbol ∇. When applied to a function defined on a one-dimensional domain, it denotes the standard derivative of the function as defined in calculus. ∇ is defined as follows:

$$\nabla = i\delta/\delta x + j\delta/\delta y + k\delta/\delta y$$

T OPERATES ON A VECTOR TO GIVE A DERIVATIVE

Vector analysis is expanded into equations of change to describe continuity, and motion including multiterms in vector form as a shorthand form for presentation compared with the original equations for three directions:

$$\delta/\delta t. \, \rho.v = -(\nabla. \, \rho.v.v) - \nabla p - \nabla \tau + \rho g$$

Navier Stokes equations apply to the conditions of constant density and viscosity and for Newtonian fluid of constant density. These can be expanded by starting from the equations of continuity and motion:

$$\nabla. \, V = 0, \text{ and } \rho Dv/Dt = -\nabla p - \nabla \tau + \rho g$$

So, the equation of motion for conditions of constant density and viscosity is

$$\rho Dv/Dt = -\nabla p - \mu.\nabla^2.v + \rho g$$

All other vector equations in transport phenomena can be trended likewise.

Chemical engineering is concerned with fluid profiles and mechanisms at the wall. Heat transfer surface, mass transfer stages, adsorption columns, and Deissler's equation model profile near the wall in transport phenomena, e.g., shear stress, and vector profile, might be parabolic toward the wall, eddy current in turbulent flow, and dependent on eddy viscosity.

Equations of change describe continuity, motion, and energy. Continuity depends on the conservation of mass, motion depends on the conservation of momentum, and energy equations are based on energy transformations from heat, kinetic, or potential energy conservation. Therefore, the application of equations of change extends to fluid flow in pipes and other unit operations equipment.

Equations of change are general equations describing the conservation of mass, momentum, and energy in a system, although, based on the differential element approach, the equations can be applied to a problem without resorting to the "thin shell" techniques.

Partial time derivatives, e.g., dc/dt, include a change in concentration with respect to time in a flowing system observed from a fixed point. Total time derivatives include changes in concentration in any direction, allowing for the velocity of the observation point. Substantial time derivatives include change of concentration with respect to time observed from a point moving at the same velocity, v, as the system fluid flow.

The equation of continuity is based on the conservation of mass of an element of volume $\Delta x, \Delta y, \Delta z$ in a flowing fluid, i.e.,

The rate of accumulation of mass = Rate of mass entering the element −
Rate of mass leaving the element.

The equation of motion is based on the conservation of momentum in an element of volume, i.e.,

Rate of accumulation of momentum = Rate of momentum entering the element −
Rate of momentum leaving the element + Sum of forces acting.

Considering the x-component and noting that the x-component of momentum by both bulk flow and molecular transport can enter the volume via three faces and leave via the other three, then the net rate of momentum flow in the x-direction by bulk flow and net x-component by molecular transport can be described mathematically and correlated to normal stress and tangential stress or shear stress in the x-direction. If the external forces acting on the element area are due to fluid pressure and the effect of gravity, then their result in the x-direction is added to the correlation.

These equations apply to any continuous medium but can only be meaningful if the stress function is expressed in terms of velocity gradients using the generalized statements of Newton's law of viscosity. These equations in the x-direction together with the equivalent equations in the y and z directions, the density and viscosity dependence, boundary, and initial conditions, determine pressure, density, and velocity components in a flowing isothermal fluid.

The equation of energy is based on the conservation of energy,

Rate of accumulation of internal and kinetic energy = Net rate of flow of internal and kinetic energy by bulk flow + Net rate of flow of heat energy by conduction − Net rate of work by the system on the surroundings.

Internal energy is molecular energy associated with the movement and interaction between molecules. Kinetic energy is $\frac{1}{2}.\rho.v^2$ per unit volume. The work term includes potential energy effects. The work is done against volume forces (gravity) and surface forces (static pressure and viscous forces), w = force × velocity in the direction of force. Note that the rate of gain of energy includes magnetic and electrical forces as well.

The equations of motion and energy for a binary system can be developed in the same way and will be analogous to those obtained for a pure system. Extension of the equations of change to the multicomponent system, similarly, the arguments applied to a binary system can equally well be applied to each component of an n-component mixture.

9 Chemical Reaction Engineering

Continuous stirred tank reactor for nth-order reaction might be described by

$$R = k\ C_n$$

For plug flow (tubular reactor),

U. C. df = r. dv	Homogenous
U. C. df = r. w	Heterogenous

The following mass balance applies:

$$dQ - r.\ dv.\ \Delta H = \mathcal{E}\ m.\ C_p.\ dT$$
$$\text{If adiabatic, } dQ = 0$$
$$\text{Non-adiabatic,} \quad dQ = U.\ \rho.\ dZ\ \Delta T$$

And

$$U.\ \rho.\ dZ\ \Delta T - r.\ dv.\ \Delta H = \mathcal{E}\ m.\ C_p.\ dT$$

For reversible endothermic reaction, temperature increases, and equilibrium constant, k, increases. For reversible exothermic, k decreases with increased temperature, and the reaction rate increases with temperature.

PURGE

In a reactor where C_2H_4 + Air $\longrightarrow$ $C_2H_4O + N_2$

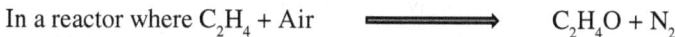

We recycle through a separator to purge N_2 and so there will be no accumulation inside the reactor due to consecutive pumping of N_2, which results in an unsteady state:

Assume purge (P) is   7.9 nitrogen, 1.6 oxygen, and ethane 0.11 = 9.61
Mass balance for ethane, 0.11/9.61 R + source 1.11
Out of reactor
0.75 (efficiency) × (0.11/9.61 R + source 1.11) = 0.11/9.61 R + 0.11
Efficiency is 1.0 − pass efficiency 0.25 = 0.75
R = 252 moles   P/R = 252/9.61 = 26.2

DOI: 10.1201/9781003384199-10

REACTORS

1. Batch
2. Continuous
   - Stirred
   - Tubular
   - Diffusion model
   - Packed bed

For zero-order format,

$U.C_0 = UC_1 + KV$
For second-order format,
$U. C_0 = U C_1 + kV C_1^2$
For a cascade of CSTR first-order format reaction,
$C_n/C_0 = 1/(1 + kV/U)^n$ U in m^3/h

BATCH REACTOR

f = Mass of reactant, reacted/initial mass of reactant,

$mR. df = r. V. dt$ V = Batch volume t = Batch time r = Reaction rate

If r is nth order forward and reaction will take place at varying temperatures, e.g., adiabatic. If isothermal at constant volume, nth order forward,

Average production rate = V/t m^3/h

The rate-limiting step is one of the rate processes whose rate coefficient determines the rate of overall conversion. If the chemical reaction rate is limiting, then we have kinetics control. If the transport mass transfer process rate is limiting, then we have diffusion control. $1/k$ is the resistance to chemical reaction, and $1/h_D$ is the resistance to diffusion and mass transfer.

The selectivity of porous catalysts, dual functional catalysts, and catalyst poisoning play a major role in reaction conversion rates and catalyst regeneration. Chemical kinetics, operating conditions, and catalyst preparation have pronounced effects on mixing, dispersion, drying, adsorption, filtration, and precipitation unit operations. For the estimation of surface area in gas permeability, pore volume and diameter are important factors. Dimensionless models of pore structure cover

- 3D array of cubes
- Pores of different radii occurrence at random
- Vertical pores or at-angle pores
- Pore constriction allowances

Surface diffusion plays a major part in transport, and hot spots in chemical reactors cause catalyst decay. In heterogeneous catalytic reactions, the particles of the solid break the velocity profile in this order:

1. Transport of reactant from bulk fluid to solid
2. Intra particle transport (diffusion) of reactant (transport inside the particle)
3. Chemisorption (Chemically attached to solids)
4. Surface reaction to produce the product
5. Desorption of product
6. Intraparticle transport of product
7. Transport of product from solid to bulk fluid

All concentration variables in the aforementioned steps are independent variables. If one of the steps is faster or slower than another, then we have not archived a steady state. One of these steps is the rate-determining step of the overall process. Only in batch reactors, one step is slower, which is defined as the rate-determining step. Usually, in continuous reactors, we cannot define the rate-determining step as the slower step. If we lump the chemical steps and physical steps, then the rate of mass transfer and chemical reaction are competing based on

The flux mass per unit area per unit time, N_a

$$N_a = h_D \cdot a \cdot (C - C_i) \qquad a = \text{Unit area/Unit volume}$$
$$\text{Reaction rate, R}$$
$$R = k \cdot C_i \qquad k = \text{Chemical rate coefficient}$$

The general approach for reactor design for different reactor models including catalysts is as follows:

1. Write down the differential equation for mass and heat transfer within the porous pellet.
2. Write down the differential equation for mass and heat transfer for the reactor.
3. Solve the pellet equations at the bed entrance.
4. Use the solution of the pellet equation at the bed entrance to march one step along and across.
5. Repeat from step 4 to get to the next pellet along the reactor.

Pellet equations usually depend on the number of components, and if # of components = n, then, pellet equations = n − 1 (Table 9.1).

Catalyst poisoning is when a foreign material that is adsorbed to the surface of the catalyst prevents the reacting materials from being adsorbed on the catalyst surface, e.g., weak poisons are removed by regeneration of the catalyst in a higher temperature air stream. Stronger catalyst poisons are removed by chemical treatment or might be replaced based on cost.

TABLE 9.1

Effect of Intraparticle Diffusion on Various Parameters

Rate-Determining Step	Order	E-Activation	Surface A	Pore Volume
Chemical reaction	N	E	S	Independent
Bulk diffusion	(n + 1)/2	E/2	(S)1/2	(V)1/2
Knudsen diffusion	(n + 1)/2	E/2	Independent	V

CATALYTIC PROCESSES

The significance of catalytic processes in chemical plant operations is of prime interest to chemical engineers. For example, it is necessary to use dual-function catalysts for catalytic reforming of naphtha in order to produce high-octane gasoline. Another case is fertilizer base synthesized ammonia was originally produced at atmospheric pressure with capacities of few tons per day. The use of improved catalyst in reactor design allowed for large ammonia plant conversions in the order of 1,500 tons/day in which primary reforming takes place at 30 atmospheres and 500 F. Synthesis reactions are governed by space-time yield in which space velocity is important:

Space velocity, SV = Reaction mixture volumetric flow/Catalyst volume

And

Space time yield = Moles product at reactor outlet/Moles feed entering × SV

Reactor design considerations include pressure drop across systems and cooling rates in catalytic bed process fluid cooling and heat exchanger services. Inter-bed cooling and independent fluid for cooling are design parameters that play a role in cold shot quenching injection to decrease the overall temperature in the convertor (Table 9.2).

TABLE 9.2

Catalytic Reactors

Type	Fluid	Solid
Fixed bed	PF	Batch
Fluid bed	PF	CST
Slurry	CST/PF	CST
Trickle bed	PF	Batch
Transport bed	PF	PF

REACTOR TYPES ARE BASED ON TWO MODELS

Type 1—single-phase models

- Phase
  - Two-phase, bubble phase is free of particles
  - Two-phase, dilute phase includes solids
- Flow of gas in bubble phase
  - Visible bubble flow
  - Entire flow of gas
- Back mixing in dense phase
  - Axial dispersion
  - Tanks in series
  - Downflow of dense phase
  - Transfer of gas between phases

Type 2—Bubbling bed models

- Phase
  - Two-phase bubble phase and dense phase
  - Three-phase bubble phase, cloud (bubble wake) phase, emulsion phase
- Flow of gas in buddle phase
- Back mixing in dense phase
  - Axial dispersion
- Transfer of gas between phases

10 Fuel and Power

The effect of process pressure is best described in the case of catalytic regeneration stage requirements in the catalytic reforming of gas oil to gasoline. The octane number sequence of operation targets 110 octane gasoline from a medium gas oil of high naphthenic content. Other examples of process dependency on temperature and pressure variable effects on product composition are described in the steam reforming of naphtha to lean gas.

This chapter addresses many of these questions:

1. What are the important features of the Lurgi process?
2. What are the key differences between a CANDU and Steam Generating Heavy Water Reactor (SGHWR)?—What are the advantages of each?
3. What are the important features of the catalytic rich gas (CRG) process as used in steam reforming naphtha? For example, a CRG reformer is to be used as a major gasifying unit in a plant producing substitute natural gas of calorific value 1,100 Btu/ft^3 (41 MJ/m^3). Feedstock is naphtha.
4. What are the important components of a thermal fission nuclear reactor (moderator)? What are the differences between thermal and fast fission types of nuclear reactors?
5. What are the key effects of a catalyst in catalytic cracking of medium gas oil to gasoline fraction? Why are these effects desirable?
6. What extra effects are required for a catalyst in catalytic reforming of gas oil to gasoline fraction? Why are these extra effects considered desirable? How are they achieved?
7. How are the characteristics of a typical feedstock for catalytic cracking differ from those for catalytic reforming to gasoline?
8. What is the role of fuel, coolant, moderator, and control rods in Magnox-type gas-cooled nuclear fission reactor? What does a biological concrete protective shield provide?
9. How is uranium extracted and refined from its ore? What are the advantages and disadvantages of separating U-235 and U-338 isotopes in the preparation of fuel for nuclear reactors?
10. What are the important differences and advantages of boiling water reactors (BWR), pressurized water reactors (PWR), CANDU, and SGHWR water-cooled reactors?
11. What is the importance of steam/hydrocarbon ratio to the operation of catalytic steam reforming of naphtha feedstocks?
12. Fouling—The tendency to deposit carbon or coke on solid surfaces is encountered in all chemical processing plants. The source and effects of this deposition have detrimental effects on operation efficiencies. Important

TABLE 10.1
Nuclear Reactors

Fuel	Moderator	Coolant	Reactor	Property
Uranium (U)	Graphite	Air	Normal	Not compatible
U	Graphite	CO_2/CO	Magnox	CO_2/graphite In-compatible
UO_2	Graphite	CO_2	AGR	CO_2/graphite In-compatible
UO_2/ThO_2 (ceramic)	Graphite/Be	He/N_2	HTGCR	Reduced In-compatible
UO_2 (pellet)— enriched Zirconium alloy	D_2O	H_2O	SGHWR	Steam passes through to minimize high pressure
PuO_2/UO_2	No moderator	Liq. Na or K	FBR (fast breeder reactor)	No moderator
			PFR	Prototype fast breeder reactor

periodic cleaning and other countermeasures are usually taken to counteract this tendency and minimize its effect. What are the advantages and disadvantages of these techniques and measures? How does it affect the plant operation?

13. Catalytic cracking and catalytic reforming processes may each be used in the production of motor gasoline. How are the chemical reactions of these processes different? What are the factors that may lead to the choice of one of the processes in preference to the other by considering characteristics of feedstock and resulting products?

14. Existing industrial "Coal plex" is a mega plant set up in which coal feedstock is converted to substitute natural gas (SNG), synthetic crude oil (Syn crude), and electric power generation. How important are these conversions to clean fuels? Controlled emissions?

The elements of nuclear reactors should have low neutron capture cross section and adapt to chemical reactions and requirements of the process, i.e., do not interfere in the reaction (Table 10.1).

NUCLEAR POWER PRODUCTION—FISSION AND NUCLEAR REACTOR

If certain nuclei are hit by neutrons of suitable energy then they may split into two almost equal-mass nuclei while omitting two or three high-energy neutrons and a considerable amount of thermal energy. It is this thermal energy which may be utilized as either a destructive force if the rate of release is rapid and uncontrolled or as a heat source of variable temperature for power production or direct heating if the rate of heat release and thus the rate of the continuing fission is controlled:

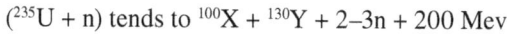

$$(^{235}U + n) \text{ tends to } {}^{100}X + {}^{130}Y + 2\text{--}3n + 200 \text{ Mev}$$

This is a neutron-induced fission. The neutron is the most likely particle to impinge on the nucleus as any positively or negatively charged particle would be more likely to be repelled by either a positively charged nucleus or a negatively charged cloud of electrons, respectively. The nature of X and Y varies widely and will be of varying stability. Generally, X and Y vary from Ba at $Z = 56$ to $Z = 62$ in the lanthanides.

It may be that the neutrons released during fission be of the correct energy they may induce further fission reactions in adjacent fissile nuclei thus making the fission reaction self-sustaining. In order that the thermal energy may be released at a rate that is practically useful as a heat source the number of fissions per unit of time must be controlled. This control may be achieved by either diluting the number of fissile nuclei with non-fissile material such that spare neutrons are less likely to encounter a fissile nucleus before leaving the mass or the neutron flux within the mass may be reduced by introducing materials of high capture cross section. In fact, both of these control methods are adopted in practical operations.

Should any array of nuclei be irradiated by any type of particle the probability of a nucleus being hit by a particle is proportional to the size of the nuclei and their separation in the array. For unidirectional radiation flux, the proportionality may be denoted by the fraction of the cross-sectional area of the array at right angles to the direction of flux that is occupied by nuclei. The probability of collision will also be affected by the size of the radiating particle.

A more useful characteristic of the material radiating particle combination is the knowledge of which of the several possible events subsequent to collision will in fact occur. For the neutron radiant flux, for instance, the probability of occurrence of neutron scattering (bouncing off) or of one of the neutron capture reactions or of fission is dependent on both the character of the nucleus and the energy of the impinging particle. The probability of each of these events when applied to a statistical population allows the statement that a certain proportion of the collision will result in one of these events. As the likelihood of collision may be represented by the relative cross-sectional area of the nuclei, then the likelihood of each of the subsequent events may be represented as a fraction of this cross-sectional area.

Cross section for collision = Section for scattering + Section for capture + Section for fission

Collision = Scattering + Capture + Fission

The balance between fission, scatter, and capture depends on the energy of the neutron and the ratio of $^{235}U : {}^{238}U$ in the fuel.

The components of a thermal nuclear reactor are as follows:

• Nuclear fissile fuel
• Moderator

- Control system
- Coolant
- Containment vessel for fuel elements (can)
- Containment vessel for coolant (pressure vessel)
- Containment for neutron flux (reflectors and biological shield)

Containment of the fuel rod (the can) is necessary to ensure against oxidation of metallic fuel elements at elevated temperatures. In addition, protection of operators from radioactive emanations of fission products that may escape into the coolant stream and protection from and ease of removal of the spent fuel pin are necessary.

General requirement of nuclear power production is the transfer of heat resulting from a fission reaction in the core to an outside power circuit via a working fluid. The character of this working fluid classifies the nuclear reactor type. Requirements for coolant are low neutron capture cross section, high specific heat, non-corrosive to fuel elements and pile internals, and stable under nuclear irradiation.

Gases such as CO, CO_2, He, and N_2 are used as coolants. Oxides of carbon cause problems with carbon deposition and oxidation at high temperatures. Inert gases have shown deficiencies such as reactions with dome fission products at high temperatures. The main disadvantage of gas coolants is their low density which gives rise to large volume flow rate requirements and the larger pressure vessel to accommodate the flow channels and the difficulty of sealing the pressure vessel, a problem which is extreme with helium as the gas tends to leak through the welds.

The benefits of liquid coolants generally are that they have high specific heats and high densities (thus low flow rate). Light and heavy water has the specific advantages of being non-corrosive and cheap and can work as a coolant and working fluid in the power generation cycle. The disadvantage of water as a coolant is the high pressure required to maintain water in a high-temperature liquid form, which requires higher pressure and more costly pressure vessels. The adoption of heavy water as a coolant (deuterium oxide) allows the possibility that the liquid coolant may act also as a moderator, thus eliminating the requirement for loid mechanical moderator lattice. Organic coolants have been used for their beneficial high specific heat or high boiling point. Hydrocarbons' disadvantage is their tendency to decompose under irradiation, which might lead to deposition and fouling on heat transfer surfaces. Liquid metal coolants have been used for the advantage of high density, high specific heat, and low pressure required. The disadvantage is that the radioactive liquid metal coolant cycle must be within the biological reactor shield.

Fusion reactors are based on two or more nuclei that may fuse together into a single more stable nucleus losing mass and emitting a large amount of energy. Achieving collision and fusion of these nuclei necessitates high kinetic energy, i.e., high temperature to overcome the proton repulsion. In addition, high density increases the probability of collision and residence time requirements to meet these conditions. The high-temperature requirements exceed any materials, which will vaporize on which it impinges and nullify this approach in practice.

CARBON-BASED FUELS

The ultimate analysis of typical crude is carbon 83–87%, hydrogen 11–14%, oxygen, nitrogen, and sulfur 7%. The types of crude by residue are paraffin waxes and asphalt bitumen. Constituents of crude and petroleum fractions include paraffinic, olefinic, isoparaffinic, diolefinic, naphthenic, and aromatic. Crude is treated through a series of separation processes:

1. Well head treatment of crude by dewatering and degassing through inclined oil gas separator
2. Separation into fractions of narrower boiling point and viscosity range in rough relationship to the product fraction properties
3. Conversion of one boiling point fraction to an alternative B.P. fraction to balance market demand through cracking
4. Conversion of one chemical type to another through reforming
5. Treatment to include final separation of materials of specific market use or adjustment of final product performance characteristics
6. Finishing by removal of harmful materials, i.e., materials detrimental to performance or sales potential (smell, color, corrosivity, and gum-forming stability) (Table 10.2)

Cracking changes molecular weight and B.P. The typical feedstock is medium gas oil (166–290°C) fraction. The typical major product is gasoline. General chemistry applied.

1. Bond breaking
2. Dehydration
3. Isomerization
4. Polymerization

BOND BREAKING IS A MAJOR REACTION IN CRACKING

Paraffins are separated into short-chain paraffins and olefins. Olefins can be separated into short chains and di-olefins. Aromatics go through dealkylation to make benzene. Naphthenes go through ring fission to make olefins.

TABLE 10.2
Breakdown of Refinery Yield

Crude Fraction	%	USA	Europe
Gasoline	23	42	18
Kerosene	10	7	4
Gas/diesel oil	10	22	26
Fuel oil	62	22	26
Other gases	5	22	14
Typical middle east yield on fractionation			

Cracking reactions could be accomplished through thermal or catalytic processes. 100% medium gas oil is cracked to long-chain olefins, naphthenes, long-chain paraffins, and saturated aromatics. By changing the temperature of the operation, we change the % of products. The reduced temperature will result in reduced cracking.

The resultants of cracking are short-chain gases, C2–C4, and di-olefins with 20–30% carbon. Short-chain olefins C6–C8, short-chain paraffins C6–C10, and benzene all have high octane values. Gasoline is 70–80%.

DISTILLATION OPERATIONS

1. Primary flash column involves the stabilization of crude in order to minimize the temperature range over which the atmospheric column has to operate and to remove permanent gases which may give rise to vigorous actions in the column, and vapor slugging in the lines.
2. Primary atmospheric fractionators involve the separation of feed into more manageable fractions roughly equivalent to market products.

Gasoline	28–200°C
Kerosene	160–250°C
Gas oil	200–360°C
Heavy fuel oil	>360°C

3. Primary flash distillate (P.F.D.) contains the dissolved gases of crude plus some low boiling liquids. It has an intrinsic value as a feedstock in the process industries either as the untreated fraction or separated into its components:
   - Permanent gases
   - Liquified petroleum gases (LPG), propane, butane
   - Gasoline

Separation of fractions above 360°C requires a separation technique that does not curtail heating to the temperatures normally required for conventional distillation. The instability of these components may result in cracking (above 400°C) in the still. To avoid these issues, vacuum and steam distillation are employed.

4. Vacuum distillation at 30 mmHg at 400°C. Fuel oil feedstock is fractionated into gas oil, wax distillate, and heavy fuel oil.
5. Steam distillation would require the addition of steam (oil/H_2O = 2:1). Steam at 170°C, and 1 bar.

The high cost of maintaining a high vacuum is balanced by the increased capital cost necessitated in columns of large diameter designed to accommodate the steam flow. Generally, an optimized hybrid of vacuum and steam distillation is adopted.

6. Specialized separation processes where a fraction having a very narrow boiling point range is required when a single chemical type or compound is required for the market, the rough separation allowed by conventional

fractionation may be inadequate. Specialized separation techniques have been developed to accommodate this requirement.

Super fractionation is useful in the production of very sharp cuts or one compound product. The column requires 100–200 trays and reflux ratios of the order 25–100: 1. For example, its use in the production of isopentane product from a de-butanized gasoline feed.

Isopentane	Overhead	82 F
n-Pentane	Bottoms	97 F

7. Molecular distillation is useful in the production of low vapor pressure lubricating oils obtained from the residue of atmospheric distillation. Operating pressure is under 10^{-3} mmHg. Process arrangement consists of two concentric cylindrical vessels having a 3–4 cm annular separation and 1 m length. The degassed feed falls over the inner heated cylinder, and low boiling components are condensed on the outer cooled cylinder.
8. Distillation with extractive solvents is used with hydrocarbons with boiling points falling in a very narrow range of 1–3°C or components forming a constant boiling mixture, which cannot be separated by fractionation.
   a. Azeotropic distillation involves an additive, generally more volatile than the feed, forming an azeotrope with one component. This azeotrope is removed, and the additive and product are separated by solvent extraction.
   b. The distex process involves a solvent additive generally of lower volatility than feed, which changes the vapor pressure characteristic of the chemical types. Feed plus solvent (aniline furfural) will be separated into paraffins and solvent + naphtha + aromatics.

Medium gas oil is typical feed for cracking. Thermally, without a catalyst often entails passing the feed through tubes in a furnace. The residence time and temperature decide the degree of cracking. Thermal cracking is less common than catalytically, in the presence of a selected catalyst (Fluid bed, fixed bed, or moving bed), and the catalyst used is neutral clay (aluminosilicate). The chemical reactions are common to both, but the catalyst increases the reactivity of selected components.

Catalytic cracking temperature is lower than thermal cracking, which makes the process more convenient. The catalyst changes the stability to cracking of components. Paraffins are highly unstable and tend to become short chains at linkage in temperature. Naphthenes are broken in catalytic cracking more considerably than in thermal cracking. The advantages of catalytic cracking are as follows:

1. Low degree of cracking and high gasoline yield
2. Production of paraffin reactivity, high aromatics, and isoparaffinic content, i.e., high-octane rating
3. Increase in olefin reactivity and low olefin content, i.e., low gum formation

4. Catalytic cracking is useful for sour crude, in hydrogen atmosphere—
dehydrogenation and hydrodesulfurization
5. Cheap catalysts, natural clays, or synthetic aluminosilicates

REFORMING

Change of organic molecular components involves thermal or catalytic reforming. Catalyst action speeds the reaction and promotes cracking. Dehydrogenation
of naphthenes and isomerization of paraffins are enhanced by selective catalysts
(Figure 10.1 and Table 10.3).

Increasing pressure in the process of thermal reforming leads to reduced volume,
i.e., larger molecules. This also increases tar polymers, increases coke deposition,
and reduces gasoline output. By contrast, in the presence of hydrogen, the olefins
double bond is saturated and more paraffins are developed with increased pressure, a
reduction of coke deposition, and increased gasoline output. Thus, at higher pressure,
catalyst deactivation is less of a problem in fixed bed operations.

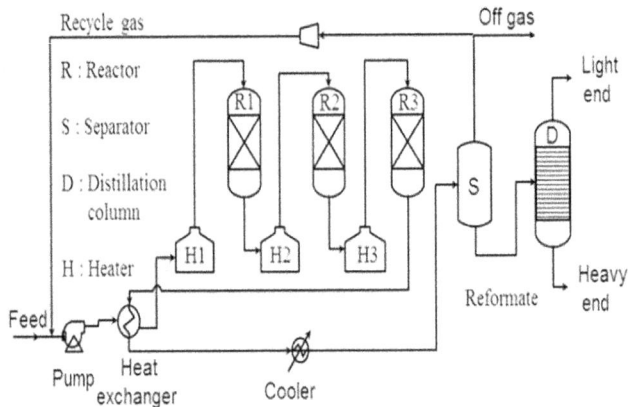

FIGURE 10.1   Catalytic naphtha reforming process.

TABLE 10.3
Process Effect on Gasoline Output

	Thermal Reforming		Catalytic Reforming	
	960°F	1,130°F	813°F	903°F
Gasoline % of feed	92.4%	68.7%	98%	91.6%
Octane	56	83	85	98.5

TABLE 10.4
Catalytic Cracking versus Catalytic Reforming

	Catalytic Cracking	Catalytic Reforming
Catalyst	Natural clays aluminosilicate	Dehydrogenation, isomerization agents
Major effect	Reaction speed	Speed reactions selectively
Major reactions	Bond breaking, high paraffins	Bond breaking, dehydrogenation
Feed	Low-cost catalyst, high paraffinic content	High-cost catalyst, high naphthenic content

Any feedstock may be either catalytically cracked or reformed. The choice is dependent on the characteristics of feedstock and desired major products (Table 10.4). Natural gas or synthetic natural gas (SNG) is measured by the Wobbe index:

$$WI = Caloric\ value/Specific\ gravity = C.V./S.G1/2$$

It is a measure of the energy throughput when supplying gas with standard pressure and a standard burner. The Wobbe index speed factor, S

$$S = Flame\ speed,\ gas/Flame\ speed\ through\ hydrogen \times 100$$

The important characteristics of liquid fuels are presented in either absolute or arbitrary values determined through standardized test procedures. For example, C.V. for any grade of oil should remain constant from batch to batch to control the heat energy content validity and cost. The carbon residue after vaporization (Conradson carbon number) is a measure of carbon-forming tendency. The carbon to hydrogen ratio, C:H, is higher for higher flame emissivity, e.g., kerosene 6:1 and heavy fuel oil 8:1. API (American Petroleum Institute), Deg. API = $(141.5/S.G.) - 131.5$ is a gravity measure related to a crude and reflective of the gasoline content of the oil. Other physical properties include vapor pressure, flash point, viscosity, cloud, and pour point.

Perhaps the knock characteristic of fuel is an important performance measure such as octane number, which is the isooctane that must be mixed with normal heptane in order to match the knock intensity of the fuel. Octane ratings may be determined using isooctane and tetra-ethyl lead (anti-knock agent). Knocking is the preignition of the fuel in an integral combustion engine caused by adiabatic compression temperature rise. This gives a characteristic to metallic knocking sound and loss of performance. Leaded fuels have been phased out in favor of higher-octane fuels.

Cetane number involves diesel knock due to delay in the ignition at the end of the compression stroke when the fuel is being injected. Aniline point is another measure of octane rating aromatic content. The diesel index reflects the content of aromatics as a function of diesel fuel performance:

$$DI = Aniline\ point\ (F).\ API\ gravity/100$$

11 Electrochemistry/ Corrosion

Corrosion is an oxidation process involving the loss of electrons from metals. This could happen in solution, dry or wet conditions. The oxide film (porous or non-porous) depends on the molar volume (MV), which is equivalent to the molecular weight divided by the density of the metal. The relative molar volume,

$$RMV = MV \text{ of oxide/MV of metal}$$

For porous film, RMV < 1. Non-porous oxide film is "protective."

Protective oxide films such as Al, Cr, Sn, Be, Si, and Ta (works against acid) could provide a layer of oxide protection for the surface of the metal. In addition, Fe is usually galvanized with a Zn casing to create a protective layer against corrosion.

Electrochemical wet corrosion involves water in immersed, atmospheric, or underground conditions. Anodic reaction (oxidation) involves the release of electrons whereby the metal becomes charged, which leads to corrosion. Cathodic reaction (reduction) involves the discharge of electrons (gained). The cathodic reaction involves oxygen absorption and hydrogen release. In the presence of oxygen and water, more corrosion will occur resulting in rust:

$$Fe(OH)_2 + O_2 \longrightarrow Fe_2O_3 . H_2O \text{ (RUST)} + H_2O$$

Atmospheric corrosion involves humidity, gaseous impurities, and metal surface conditions. The presence of solid impurities where salt might be under dirty conditions will enhance the absorption of water and promote corrosion—underground corrosion with acidic soil or differential oxygen concentration and soil conductivity due to differential salt concentration, which creates high salt soil (anodic) leading to corrosion. Stray current through the ground pipe creates anodic and cathodic parts of the pipe which leads to corrosion. Microbial corrosion is promoted by bacteria (*Vibrio desulfericous*) catalyzing reaction in the soil. This type of corrosion might be fought by high-pH media.

The Pourbaix (E/pH) diagram (Figure 11.1) correlates the electrical potential to pH as it relates to corrosion. As E increases by becoming more positive, anodic protection becomes more effective. As the region of immunity is approached with a decrease in E, cathodic protection becomes more important in preventing corrosion. The limitation of the Pourbaix relationship is that there are no kinetic data to determine the rate of reaction, which is not known. Metal water and oxygen are involved, which actually are causes of corrosion. The mechanism involving metal hydroxide

DOI: 10.1201/9781003384199-12

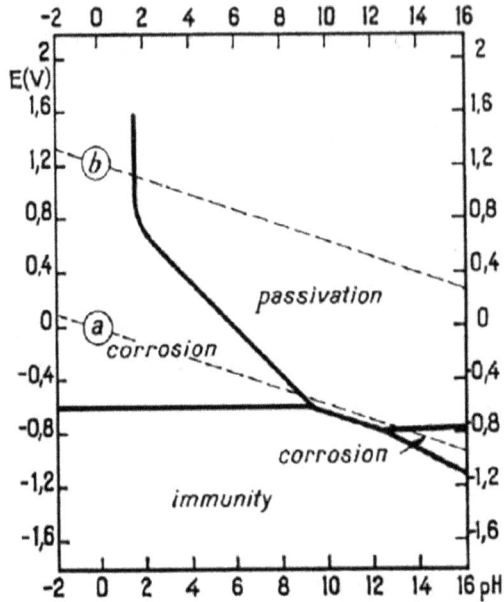

FIGURE 11.1   Pourbaix diagram for Fe in water.

TABLE 11.1
Relative Molar Volume (RMV) of Porous and Non-Porous (Protective) Oxide Films

Metal	RMV	Porous	Non-Porous
K	0.4	P	
Na	0.6	P	
Ca	0.6	P	
Ba	0.7	P	
Mg	0.8	P	
Sn	1.3		X
Cd	1.4		X
Al	1.5		X
Pb	1.5		X
Ni	1.5		X
Zn	1.6		X
Ag	1.6		X
Cu	1.7		X
Fe	2.0		X
Cr	2.0		X
W	3.4		X

as a precipitating ion that creates a film might be replaced by other ions which may not protect.

Evans, U.R. et al. (1975) *An Introduction to Metallic Corrosion*. Edward Arnold.
Scully, J.C. (1966) *The Fundamentals of Corrosion*. Pergamon.

In general, dissimilar metals promote corrosion, e.g., coatings and bulk metal. Also, single metal wet corrosion increases when the cathode is widely spread, and anode current creates tiny pitting in metal sheets, for example, corrosion of iron at break in mill scale in the presence of an electrolyte medium. Pitting will occur as the mill scale is cathodic, while the iron substrate is anodic. In addition, metal roughness affects the surface and might create anodic bumps and cathodic indentations that would promote corrosion. Concentration changes in electrolyte and differential aeration, which might affect oxygen concentration, would attack crevices in boilers and increase corrosion levels.

Atmospheric corrosion is increased as humidity increases and would become catastrophic for Fe when RH > 75%. Gaseous impurities occur sometimes when the atmosphere contains sulfur dioxide or hydrogen sulfide. When water condenses on the surface and these gases' affinity for water increases, at elevated dew point temperature, acidity increases causing corrosion. These gases are absorbed in water film, which creates a sulfurous acid, and the metal surface catalyzes the reaction and transforms to sulfuric acid, which causes severe corrosion to the metal substrate and similarly for carbon dioxide in the creation of carbonic acid. Solid impurities such as dust particles might be chemically inert, e.g., sand but may have an anodic area under pile which will cause corrosion. Chemically reactive salt carried over by wind from the oceans is of great concern for corrosion. Chemically inert, but absorptive water would allow for increased rates of corrosion on metal.

The presence of H^+ and dissolved oxygen allows cathodic reactions to take place:

$$M \implies M^{++} + 2e-$$

TABLE 11.2
Effects of Locality on Atmospheric Corrosion

Place	Description	Mild Steel Corrosion Rate, $\times 10^{-3}$ inches/year
Khartoum	Hot, dry	0.02
Singapore	Hot humid	0.46
Llanwrtyd well (UK)	Cold, wet	0.92
Gongell (S.A.)	Warm, humid	1.44
Woolwich (London)	Warm, humid, industrial	1.84
Mother well (Scotland)	Cold, wet, industrial	2.02
Sheffield	Warm, wet, industrial	4.25

This type of chemical reaction is dependent upon the electrode potential of metal with respect to the solution. In solution, these reactions involve electrons and also H^+ and OH^- and therefore are pH dependent. These parameters can act independently or together.

Cathodic protection is an important technique to connect a metal to a base to pull the potential down. For example, a sacrificial anode like an Mg anode is used to provide electrons to the underground pipe, which acts as the cathode, as the pipe is losing electrons to the surrounding soil. In addition, galvanized Zn valves are used as a zinc anode to protect a Fe-based heat exchanger from corrosion through a cathodic reaction. Impressed voltage involves the use of a carbon anode to protect the pipe (cathode). Pt can be used but is expensive. This can be accomplished by using a Ti sheet with Pt applied to it. Similarly, stainless steel or Pb can also be used for cathode protection.

Anodic protection is usually used to protect vessels by moving the potential to the passivity region. The wall of the vessel will act as an anode that is wired to a cathode dipped into the solution inside the vessel to produce a metal oxide or hydroxide passive film, for example, the storage of acid in mild steel using anodic protection.

Inhibition involves the use of some chemicals to stand against the chemical reaction that causes corrosion. In anodic and cathodic protection, we employ the potential and electrical interference against the corrosion mechanism and process. For example, in anodic inhibition, a chromate such as $FeCrO_4$ would act as a stifling agent at break in mill scale. Cathodic inhibition might involve the use of calcium hydroxide to form a film on the cathodic area to prevent oxygen from reaching it, thus ceasing the reaction.

Adsorption inhibitors are adsorbed at the surface of materials. These inhibitors are usually organic in nature, e.g., pyridine and quinoline glue.

Protective coatings require surface preparation such that the coating must adhere tightly to the metal surface. Degreasing by taking off oil film by using solvents to dissolve oily film is common practice in surface preparation. Rubbing, acid pickling, or blasting techniques are also used for the removal of the oxide film. Surface roughness is changed in the case of plantings by polishing first.

Organic coatings are applied following surface preparation. For example, in the case of paint application to steel, phosphoric acid is used to convert FeO to $FePO_4$ before applying paint. Similarly, to wet the surface and cause adhesion, chromate or lead primers are used. An undercoat deals with pigments and supplies the color to make a thick coating. Topcoats contain pigment and have a varnish to provide a shiny surface. The main job of the under and topcoats is to provide a barrier against oxidation and thereby corrosion. A typical film thickness is 1/8 mm. For thicker film applications, plastic coatings of plastic melt might produce a coating of 2–6 mm. Metal is heated or coated in the fluid bed of plastic under vacuum to prevent air bubbles. For example, polyethylene, nylon, PVC, and PTFE are all used in many applications. Metallic coatings such as noble metals are very resistant to corrosion, e.g., Pt and Au. The most popular metal coatings are Zn, Cu, Pb, Ni, Cr, Cd, and Sn as they are self-protecting against corrosion. Zn is useful to protect materials that are higher in electrochemical series. Metal coatings employ dipping and rolling, electroplating, flame spraying, cladding, and diffusion coating at higher temperatures to make a coating like an alloy (Table 11.3).

TABLE 11.3
Comparison of Protective Coatings

Type	Advantage	Disadvantage
Organic	Flexible, easily applied, cheap	Oxidize, relatively soft, limited temperature, soluble in organics
Metallic	Deformable with base metal, Insoluble in organics, conducting	Establishes galvanic cell if damaged
Ceramic	Temperature resistant, hard, does not produce galvanic cell, resistant to acids	Brittle, insulating, liable to thermal shock

CORROSION TESTING

Tests have a rate and might be accelerated by changing temperature or concentration. For atmospheric exposure, tests are conducted in a box sheltered from the rain. To simulate a marine environment, materials are immersed underneath the water at varying temperatures. Laboratory testing involves acid testing to initiate atmospheric conditions. In addition, tests are conducted to reflect on chemical plant exposure, Aeration, to measure how much oxygen is involved in the system, liquid agitation, and other process features are simulated. Need to avoid setting up a galvanic cell effect by setting up the right insulation. Worst-case conditions should be considered including temperature and chemicals in use to evaluate to what extend may corrosion take place.

Corrosion penetration is described (Tables 11.4 and 11.5):

$$\text{Penetration} = 365 \times \text{wt. loss per day/density} \times \text{surface area} = \text{mm/year}$$

Tests include the amount of metal dissolved in liquid, which is correlated to corrosion level or measuring the amount of oxygen absorbed and knowing the amount of hydrogen formed. In this case, a hazard may arise from this situation by having an atmosphere of hydrogen around the corroded area and an explosion or fire may take place. In addition, a wire of material might be measured as it correlates to its increase in resistance.

TABLE 11.4
Corrosion Penetration

Penetration, mm/year	Corrosion Level
<0.1	Resistant
<1.0	Fairly resistant
>1.0	Not resistant

TABLE 11.5
Corrosion Number

Penetration	Corrosion # = Log_{10} (mm/year × 10³)	Corrosion Resistance
<0.001	0	Immune
0.01	1	Exceptionally good resistance
0.10	2	Good resistance
1.0	3	Sufficient resistance (monitor)
3.0	3.5	Limited resistance (caution)
10.0	4	Poor resistance (unusable)
>10.0	4+	No resistance

TABLE 11.6
Cost Index of Primary Metals

Material	Clad	Solid
Mild steel	1.0	1.0
304 Stainless steel (18/8)	2.3	3.5
316 Stainless steel Cr 18, Ni 12, Mo 3%)	2.6	4.3
Monel (Ni 66, Cu 33%)	4.5	9.8
Ti	4.9	10.6

TABLE 11.7
Selection of Materials

	Inorganic			Organic	
Acidic	Glass, ceramic, carbon	Oxidizing (metals with protective oxide film), non-oxidizing (rubber, plastic)	Glass, ceramic, carbon, not rubber, plastic	Acidic	Metals except Al, Pb, Sn, and brass
Neutral	Except mild steel			Neutral	Metals
Alkaline	Metals except Al, Pb, Sn, and brass. Rubber and plastic are acceptable, but not glass or ceramic				

Corrosion is usually expressed in penetration; nonetheless, it is important to evaluate how the material profile looks like the corrosion might have covered the whole surface. Following testing, a mathematical model of the system should include corrosion allowance based on thickness measurements. This allowance is dependent on the materials in use and is connected to an economic scale.

The selection of materials for construction would consider the physical environment, cost, and chemical environment targeting strength, temperature, transmission, and surface properties. Transmission properties may involve radiation, electricity, or heat. Surface properties include roughness, softness, hardness, and resistance to grinding or impact. Cost impacts building, operation, production, and maintenance of the plant (Table 11.6).

In summary, the galvanic series must consider oxide film dissolved (active) and oxide film present (passive). In all cases, less noble metals corrode, and there is a dire need to avoid galvanic cells (Table 11.7).

12 Unit Operations

SOLVENT EXTRACTION

Feed is introduced into a separator, which is combined with a solvent in stages that include stripping and then enriching. The output streams include solvent, extract, and raffinate. For a given extract reflux ratio, the locations of the different stages are independent of any raffinate reflux. Consequently, the total number of ideal stages must be the same both with and without raffinate reflux. This is expressed as the Reflux ratio mass balance:

$$R_0 = E-Q \qquad Q = Pe + Be$$

Back mixing and dispersion are correlated and are dependent on the number of stages in the extraction column. The efficiency of extraction depends on the number of stages, interfacial area, and mass transfer coefficient. Therefore,

$$h = n \text{ stages. (H.E.T.S)} \qquad h = \text{hold up}0$$
$$n = K. \text{ a. } \Delta C$$

In the extraction process, one phase is dispersed in the other in the form of droplets. Mass transfer depends on the method of formation and density of the solution around the drop. Drops may be prolate, spherical, or oblate. Mass transfer due to circularity is several times greater than stagnant drops.

ΔC is dependent on mass balance and equilibrium of the system. The mass transfer coefficient, K, depends on drop size and Reynolds number. Interfacial area, a, depends on equipment geometry, drop treatment, and formation in equipment. The drop volume is correlated to nozzle velocity. Holdup is described as the fraction of the column occupied by the dispersed phase, e.g., in the spray column at drop holdup less than 8%, no coalescence occurs. The residence time drops,

$$\text{Residence time (drop) } t = \text{Height of column/Drop velocity}$$

The design of equipment for simultaneous mass transfer and chemical reaction depends on the reaction rate and the reaction of gas A and liquid B taking place at the surface of the liquid, i.e., at the interface.

For example, in oil refining, the objective is to extract groups of compounds having remarkably similar physical properties, e.g., in the extraction of lubricant oil, there may be a number of compounds ranging from 50 to 500 that need to be extracted. In aviation fuel, all aromatics have to be extracted to eliminate smoke and inconsistent burning. Consequently, an analysis cannot be made in terms of specific components

DOI: 10.1201/9781003384199-13

and an alternative index must be obtained. Generally, the index implies specific gravity (s.g.), and viscosity gravity constant (v.g.c.).

Mixing crude and a solvent for extraction is represented by the joining line on the phase diagram. The analysis of any extraction process depends on the following:

1. The phase equilibrium diagram
2. Mass balance and stagewise calculations

The stagewise analysis is characteristic of the process and depends on mass balance and the concept of net flow irrespective of the unit. The phase diagram can be described in any set of units, e.g., specific gravity or v.g.c. For example, the phase diagram can be expressed in terms of s.g. as follows.

> Consider crude kerosene (primary distillation) if mixed with desired solvent in a known ratio of oil to solvent and agitated for 4 hours, until equilibrium is established. Then, phase is separated and volume to oil ration is estimated for each phase. The solvent is separated and s.g. of the oil fraction is determined. On the phase diagram, this will give points A and B. Since the solvent-to-oil ratio of each phase is known, then points P and Q can be located on the diagram. The straight-line PQ is the tie line and should pass through the total mixture M. The linearity PMQ is a measure of the accuracy of the experiment. P and Q are two points on the equilibrium curve. Therefore, this type of experiment must be repeated at least four times in order to establish an equilibrium curve and to produce sufficient data to apply one of the tie line correlations procedures. The stagewise analysis is thereafter carried out in the normal way.

The function of the designer is to evaluate the mass transfer coefficient and residence time of dispersion in the particular extraction equipment. The mean value of the driving force is the mean perpendicular distance between the equilibrium curve (distribution curve) and the correlated operating line. This is best achieved by using Simpson's rule:

$$\text{Divide } X_n \text{ into } X_0 \text{ (2n intervals),}$$

Then,

$$\Delta C_m = \rho . \Delta y_m$$

The actual extent of mass transfer during drop formation depends on the method of formation and to a large extent on the densities of solution around the drop. Consequently, in current modern extraction equipment, a dispersion is formed as the phase enters the equipment and is not intentionally coalesced until the phase is about to leave the equipment. Common types of extraction equipment are based on the rotating disc column (R.D.C.) extractor which has the dispersion phase at the bottom of the column, and the continuous phase of the column fitted with staters, rotors, turbines, and baffles based on the design requirements, and the coalescence phase at the top of the column.

FLUIDIZATION

Pressure drop in a fluidized bed is dependent on the weight of particles, weight of bed per unit area, and flow as a function of drag force on the particle:

$$\Delta P = (\rho_s - \rho).g.Z(1-e) \quad \rho = \text{Density} \quad Z = \text{Bed height}$$
$$C = 1-e = \text{Solids volumetric concentration} \quad e = \text{Void fraction}$$
$$V_s = A.Z.C \quad A = \text{Area}$$

The degree of fluidization, n, is dependent on the fluid velocity and in a fluidized bed n-top > n-bottom

$$N = U/Uo \quad U = \text{Fluid velocity}$$

SEDIMENTATION

In continuous operation, the sedimentation rate increases based on the characteristic of the sludge and by adding an electrolyte that enhances flocking. As we introduce heat, the viscosity of the fluid is decreased and smaller particles will grow. Thixotropic sediment is usually developed through slow stirring. Sedimentation capacity is dependent on the design of the clarifier and the thickening process before the discharge point.

$$\text{Discharge flux} = \text{Discharge rate} \times \text{Discharge concentration}$$

CRYSTALLIZATION

It is important to remember that crystallization is a separation process, but not a purification one. Crystallization is dependent on temperature, concentration, and impurities in the solution. The solubility curve is proportional to concentration and temperature, and differential concentration is the driving force for crystal growth. Temperature affects crystal size, and impurities affect the rate of crystallization and freezing point depression temperature. The super solubility curve defines crystals; at that temperature point, cooling can be stopped. There are two processes of nucleation and crystal growth involved. Nucleation could be homogenous or heterogenous. The nucleation rate is dependent on solubility and applies to both homogenous and heterogenous nucleation. If the cooling rate is reduced, it is likely to initiate the crystal growth. In heterogeneous nucleation, some crystals are formed before crystallization starts, and this same material enhances crystallization. Usually, secondary (heterogeneous) nucleation occurs at a lower supersaturation than primary nucleation (homogeneous).

Mass transfer effects during crystallization are dependent on reaction, diffusion, and bulk fluid concentration. Crystal formation is driven by lowering the temperature and interface reaction zone. The reaction rate could be first or second order, and the interface reaction rate is proportional to the crystal growth rate. Growth rate and kinetic control are correlated to supersaturation. In practice, there might be discontinuity due to the presence of impurities, which would cause occlusions that are cavities filled with mother liquor creating faulty crystals.

In summary, the determination of crystal growth varies with temperature, speed of formation, and concentration. Change in mass of a crystal suspension and indirect usage of mixed suspensions affect the mixed product removal and are proportional to growth and nucleation rates.

LEACHING

This technique is used to remove a soluble component of a solid system. For example, to take the sugar out of sugar beet by leaching it in water:

1. Put solids in solvent
2. Provide a contact area
3. Dissolve the soluble part of the solid in a solvent
4. Solute dissolution
5. Solute diffusion through the solution within the total solid
6. Transfer of solute into bulk solution outside solid

A leaching model example is described for the manufacturing of caustic soda, NaOH. Caustic soda is manufactured by lime-soda (sodium carbonate) processing by treating this solution in water. After the reaction is complete, calcium carbonate sludge containing one part calcium carbonate to nine parts of water by weight is fed continuously to thickening units in series and is washed counter currently. The solid discharged from each thickener contains one part by weight of calcium carbonate to three parts of water. On drying, x% of caustic soda is recovered.

DISTILLATION

Macabe Thiele postulates the concept of constant molal overflow and adiabatic conditions during the separation process (distillation). The upper operating limit (U.O.L.), $y = R/R + 1 \cdot x + x_D/R + 1$

The lower limit (L.O.L.), $y = (Lo + q_F) \cdot x/(Lo + q_F - w) - W \cdot x_w/(Lo + q_F - w)$
                q line, $y = (q/q - 1) \cdot x - x_F/q - 1$

The steps involved in constructing distillation mapping plots are as follows:

1. Plot vapor–liquid equilibrium
2. Plot X_D, X_F, and X_w
3. Plot $x_D/R + 1$
4. Plot U.O.L.
5. Calculate q
6. Draw q line through X_F
7. Plot L.O.L. through X_w, cross U.O.L. with q line

8. Step: Count # of theoretical plates = step − 1
9. Efficiency = η = # T.P/# A.P
10. Locate the feed plate

In gas absorption, the height of packing, molar density, equation of operating line, and equilibrium constant mass transfer coefficient for gas and liquid phase are all described in mathematical models that are correlated to predict the number of theoretical stages and predicting the ratio of gas flow rate per unit area to liquid flow rate per unit area, which is graphically determined as the crossing of the operating line and equilibrium line (pinch point).

Distillation is based on mass and heat transfer calculation relating number of plates in the distillation column, and the connection from the reboiler to the condenser. The relative volatility of the fraction components is key to the separation of these components through the distillation process. The vapor–liquid equilibrium is dependent on the liquid leaving a tray and the hold-up of liquid on each tray. Distillation may also occur with chemical reactions. The mass balance for the vapor–liquid correlation will be dependent on the rate of reaction and plate hold-up volume. Back mixing of liquid on the tray is splashed by the action of gas, a portion of which is thrown backward in the direction from which it enters into the tray. Tray efficiency has limitations with respect to numerical computations based on a small element of froth over the tray. This is determined by the concentration of vapor leaving and the concentration of vapor in equilibrium with the liquid in the element of the froth. The flow pattern describing bubble, cellular foam, froth, or spray is dependent on gas supercritical velocity against hold-up. The dispersion of bubble flow v implies that heat transfer plus mass transfer led to higher efficiency. Velocity increases as hold-up increases. Froth with very mobile dispersion, smaller bubbles with a wide size distribution, and vigorous liquid circulation increase in velocity are related to a decrease in hold-up. Higher velocities are correlated with liquid dispersion in the continuous gas phase, e.g., spray transitioning of bubbles.

In thermal distillation, the temperature gradient heat transfer effect is when the dew point of vapor is greater than the boiling point of the liquid. The heat transfer from vapor to liquid, which helps mass transfer in distillation, will lead to higher efficiencies.

Gilliland empirical equation is used to estimate the number of stages required in a distillation column. It uses the minimum number of stages as well as minimum and actual reflux ratios in the calculation. Gilliland correlation is an approximation method for distillation-column calculations, which correlates the reflux ratio and the number of plates for the column as functions of minimum reflux and minimum plates. The mathematical model of the column is obtained by performing a total mass balance and partial mass balance to the component.

Liquid surfaces with a high surface tension contract when contacted with a surface of lower surface tension. For column separation unit operations, the selection factors for the choice of a tray or packed column depend on the physical and chemical properties of the system, properties of the fluid, fluid flow, physical properties of the column, and equipment availability (Table 12.1).

Flooding occurs in any device where liquid and vapor flow counter currently through the same space. When the energy potentials causing the flow upward and

TABLE 12.1

Packed Columns against Tray Columns

Packed Column	Tray Column
Preferred for foaming systems as packing tends to break up in foam formation	Sludge or solids
Corrosive system with ceramic materials	Vapor is dispersed, and liquid is turbulent. Mass transfer efficiency-controlled system good for liquids
Lower pressure than tray; use for temperature-sensitive systems	Engineering construction favors internals
Vapor turbulence; liquid dispersion. Mass transfer efficiency-controlled system good for vapor	
Heavier	

downward are balanced, the point of flooding is reached. When the energy potential of vapor is greater than that of the liquid, the liquid backs up and the column floods. The design of packed towers defines packing and diameter, permissible pressure drop, reflux ratio, height (packing and column), and maximum throughput with a minimum height of packing. Column calculations include the following:

- Feed, reflux ratio, select packing
- L/V $(\rho_L/\rho_V)0.5$
- ΔP
- Cross-section diameter
- Height theoretical unit (HTU)
- Number of theoretical units (NTU)
- HETP
- Z, height of liquid packing

Stage-to-stage analysis requires the liquid phase to be well mixed, the vapor will flow through the liquid in plug form, and the mass transfer process is gas-phase controlled. Part of the counter-current cascade calculations is made based on the direction of vapor flow from stage to stage. The equilibrium line is when $E_{MV} = E_{ML}$. Equilibrium value is the composition obtained if the feeds to a stage achieve complete equilibrium. Overall E = NTP/NAP

NTP—number of equilibrium stages at a specified reflux ratio

NAP—number of actual stages at a specified reflux ratio

The multicomponent distillation design method approach is when the minimum number of plates, minimum reflux ratio, optimum reflux ratio, and feed plate location are defined. In addition, tray-to-tray analysis and tray efficiency calculations to

translate the theoretical plate into actual plates are prerequisites for column mechanical design.

Considerations for physical separation or if a chemical reaction is present during the
vapor–liquid heats of mixing, latent heat calculations. The calculation of Nm, the minimum number of theoretical trays at the total reflux operating line, relates to vapor–
liquid composition passing between trays. The composition of vapor leaving the tray is
equal to the composition of the liquid leaving the tray above. The vapor-to-liquid ratio
is based on the boundary situation starting at the still and correlated at total reflux for
the first plate and then calculated for partial condenser and total condenser.

Multicomponent extraction is dependent on the feed composition with designated light
and heavy keys and residue. To achieve sharp separation, the unit operation must have
constant relative volatilities and constant molal overflow at a mean column temperature.

The Fenske equation in continuous fractional distillation is an equation used for
calculating the minimum number of theoretical plates required for the separation of
a binary feed stream by a fractionation column that is being operated at total reflux
(i.e., which means that no overhead product distillate is being withdrawn from the
column) (Figure 12.1).

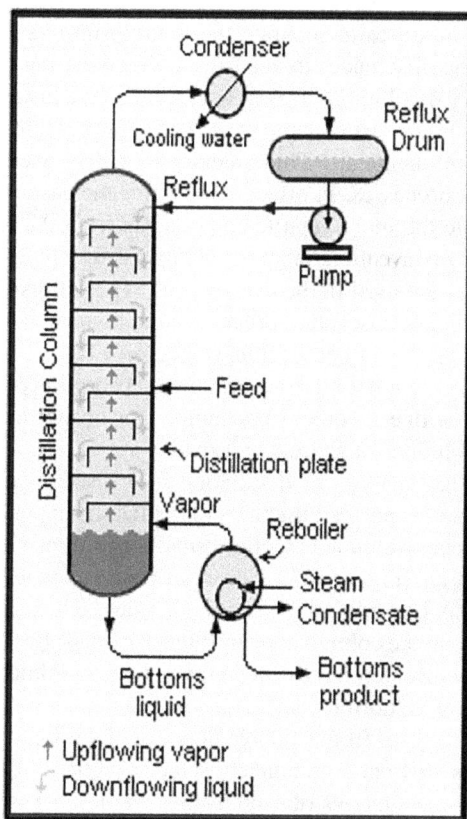

FIGURE 12.1   Separation by distillation.

When designing large-scale, continuous industrial distillation towers, it is particularly useful to first calculate the minimum number of theoretical plates required to obtain the desired overhead product composition.

Fractional distillation is the separation of a mixture into its component parts or fractions. Chemical compounds are separated by heating them to a temperature at which one or more fractions of the mixture will vaporize. It uses distillation to fractionate. Generally, the component parts have boiling points that differ by less than 25°C (45°F) from each other under a pressure of 1 atmosphere. If the difference in boiling points is greater than 25°C, a simple distillation is typically used. It is used to refine crude oil.

A theoretical plate in many separation processes is a hypothetical zone or stage in which two phases, such as the liquid and vapor phases of a substance, establish an equilibrium with each other. Such equilibrium stages may also be referred to as an equilibrium stage, an ideal stage, or a theoretical tray. The performance of many separation processes depends on having a series of equilibrium stages and is enhanced by providing more such stages. In other words, having more theoretical plates increases the efficiency of the separation process be it either a distillation, absorption, chromatographic, adsorption, or similar process.

A fractionating column or fractional column is an essential item used in the distillation of liquid mixtures to separate the mixture into its component parts, or fractions, based on the differences in volatilities. Fractionating columns are used in small-scale laboratory distillations as well as large-scale industrial distillations.

Relative volatility is a measure comparing the vapor pressures of the components in a liquid mixture of chemicals. This quantity is widely used in designing large industrial distillation processes. In effect, it indicates the ease or difficulty of using distillation to separate the more volatile components from the less volatile components in a mixture. By convention, relative volatility is usually denoted as alpha.

Relative volatilities are used in the design of all types of distillation processes as well as other separation or absorption processes that involve the contacting of vapor and liquid phases in a series of equilibrium stages.

Relative volatilities are not used in separation or absorption processes that involve components reacting with each other (for example, the absorption of gaseous carbon dioxide in aqueous solutions of sodium hydroxide).

Stage (tray) to stage analysis of distillation in a distillation column takes into consideration liquid and vapor mol fractions to predict the upper and lower operating lines, which assume constant molal overflow, small heats of mixing, and Raoult's and Dalton's laws application (Figure 12.2). The equilibrium method utilizing bubble and dew point calculations is used to calculate the k value, which is a function of temperature and pressure in the column at operating conditions. Liquid and vapor phases are determined based on column pressure, temperature, and fluid composition, which is a function of relative volatility.

As the flow rate increases from a bubble flow to cellular foam into froth and spray flow pattern, there is still not a completely reliable method of predicting tray efficiency. Although the calculations of establishing the number of theoretical trays in a given separation are achievable, the conversion of these trays into actual stages by a tray efficiency factor remains unsolved. In the process of dispersing vapor through

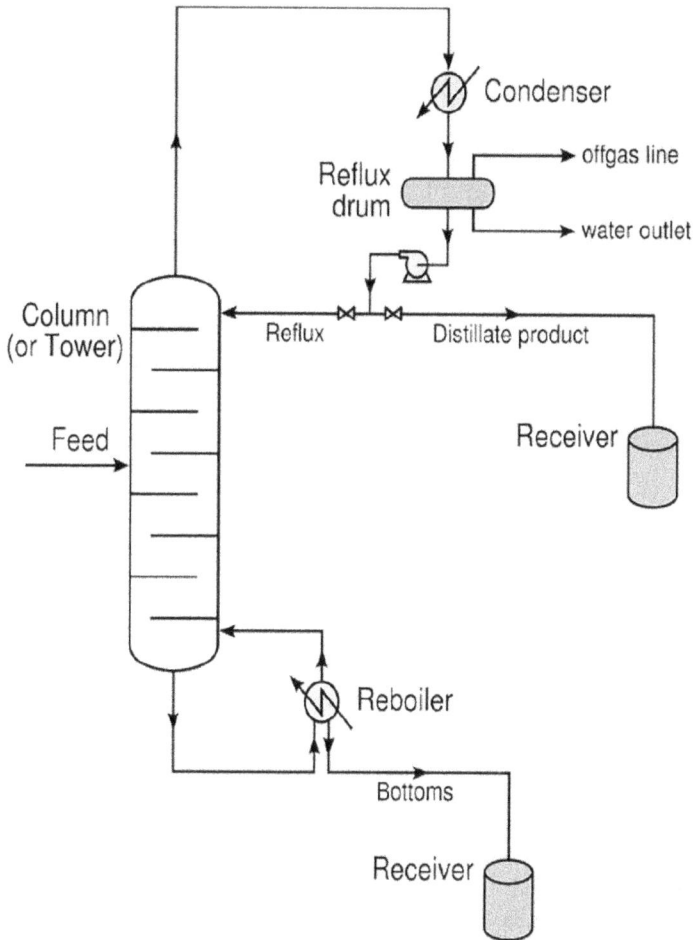

FIGURE 12.2   Distillation process.

liquid on a perforated tray, different flow patterns arise according to flow conditions, systems properties, and plate geometry. These patterns are different, and it is not possible to describe plate operations by a combined mathematical model. It is necessary to determine the flow pattern occurrence and then describe the appropriate two-phase flow hydrodynamics and mass transfer characteristics.

Liquid surfaces with higher surface tension tend to contract when contacted with a lower surface tension. Therefore, surface tension and heat transfer effects are exacerbated in distillation columns due to temperature and concentration gradients, which create a channeling wetting effect. The liquid running down the column is not of a uniform thickness, and hence local thin film becomes saturated with heavier components. If the volatile component has a higher surface tension than the less volatile component, the thin film areas with lower surface tension will break up. When the surface tension of a mixture is greater at the top of the column than at the bottom, i.e., surface tension decreasing down the column, a negative system is created. The converse of this is a positive system, where surface tension increases down the column. In a positive system, the thin film has a higher surface tension, which stabilizes the film. For both systems, the relative volatility value of the system must be higher to allow for the development of surface tension gradients. Surface tension effects also play a role in vapor–liquid contacting on bubble trays. Where vapor is dispersed in the liquid, the bubbles formed will be stabilized if the system is positive, because, at the thin films between bubbles, the less volatile component in the films has higher surface tension leading to gas stabilization of foam or froth. The converse effect will take place if the system is negative as bubbles are destroyed resulting in droplet formation.

In a distillation column, a vertical temperature gradient is developed. The dew point of the rising vapor is always higher than the boiling point of the descending liquid. Inter-diffusion of components by mass transfer involves the transfer of heat from vapor to liquid, which results in condensation sand vaporization at the interface. This thermal distillation augments the mass transfer separation effect. This implies that plate efficiencies are higher than by mass transfer alone. Both surface tension and thermal distillation due to temperature gradients play a lesser role in difficult separations where a large number of stages are required.

Liquid mixing on bubble plates is a complex process that, under steady-state conditions, any concentration gradient that is present in the liquid on the plate as a result of the difference between the rate of removal of volatile components by stripping gas and the rate of replenishment by the inflowing liquid is because of the distribution effect due to liquid mixing. Predicting efficiencies for commercial bubble trays used in multicomponent fractionation is dependent upon the effects of tray design, operating variables, and system properties of the gas and liquid film efficiencies and is the summation of these that calculate the overall tray efficiency. With foaming systems, packed columns are usually preferred since the packing tends to break up foam formation. For systems that contain sludges or solids, a tray column is usually a better choice selection. For corrosive systems, packed columns are often chosen because of the ability to use ceramic materials. For temperature-sensitive systems, packed columns are often favored since they are expected to operate at a lower pressure drop than a tray column. In packed columns, the liquid phase is dispersed, and the vapor phase is in turbulence, resulting in effective mass transfer for vapor control systems. The opposite occurs in plate columns where the vapor is dispersed and the liquid is in turbulence, resulting in good mass transfer efficiency for liquid phase-controlled systems.

There are three possible ways of bringing vapor and liquid together for mass transfer purposes:

1. Both phases may be continuous
2. Gas phase continuous, liquid phase stationary
3. Liquid phase continuous, gas phase stationary

Packed towers are phase contacting equipment that have limitations in their capacity to handle liquid and vapor loads. In distillation, the ratio of vapor and liquid is fixed by the reflux ratio. The operating limits, therefore, lie between the minimum and total reflux ratio. For a given product rate, the actual quantities of vapor and liquid are fixed in relation to the minimum or total reflux. This limitation is based on producing a specified product. In addition to the operating limits imposed by the separation of desired components, there are other limits to capacity. For higher flow rates of liquid and vapor, column flooding conditions might be encountered. In this situation, liquid backs up and fills all space in packing. On the other hand, at lower flow rates, there is an insufficient flow to cover the packing with liquid and vapor bypassing resulting in lower efficiency of mass transfer. Nonetheless, at some higher flow rates, excessive pressure drop might occur.

When the energy potentials causing the flow upward and downward are balanced, flooding will occur. When the energy potential of vapor is greater than that of liquid, then liquid backs up and the column floods. The design of packed towers involves the selection of packing, and tower diameter to get reasonable flow rates, good contact of liquid and vapor phases, with permissible pressure drop, the selection of reflux ratio, and height of packed column section. Optimization of these variables involves the greatest throughput of a given product with the least height of packing. When the operating lines and equilibrium line in the graphical representation of mass transfer units are linear and parallel, then this unit will be equal to a theoretical plate. Selection of packing as a guide is for the largest size packing consistent with tower diameter (Table 12.2). The height of packing, Z

$$Z = N.T.U \times H.T.U, \quad \text{where}$$
N.T.U = Number of transfer units
H.T.U = Height of transfer units

Distillation can be coupled with chemical reaction. The process is described by a set of equations in mass and enthalpy similar to conventional distillation, but with an additional term to allow for the effects of chemical reaction. For example,

$$L_{m+1} + V_{m-1} = L_m + V_m + \Delta R_m Z_m, \quad \text{where}$$
L = Liquid;   V = Vapor;   R = Rate of disappearance of a reactant;
Z = plate holdup.

The overall component molar balance correlation should describe reaction rate concentrations, mol fraction, and rate of formation of the product.

TABLE 12.2
Packing Size

Packing Type	Column Diameter D, ft.
Raschig rings	<D/30
Berl saddles	<D/15
Pall rings	<D/10

Application of computerized methods to distillation allows for stagewise calculations. For example, in a McCabe-Thiele model, the assumptions are for a binary (two components) system with 100% plate (tray) efficiency, and the mass balance at equilibrium involves four streams plus any inputs and outputs:

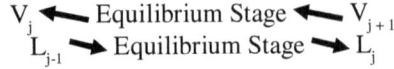

$$V_j \longleftarrow \text{Equilibrium Stage} \longleftarrow V_{j+1}$$
$$L_{j-1} \longrightarrow \text{Equilibrium Stage} \longrightarrow L_j$$

The MESH equations represent <u>M</u>ass balance, <u>E</u>quilibrium relation, <u>S</u>ummation, and <u>H</u>eat balance equation. Decisions to be made in this approach are whether to solve equations stage by stage, or by type of equation. If by type, the usual order is to calculate mass and energy balances but leave out enthalpy, if the system assumptions allow for that. If repetitive iterations are assumed, then how do we correlate variables found in each equation, and how that relates to the mol concentration of desired components resulting in overall efficiency calculations of the separation column. If we combine M and E equations and then perform the S equation, then we use this to find L and V of the fluid.

The concept of separation efficiency of plate columns is adopted as a substitute for the exact direct calculation of the amount and extent of mass and heat transfer between phases in contact on actual plates in the actual column. Practically, this is compared to an ideal column where the phases leaving each plate are in thermodynamic phase equilibrium. Based on materials and enthalpy balances, the use of separation efficiency presents a rapid method of determining column actual performance. Therefore, plate efficiency reflects the actual mechanism of separation transfer processes occurring on the plate. Thus, efficiency can be predicted in terms of rates of mass and heat transfer, interfacial area, contact times, and extent of mixing of phases.

A number of different expressions for plate efficiency have been proposed. It is imperative that the defined efficiency should be useable in the computation sequence for the separation device under consideration with the intent to minimize complexity and iteration in calculation modeling. For example, Murphree efficiency, E_M, is based on the following criteria:

Vapor E_{MV}

- Liquid phase is well mixed.
- Vapor flows through liquid in a plug-flow pattern
- Mass transfer process is gas-phase controlled.
- Stages are part of the countercurrent cascade.
- Calculations are made in the direction of vapor flow from stage to stage.

Liquid E_{ML}

- Vapor phase is well mixed.
- Liquid flows through vapor in a plug-flow pattern
- Mass transfer process is liquid-phase controlled.
- Stages are part of the countercurrent cascade.

- Calculations are made in the direction of liquid flow from stage to stage.
- In general, $E_{MV} \neq E_{ML}$. Only when the operating and equilibrium lines are parallel, then the two Murphree efficiencies are equal. The overall efficiency E_O is defined as follows:

E_O = Number of equilibrium stages for a specified quantity of separation at a specified reflux ratio/Number of actual stages at a specified reflux ratio

Obviously, the efficiency of each plate in a plate column need not be solely based on mass transfer. In heat transfer cases, Carey introduced temperature dependency for thermal efficiencies. Although Murphree plate efficiencies are widely recognized, it has its limitations in having liquid and vapor efficiencies for each plate to describe the same process. This is only readily usable for calculations of vapor efficiency up-the-column and liquid efficiency for down-the-column. However, these efficiency calculations are limited to saturated compositions and cannot be readily extended to cover unsaturated phases. Even in multicomponent distillation, the assumption that streams leaving a plate are saturated is not accurate although the deviations are small. The Holland vaporization efficiency calculations are more finite for reflecting the efficiency of the transfer processes occurring. Therefore, the vaporization efficiency should be considered the basis of efficiency correlations as it describes the degree of completion of the equilibration transfer process.

Mathematical modeling and techniques are critical to the understanding of mass and heat transfer process calculations. Many differential equations are used to express definitions and notations used in process representation. Many graphical solutions are correlated based on these models that in many cases are based on experimental studies and empirical experiences. All expressions of binomial or polynomial relationships can be covered by a functional equation that correlates process parameters and systems design geometrical values.

One of the most uses of partial differentials is in solving differential equations. Parametric differentiation is based on the premise that some graphs are complicated to express by a relationship between x and y. So, in these cases, it is easier to introduce another variable (parameter) that adds value to the process conditions variations and significant interactions. Moreover, an expression of the $y = f(x)$ known as an explicit function is easy to differentiate. Other expressions that cannot be correlated in this format are implicit functions. In many cases, the use of logarithmic functions is used to simplify implicit functions to solve parametric equations.

Stationary points in a mathematical correlation are plotted in graphical forms to show variations as conditions change. This is important to observe trends and rate of change. These types of representations can quickly show and demonstrate maximum and minimum process conditions that are important for design limitations of process unit operations and equipment in service. Similarly, they can show process inflections or deflections as variables are changing.

13 Process Dynamics

Mass and energy balances are critical to the understanding of process dynamics and measurements in steady and transient states. With chemical reaction, a balanced equation can be expressed:

In = Out + Reacted + Accumulated
For an unsteady state without reaction, the equation degeneration:
In = Out + Accumulated
For a steady state with reaction:
In = Out + Reacted:
Without reaction
In = Out

A basis for the calculation has to be specified:

- Complete change in a batch operation
- Continuous operation (an arrow in and an arrow out)—either time basis, e.g., 1 hour, 1 day, 1 year, and so on, or mass basis, e.g., 100 kg, 100 kg moles, and so on

Considerations in establishing balances in plant design start with materials flow incoming and outgoing. Typically, a graphical flow sheet describing a reaction in terms of process knowledge can be followed (Figure 13.1):

$$A \rightleftharpoons B + C$$

QUANTIFYING FLOW SHEETS

In terms of mass balances, quantities of each compound are reflected in total composition. For energy balances, enthalpies of the total streams and temperatures reflect heat transferred in the plant. This will be in preparation for chemical engineering design calculations. For example, how high is a packed tower? How many plates are in a tower? How many stages are in an extractor? What is the diameter of a tower to allow for flow to take place? How much heat transfer surface is needed?

In terms of mechanical engineering design, e.g., how thick is the vessel wall, schedule of pipes, and materials of construction? This is followed by plant layout and instrumentation for control loops (Figure 13.2).

$$\text{Steady State (SS)} \quad mC_pT_0 + Q = mC_pT_1$$
$$\text{Unsteady State USS)} \quad m\,C_pT_0(t) + Q + q(t) = mC_pT_1(t)$$

DOI: 10.1201/9781003384199-14

2-Phase extraction by solvent

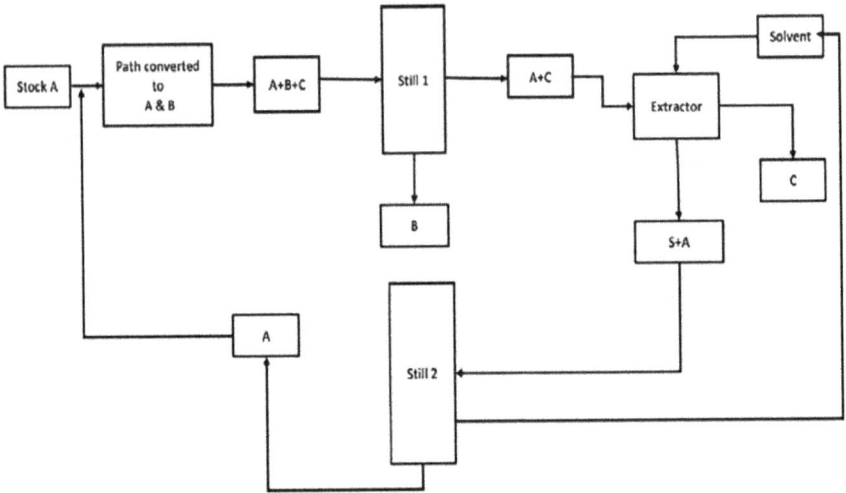

FIGURE 13.1   Solvent extraction.

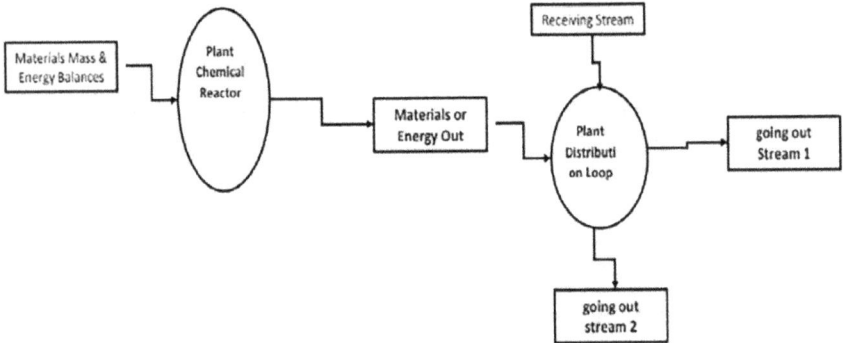

FIGURE 13.2   Mass and energy flow diagram.

When there is perturbation (P) in the system, then P = USS − SS. If there is no perturbation, then the consideration is for transient effects and conditions during USS. If there is no chemical reaction, then the system might be described by

In = Out + Accumulated
If there is a chemical reaction, then
In = Out + Reacted + Accumulated

For example, in a well-stirred reactor under transient conditions, the typical reaction rate may be described as a first-order reaction, $r = kC$.

Second-order reactions, $r = kC^2$, would need linearization to describe a reaction model. For non-linear flow, we use binomial theorem including perturbation parameter description in mathematical correlation. For mass and energy balances, we account for enthalpy and quantity of heat transferred. In addition, for mass transfer considerations, the head driving force is accounted for in the mathematical model. Under conditions where all input is accumulated and output is zero, a dead-end condition is described.

Process attenuation is directly proportional to amplitude,

Attenuation, α_w = Input amplitude/Output amplitude
Gain, or amplitude ratio = $1/\alpha_w$

Gain is usually described as the coefficient of transformation term. Bode plots show the frequency response, that is, the changes in magnitude and phase as a function of frequency. This is done on two semi-log scale plots. The top plot is typically magnitude or "gain" in dB. The bottom plot is phase, most commonly in degrees. If both the gain margin GM and the phase margin PM are positive, then the control system is stable. If both the gain margin GM and the phase margin PM are equal to zero, then the control system is marginally stable. If the gain margin GM and/or the phase margin PM are/is negative, then the control system is unstable. A positive gain margin means how much the control system gain can be increased, while a negative gain margin means how much the control system gain can be reduced. Therefore, in response to various uncertainties, the control system should satisfy negative and positive gain margins and phase margins. A gain margin of infinity means that no matter how much you increase the gain, the system will always be stable. For example, a state of equilibrium of a body (as a pendulum standing directly upward from its point of support) such that when the body is slightly displaced, it departs further from the original position.

The Laplace transform plays an important role in control theory. It appears in the description of linear time-invariant systems, where it changes convolution operators into multiplication operators and allows defining of the transfer function of a system. The key advantage of transfer functions is that they allow engineers to use simple algebraic equations instead of complex differential equations for analyzing and designing systems. The main limitation of transfer functions is that they can only be used for linear systems. While many of the concepts for state-space modeling and analysis extend to nonlinear systems, there is no such analog for transfer functions and there are only limited extensions of many of the ideas to nonlinear systems. Partial fractions of Laplace transforms include single roots, repeated roots, and quadratic roots. This is rationalized by comparing coefficients to calculate underdamped or overdamped oscillations.

A Nyquist plot is a parametric plot of a frequency response used in automatic control and signal processing. The most common use of Nyquist plots is for assessing the stability of a system with feedback. In brief, Bode plots show the frequency response of a system. There are two Bode plots—one for gain (or magnitude) and one for phase. The amplitude response curves are examples of the Bode gain plot. The Nyquist plot combines gain and phase into one plot in the complex plane.

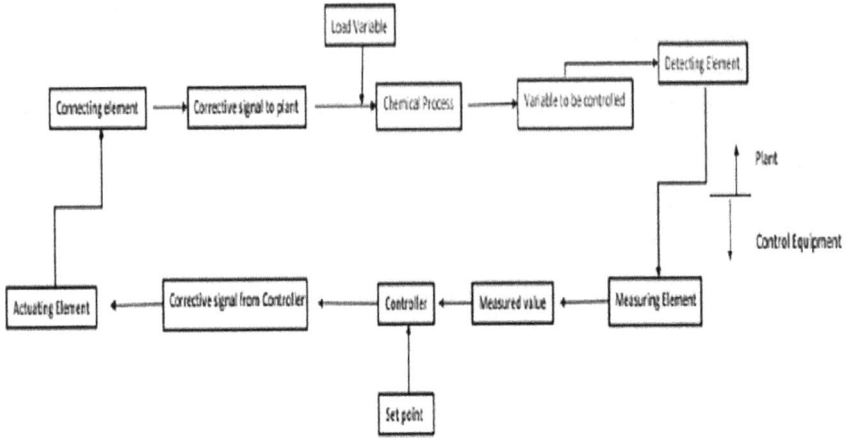

FIGURE 13.3 Feedback control loop.

Feedback factor is the fraction of the amplifier output signal fed back to the amplifier input. A system is said to be stable if its output is under control. Otherwise, it is said to be unstable. A stable system produces a bounded output for a given bounded input. The Nyquist criterion states that the amplitude ratio at 180° lag time must be < 1 for stability.

Feedback control involves input (forcing variable), error signal, corrective signal from the controller, corrective signal from value, process signal, disturbance (forcing variable), output signal, measured signal, controller item, value item, process item, measurement item, and transfer function (Figure 13.3). The fixed set points accommodate the variable load. In addition, if the load is fixed, then adjustments allow for a variable set point.

The corrective signal from the controller is directly proportional to the error signal and expressed in an equation that uses three coefficients: (1) k1 = proportional action factor, (2) integral factor, and (3) derivative factor.

The output signal is a function of time and is expressed as an exponential function that becomes asymptotic following offset and will show a higher error (Pot value) with no control (Figure 13.4).

The transfer function can be a representation of a simple linear ramp (straight line function), a decaying exponential, or a rising exponential. In addition, the physical condition can also be graphed as a sustained sine or cosine wave.

SYSTEM STABILITY DETERMINATION

The characteristic equation

$1 + G_c G_v G_{p1} G_{p2} \ldots \ldots G_{pn} G_{m=0}$, can be expanded to generate a polynomial, $P_0 S^n + P_1 S^{n-1} + P_2 S^{n-2} \ldots \ldots P_n = 0$, where P_0 is assumed to have a positive value.

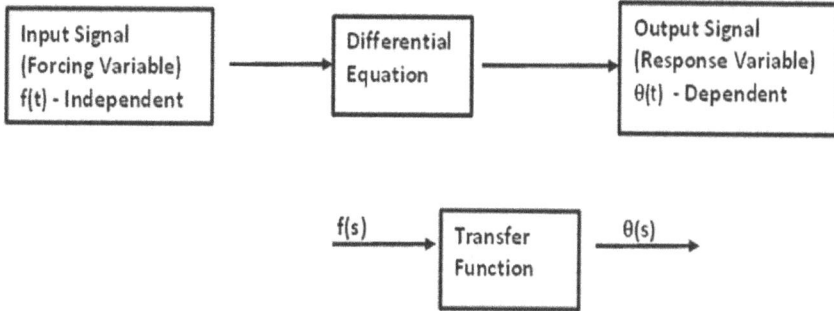

FIGURE 13.4 Input/output functional expression—Laplace transforms of differential equations.

If the coefficients vary in sign, or if one or more is zero, then the system response is unstable. If the coefficients are all of the same sign and no coefficient is missing, then the system response might be stable these possibilities may be further investigated based on the Routh criterion.

The coefficients of the polynomial are arranged in two rows, and then additional rows a1, a2, ..., b1, b2, ..., c1, c2, ..., are calculated, such as:

$$
\begin{array}{llll}
P_0 & P_2 & P_4 & P_6 \cdots \\
P_1 & P_3 & P_5 & P_7 \cdots \\
a_1 & a_2 & a_3 \cdots \cdots \text{until zero is obtained} \\
b_1 & b_2 & b_3 \cdots \cdots \text{until zero is obtained} \\
c_1 & c_2 & c_3 \cdots \cdots \text{until zero is obtained} \\
d_1 & \cdots \text{until zero is obtained in the first column,} & \text{where,}
\end{array}
$$

$$a_1 = (P_1 P_2 - P_0 P_3)/P_1 \qquad a_2 = (P_1 P_4 - P_0 P_5)/P_1 \cdots$$
$$b_1 = (a_1 P_3 - a_2 P_1)/a_1 \qquad b_2 = (a_1 P_5 - a_3 P_1)/a_1 \cdots$$
$$c_1 = (b_1 a_2 - b_2 a_1)/b_1 \qquad \cdots$$

The number of changes of sign in the left-hand column is the number of roots with positive real parts, e.g.,

$$s^3 + 6S^2 + 12S + 16 = 0 \qquad \text{gives}$$

$$
\begin{array}{lll}
1 & 12 & 0 \\
6 & 16 & 0 \\
16 & 0 & \\
0 & &
\end{array}
$$

The system is stable.

14 Mathematics

Differential equations are essential to mathematical modeling in chemical engineering. For example, a stone thrown vertically against air resistance of k × velocity is described by the equation:

$$dv/dt = -(g + kv)$$

In electronics, when a force of E volts is applied to a circuit comprising of an inductance and resistance R ohms in series, it is described in a similar equation.

A differential equation is a relationship between x, an independent variable, and y, a dependent variable, $y = f(x)$

Ordinary differential equation is an equation that contains only the total derivatives, i.e., dy/dx d^2y/dx^2

Partial differential equation is an equation that contains only partial derivatives, i.e., $\delta y/\delta x$ $\delta^2 y/\delta x^2$

The order of the differential equation is that of the highest-order derivative it contains. The degree is that power that the highest order is raised to, e.g.,

$$X^2(d^3y/dx^3)^2 + y^4 = x \qquad \text{order} = 3 \qquad \text{degree} = 2$$

General linear differential equation, $a_n.d^ny/dx^n + a_{n-1}.d^{n-1}/dx^{n-1} +. . .$

a is either function of x or a constant, and there are no terms of multiple derivates like dy/dx. d^2y/dx^2

If $f(x) = 0$, then the equation is homogenous, otherwise non-homogenous.

To find solutions of differential equations, they must be expressed in the form $y = G(x) = g(x)$, i.e., containing no differentials.

If we consider the nth-order differential equation, then the solution in this problem contains n integration. Therefore, n constants of integration (n unknowns). This is known as the general solution. If these calculations in some cases spot and identify a solution, then this is known as particular integral (P.I.)

Given an nth-order differential equation, then the solution to the homogenous equation is called complementary function (C.F.), and it will contain n constants.

The general solution of an nth-order differential equation is the sum of the P.I. and C.F., which will contain constants. A differential equation is homogenous if the power of the terms of the equation is constant (dy/dx to be of power zero).

15 Numerical Methods— Computation

There are several numerical methods that are utilized in approximations used in mathematical modeling to describe physical experimental and empirical data in chemical engineering. As we recognize that the chemical engineering discipline is a highly mathematical discipline that requires advanced and complex calculations, these numerical methods are helpful tools for these types of calculations that might be very extensive and with many detailed subroutines connected to the main program. For example, variations of physical properties as a function of process parameters and to cover systems profile from start point to end point would require subroutines to calculate these properties and call upon the subroutine in the main program to capture the physical value as required at a point in time representing the values as it relates to the variables in the process. As such, interval halving, Regula Falsi, secant method, Newton's method, Newton Gregory, Lagrange, trapezoidal, Romberg, Simpson's, and Gaussian are all mathematical models that use numerical method calculations. Each of these methods provides for certain approximations that provide for differential values, improved values, and more accurate values.

These computations involve compilation errors as in error analysis to identify expected but not found errors. In addition, these will look for in-line errors against the primary compilation in format, read, and write messages. The applied numerical analysis covers solutions of non-linear equations, interpolation, numerical integration, numerical differentiation, and numerical solution of ordinary differential equations, for example, finding the roots of a non-linear equation. To evaluate the $y(x)$ relationship, points are on either side of the root. They might have opposite signs of function (Regula Falsi method); if you have two guesses, the third point will be calculated. Two points closest to the root and a small value of function would be estimated by a secant method (Figure 15.1).

DO-LOOP

A DO statement is used to execute a piece of the program repeatedly, with an integer variable, called the loop variable, taking a succession of increasing values:

n—continue is a statement number that must occur later in the program
i—m1, m2, m3 is an integer variable NOT changed between D(0) and end of loop
m1, m2, m3 are integer variables or positive constants with no sign

The program from the DO statement number n (which must follow) is executed first with I = m1, then with i = m1 + m3, i = m1 + 2m3, . . . , until the value of

DOI: 10.1201/9781003384199-16

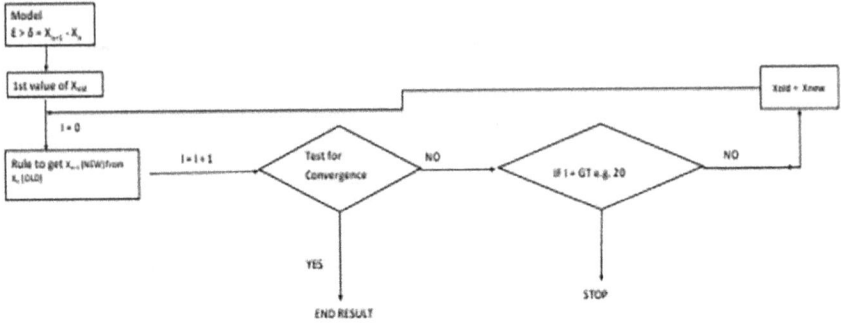

FIGURE 15.1   Operational validity.

I exceeds m2. When this occurs, the execution proceeds beyond n. Loops can be nested. If a loop is opened within another loop, it must be closed within the first loop, or at the same statement numbers. Nested loops must have different loop variables:

1. The inner loop must be completely within the outer loop
   For example,
   D(0) 10   I = 1,N
   D(0) 20 J = 1,N
   20   Continue
   10   Continue
   The aforementioned four-step loop is acceptable—Yes
   The following four-step loop is not acceptable—No
   D(0) 10 I = 1,N
   D(0) 20 J = 1,N
   10 Continue
   20 Continue, hence
   D(0) 10 I = 1,N
   D(0) 10 J = 1,N
   10   Continue is Acceptable—Yes
2. The last statement of a loop cannot be
   Go To   IF   Stop D(0)
                          Return
   D(0) 10 I= 1,N
   IF (I.EQ.2) Go to 3
   10 Continue
3. No statement inside a Do loop can alter the loop variable or the parameters
   I = M, for example,
   D(0) 10 I = M,N,K
   D(0) 20 I = 1,2
   20 Continue
   10 Continue

4. Transfer: A jump may be made out of the range of a D(0) loop or within it, but not into its range. You cannot jump in from 1 to 3, for example,

   D(0) 10 I = 1,N
   3 Continue
   10 Continue

Subscripted variables imply that it is possible to define an array of variables with one or more subscripts, e.g., X_j with j running from 1 to some positive integer n. For example,

   DIMENSION   X(10), Y(10), Z(10,5), K(10)

Will set aside space for 10 variables for X, 10 variables for Y, an array of five columns for each of 10 variables for Z, and 10 integer variables for K.

In most situations, a subscripted variable may be used where an un-subscripted variable is allowed. The subscript is an integer constant or variable. For example,

$$Y(1) = X(1) + Z(2,3)$$
$$K(2) = 1$$
$$Z(1,2) = X(10) + Y(9)$$

The subscript in brackets can be an integer variable or expression involving one. This means that the loop variable of a DO loop can be used to scan a whole array. For example,

   Do 1   I = 1,10
       X(I) = 0
     1 Continue

This will set all the variables X_i equal to zero. Also,

   Do 2   I = 1,N
   READ (1,100)   X(I), Y(I)
     2   Continue
   100   Format (2F10.0)

This will read N lines each with a pair of values.

LEAST SQUARE REGRESSION

This involves fitting a straight line to data:

$$Y = ax + b$$

The consideration of varying a and b to minimize standard deviation, S, for given X_i, Y_i, where S

$$S = \Sigma(y_i - ax_i - b)2$$

TABLE 15.1
Functional Relationships to the
Number of Points on the Graph

Number of Points	Expression
2	Line
3	Quadratic
4	Cubic
n	n−1 Polynomial
n + 1	n Polynomial

If testing for a curve, then assume quadratic, i.e.,

$$y_i = ax_i + b + cx_i 2$$

Regression least squares do not exactly go through the points on the graph. By contrast, interpolation is a method to exactly fit a function through data points (Table 15.1).

Interpolation Expression (equally spaced points):

$$y(x_s) = y_0 + s\Delta y_0 + s(s-1)/2.\ \Delta^2 y_0 + s(s-1)(s-2)/3!.\ \Delta^3 y_0 + s(s-1)(s-2)(s-3)/4!.$$
$$\Delta^4 y_0 +.\ .\ .$$

If the points are not equally spaced, then we can no longer apply the aforementioned expression formula. There is a need for a new polynomial to go exactly through unequally spaced points. Therefore, the simplest formula for Lagrange interpolation for unequally spaced points

$$y = (x-x_1)/(x_0-x_1)\ y_{0 +} (x-x_0)/(x_1-x_0)\ y_1\ \cdots$$

NUMERICAL INTEGRATION

Pressure, bar	Enthalpy—h, kJ/kg
50	2,794
60	2,784
70	2,772
80	2,758
90	2,743
100	2,725

Integral from 50 to 90 bar,

$$\int E.dp = h/2\ (.\ .\ .) = 10/2\ (f50 + 2f60 + 2f70 + 2f80 + 2f90)$$
$$= 5\ (2{,}794 + 2(2{,}784 + 2{,}772 + 2{,}758) + 2{,}743)$$

$$= 1,105,585 = A1,$$
$$A2 = 20/2 \ (f50 + 2f70 + f90) = 110,810,$$

Apparently, A1 is more accurate than A2,
Romberg method correlates error $\propto h2$

$$A1 = \text{Actual value with step h}$$
$$A2 = \text{Actual value with step 2h}$$
$$\text{Exact value of } \int E = A1 + c \ h^2$$
$$\int E = A2 + c \ (2h)2$$
$$E = A1 + 1/3 \ (A1{-}A2) = 110,585 + 1/3(-225) = 110,510 \text{ as an improved}$$
estimate based on error analysis.

The numerical integration trapezoidal rule applies if all intervals are equal to h. The Romberg method can be used for equal intervals. Trapezoid is equivalent to a straight line between points. Three points are needed to fit a quadratic equation. Practically, it needs to apply to all quadratic or polynomials of lower order.

Simpson's rule is approximated:

$$I = h/3 \ (f(x_0 - h) + 4f(x_0) + f(x_0 + h))$$

Trapezium rule and Simpson's rule use fixed x, which describes a polynomial of order n fitted using n + 1 points. Gaussian quadrature allows x1 and w1 to be chosen to give the highest possible order to fit

$$\int f(x)dx = w1 \ f(x1) + w2 \ f(x2) \ldots$$

No. of Points	No. of Parameters	Order
2	4	3
3	6	5
n	2n	2n−1

Numerical differentiation is useful in solving differential equations by the method of finite difference. For example,

One term	$df/dx = (f1 - f0)/2h + 0(h)$
Two terms	$df/dx = (f2 - f0)/2h + 0(h^2)$

Computers' dimension purpose is to reserve space for arrays. A master dimension,

$$\text{Xyz (20), Ink (5,5)}$$

Executable changes the values in a variable, or does some operation, e.g.,

Input/Output
x = 2
Stop

Non-Executable defines the setup within which operation occurs,

- DIMENSION
- END
- FORMAT
- REAL INK
- INTEGR X,R
- = PLACE AFTER MASTER AND BEFORE ANY EXECUTABLE STATEMENTS

Numerical solutions of ordinary differential equations arise quite often in chemical engineering problems:

$$C_0U = CU + Vdc/dt - R \qquad R = VkCn$$

Analytical solutions are often difficult or almost impossible:

$dy/dx = x^2 + y^2$ analytical solution? Since the analytical solution is difficult, then use numerical methods (Taylor's or Euler's methods).

Fortran techniques for problem-solving using computer language by using computer programming and iterations to solve complex and lengthy calculations are highly dependent on machine code. Fortran is a precise language that requires exact sequential codes to initialize a problem and input desired variables that are related in subroutines, which are called upon in the main program including do-loops to perform the calculations based on mathematical models describing the process. The ways of solving the problem using computers, one has to define the problem, decide if a computer is needed, employ the required math formulations, and develop a process flow chart to follow the logic behind process flow (Figure 15.2).

The aim of this section is to enable the reader to understand the basic principles of numerical solutions of ordinary differential equations. The methods described use Euler and enhanced methods including subroutines. The general solution of a differential equation contains one or more arbitrary constants and when plotted on a graph would give a family of several curves. When a differential equation is solved numerically, it is only possible at one time to obtain one particular solution, satisfying particular initial or boundary conditions. A differential equation with sufficient conditions given at one point is called an initial value problem. A differential equation can be of first order or nth order. A graphical solution is equivalent to finding y as a function of x. Euler's method is the numerical equivalent of the graphical process, and the coordinates of the successive point are calculated instead of being obtained graphically.

Before writing a program, it may be helpful to emphasize the main steps of the calculation by means of a flow diagram. It is usually necessary to experiment with different steps, besides calculating the various quantities. In addition, the program needs to include a printout of results, a test to see when the desired range of x values has been covered before repeating the loop. The actual calculation of a new point from a previous one is referred to as the integration of one step (Figure 15.3).

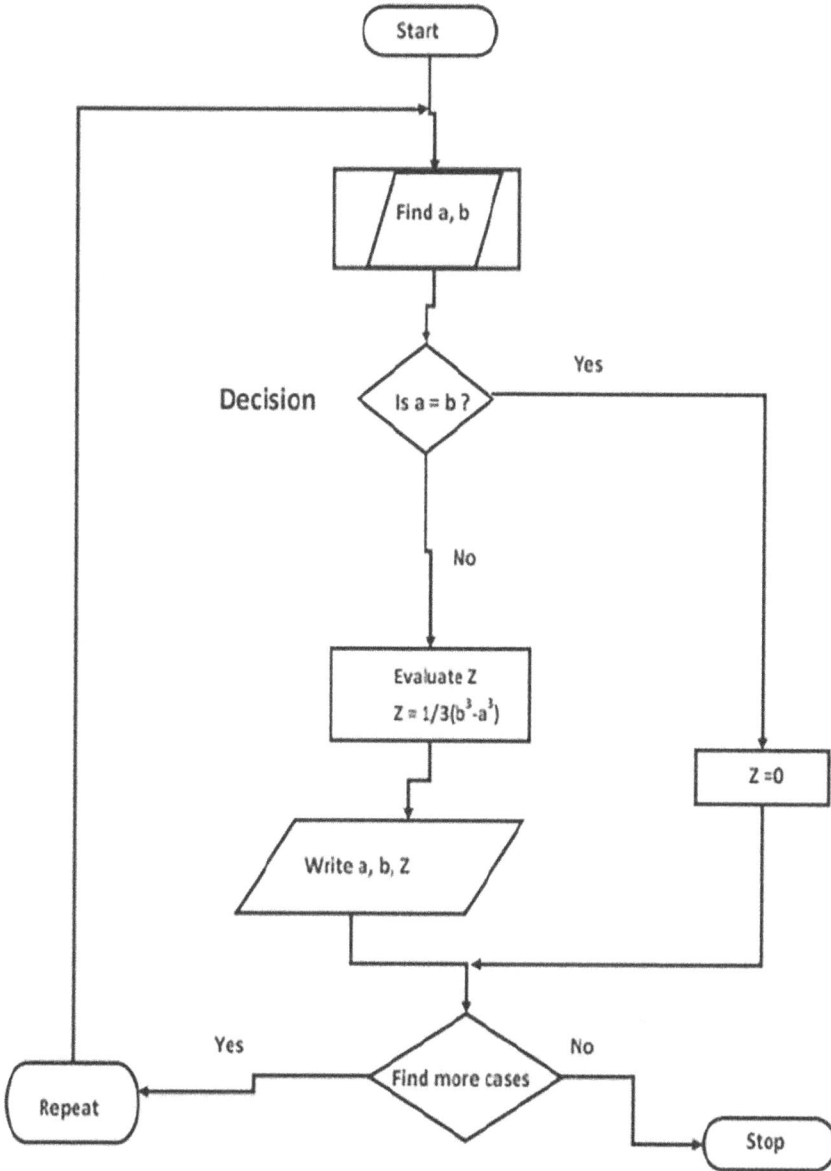

FIGURE 15.2   Computer program—flow chart.

At many computer installations, standard subroutines are provided for the solution of sets of simultaneous differential equations by a variety of numerical methods. A standard notation is usually adopted in which a dependent variable instead of being denoted by different letters is represented in the computer as the elements of a single array. A general subroutine usually confines itself to the integration part

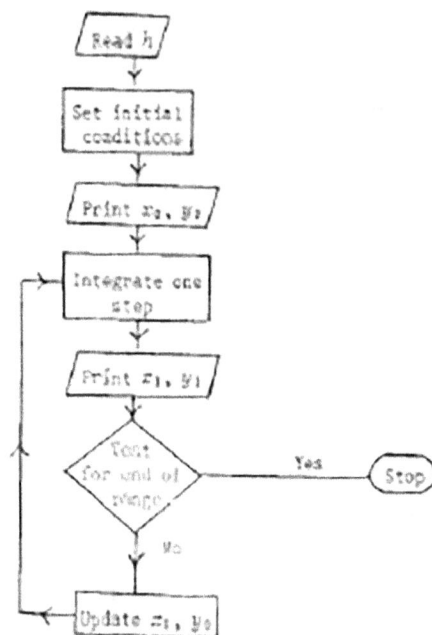

Program

```
C  EULER'S METHOD
C
      READ, H
C
C     SET INITIAL CONDITIONS
          X0 = 0.0
          Y0 = 1.0
C
      PRINT, X0, Y0
C
C     INTEGRATION PROCESS
C                  (DERIVATIVE DENOTED F0)
   10     F0 = X0 + Y0
          X1 = X0 + H
          Y1 = Y0 + H*F0
C
      PRINT, X1, Y1
C
      IF (X1 .GT. 0.9999) STOP
C
C     UPDATE X0, Y0 AND REPEAT INTEGRATION
          X0 = X1
          Y0 = Y1
          GO TO 10
C
      END
```

FIGURE 15.3   Numerical solution flow chart.

Flow diagram for main program

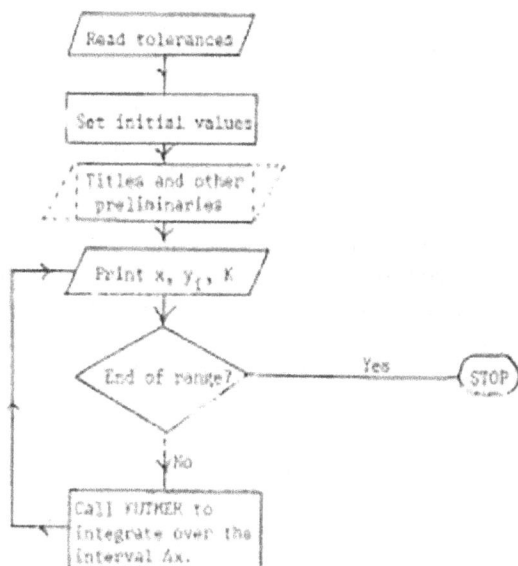

Main program

```
C   SOLUTION OF SIMULTANEOUS DIFFERENTIAL EQUATIONS BY KUTTA-MERSON
C
      DOUBLE PRECISION X,Y(2),HEST,TOL(2)
C
      READ(5,500)TOL(1)
          TOL(2) = TOL(1)
C
      X = 0.0D0
      Y(1) = 0.0D0
      Y(2) = 1.0D0
C
C     TITLES & PRELIMINARIES
          WRITE (6,610)TOL(1)
          HEST = 0.2 D0
          X = 0
C
   10 WRITE (6,620) X,Y(1),Y(2),K
C
      IF(X.GT.2.9999 D0) STOP
C
      CALL KUTMER(2,X,Y,0.2 D0,HEST,TOL,K)
      GOTO 10
C
  500 FORMAT(F10.0)
  610 FORMAT(1H1,'SOLUTION OF SIMULTANEOUS EQUATIONS BY KUTTA-MERSON'///
     1          1H ,'TOLERANCE =', 1PD8.1///
     2          8X,'X',13X,'U',14X,'V',40X,'K'/)
  620 FORMAT(1H ,F10.3,2F15.5,34X,I4)
      END
```

FIGURE 15.4   Flow diagram for the main program.

of the computation, with parameters defined by the user. Often, a step is necessary to achieve the desired accuracy and verification. Subroutines normally complete the calculation steps and compute more than one step before returning control and transferring to the main program. The interval over which the integration is performed is specified by the user. General differential equation subroutines cannot calculate the associated derivatives because the forms of the functions are different. A second auxiliary subroutine is subordinate to the general subroutine, which will compute the values of the required derivatives (Figure 15.4).

Special methods exist for numerical solutions of second-order and higher differential equations. However, it is always possible to express a higher-order equation as a set of simultaneous first-order equations, which can be solved by the associated methods described. This has the advantage that the same standard subroutine can be used for all cases.

If a program is required to solve a fairly general problem in which differential equations depend on physical parameters that differ from case to case, it may be desired to read in the values of these parameters as data.

When a standard subroutine I used, the precise form of the differential equations is represented by an auxiliary subroutine, and physical parameters referenced would be required in this subprogram. However, such data would normally be read in the main program, and a method is needed for communicating these values to the auxiliary subroutine when needed. The normal way of transferring data from the main program to a subroutine is as arguments of the subroutine.

16 Chemical Plant Design

Safe plant design involves the analysis of stress as it applies to the various unit operations structures. Stress, $\sigma = F/A$, which is the force per unit cross-sectional area, can be tensile or compressive. Tensile stress makes objects longer, while compressive stress makes them shorter. These two stresses work perpendicular to the plane considered. If a force acts parallel to the plane in consideration, then it is called a shear, τ. If a stress is applied to a component, it might cause deformation (Strain). Therefore, Strain is defined as deformation/original length:

$$\text{Strain } \varepsilon = \delta l \,/\, L \qquad \delta l = \text{Deformation}$$

Elastic behavior describes the material's ability to return to its original size after being stressed. The elastic modulus, E (Young's modulus) = Stress/Strain = σ/ε.

Beyond proportional limits, materials will not come back to their original size. Another small limit, which is called the elastic limit, is just beyond the proportional limit, followed by the plastic limit, where the material is completely changed. For example, the yield point for steel is of importance in design considerations. The maximum on the ductility curve is the ultimate tensile stress before the material breaks (engineering breaking stress point). Hence, the safety factor to produce a design stress is an empirical factor for yield stress divided by 1.5 (proof stress). This applies to ductile materials only. Conversely, in brittle materials, fracture occurs, whereby breaks happen suddenly without warning due to fatigue and failure. For example, cast iron should not be used in tension.

Materials' toughness is not an indication of materials' strength. The amount of energy needed to break a specimen is equivalent to the amount of force applied in a specified distance. Thus, malleable materials have the ability to be hammered, and it is possible to go back to the ductile region by heat treatment.

Thermal strain and stress might cause unrestricted deformation upon materials' exposure to heat. This is described as follows:

$$\delta l = L.\alpha.\Delta T \quad L = \text{Original length, } \alpha = \text{Coefficient of linear expansion,}$$
$$\Delta T = \text{Change in temperature}$$

The principle of superposition states that if several causes such as temperature and heat act simultaneously on a system, and each effect such as stress or strain at a given point is proportional to its cause, then the total effect is the sum of the individual effects considered separately. Lateral strain has always an opposite sign to axial strain, thus the ratio (Poisson's ratio):

$$\mu = \text{Lateral strain/Axial strain} = -(\delta w/W)/(\delta l/L)$$
$$\text{For steel,} \qquad \mu \text{ (elastic)} = 0.3 \qquad \mu \text{ (plastic)} = 0.5$$
$$\text{For metals,} \qquad 0.25 < \mu \text{ (elastic)} < 0.33$$

DOI: 10.1201/9781003384199-17

Similarly, volumetric strain, which is reflected by a change in volume, is described as follows:

$\Delta V = V_o \cdot \varepsilon_x (1-2\mu)$. In addition, strain energy is stored in a specimen when put under strain. For example, energy is stored in a spring of a car when the car is subjected to a shock. The energy of the shock is absorbed in the spring.

When a force acts parallel to the area in consideration, shear stress, $\tau = F/A$. A double shear might occur on two acting surfaces such that $\tau = F/2A$. For example, in a pulley block, the weakest point is the pins in shear, e.g., shear strength is 300 MN/m^2, and safety factor is 5, then the maximum allowable shear stress is 300/5 = 60 MN/m^2. The area resisting double shear is equal to 2 × pin cross section. In addition, shear strain, γ, is defined as the relative movement between planes/perpendicular distance between planes. Therefore, shear modulus (modulus of rigidity), G, is described as follows:

$$G \text{ (Elastic)} = \tau / \gamma = E \text{ (Young's modulus)} / 2(1+ \mu) \qquad \mu - \text{Poisson's ratio}$$

Materials creep (creep stress relaxation) occurs when strain increases and stress decreases due to continuous constant stress over time. In addition, the endurance limit decreases as temperature increases. If not sure in what direction are materials strained, Rosette strain gauges might be used.

Biaxial stress is a common occurrence for most structural parts, so strain gauge rosettes are routinely used for experimental stress analysis. There are two common configurations of strain gauge rosettes: Rectangular and delta. A strain gauge rosette is a term for an arrangement of two or more strain gauges that are positioned closely to measure strains along different directions. Strain gauges are used within load cells; the measurement of the strain and stress the load cell is under to determine weight and quantities. Strain gauges work by measuring the change in electrical resistance across a thin conductive foil. The gauge factor is the sensitivity of the strain gauge. It converts the change in resistance to the change in length. The strain gauge is one of the most important tools of the electrical measurement technique applied to the measurement of mechanical quantities. Strain gauges convert the applied force, pressure, torque, and so on into an electrical signal which can be measured. Force causes strain, which is then measured with the strain gauge by way of a change in electrical resistance.

The three types of strain gauge configurations, quarter-, half-, and full-bridge, are determined by the number of active elements in the Wheatstone bridge, the orientation of the strain gauges, and the type of strain being measured. Each strain gauge has its limitations in terms of temperature, fatigue, the amount of strain, and the measurement environment. These limitations must be examined before a strain gauge is used.

Metal strain gauges are also commonly used. They are typically a winding pattern of an etched metal wire on a flexible polyimide film. The copper–nickel alloy is among the most commonly used materials. Unlike semiconductor strain gauges, metal strain gauges change their resistance due to geometry changes.

17 Experimentation

Evaluating thermal characteristics of vertical unlagged plain against annular finned steel pipe while steam is condensing will show the importance of extended heat surface transfer effects to increased efficiency in plant operations. In addition, in shell and tube heat exchangers, thermal performance is determined using a Wilson plot to measure the overall fouling resistance of this exchanger while the heating medium is condensing steam on the shell side. Understanding fouling effects is significant to plant cleaning schedule and economic cost impact.

Simultaneous heat and mass transfer might occur during operations. This can be simulated and measured by determining heat and mass transfer coefficients and j-factors while evaporation is taking place from a water surface in a wind tunnel with heated air.

Heat pumps are economic systems for heating and cooling that can operate by system reversal to achieve the required temperature zones for space HVAC requirements (Figure 17.1). Refrigerant (Freon 12), R22, might be used as the working fluids and local conditioned water supply as the fluid for the evaporator and condenser. The pressure–enthalpy diagram is developed for a given refrigerant with optimum condenser and evaporator water flow rate.

Pneumatic conveying of solids in horizontal and vertical pipes requires using gas at certain operating velocities. The rate of solids transport is dependent on gas velocity and pressure drop across the distance of travel including pipe length and fittings effects on pressure. The size and density of solids are important factors in this calculation.

Filtration of solids from bulk solution or slurry can be accomplished by using a plate and frame filter press. Different size cartridge filters have been recently developed for commercial applications for sterilizing applications, e.g., filtering through a 0.2 μm pore size is considered necessary to generate a sterile filtrate. Solutions for filtration can range in volume from just a few milliliters to many liters. In addition, microfiltration and ultrafiltration media has been optimized to achieve protein separations and cold pasteurization.

Packed bed columns are largely employed for absorption, desorption, rectification, and direct heat transfer processes in the chemical and food industry, environmental protection, and processes in thermal power stations like water purification, flue gas heat utilization, and SO_2 removal. Packed beds are typically columns filled with a packing material that allows fluids to flow from one end to the other. Frequently used as a means of increasing contact between a liquid and gas. A packed column refers to a column that contains a fully packed stationary phase made up of fine particles. In contrast, a capillary column refers to a column whose stationary phase is coated on the inner surface. Plate column is used for relatively large diameters and many stages. A packed column is suitable for low-capacity operation. Plate column

DOI: 10.1201/9781003384199-18

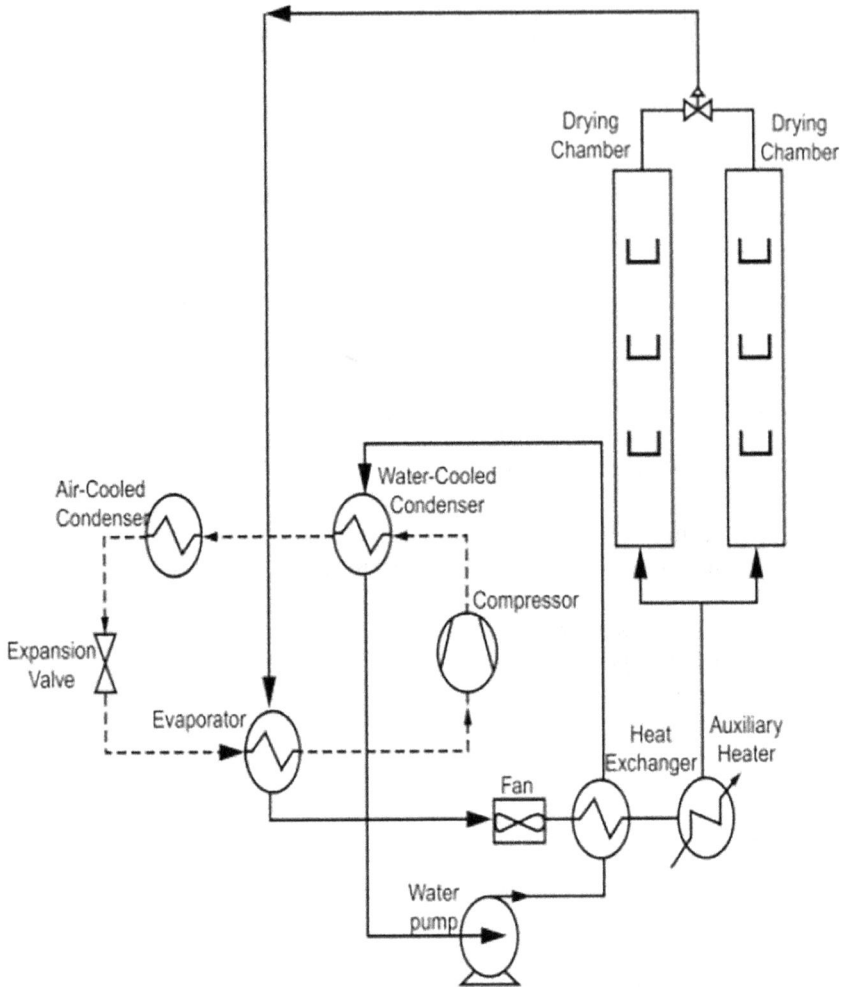

FIGURE 17.1   Heat pump process.

can handle a wider range of gas and liquid flow rates. Alumina of activity II or III, 150 mesh, is most employed. The techniques for packing a column described in the following use silica as the stationary phase but are equally suitable for use with alumina. Excellent mixing of reactants in the fluidized bed helps to minimize temperature variations and renders this system attractive for carrying out gas–solid reactions. The fluidized bed reactors show better conversion of reactants in comparison with packed beds when solid reactants are used in the form of small pellets. Packed towers offer a lower pressure drop. Tray towers are better at handling solids or other sources of fouling. Tray towers are better at handling lower liquid rates. Packed towers are better for corrosive liquids. There exist two methods to fill columns: Dry packing and slurry packing (filtration technique). The dry filling method is easy to perform, but the slurry method gives, especially for smaller particles, more efficient and reproducible columns. In standard packed absorption columns, a gas mixture travels up through a gas absorption tower and the solute is transferred to the liquid phase and thus gradually removed from the gas. The liquid accumulates solute on its way down through the column.

BATCH DISTILLATION AT CONSTANT REFLUX RATIO IN PACKED COLUMN

1. The reboiler is charged with a mixture that requires separation.
2. The reflux timer is set to desired reflux ratio.
3. Cooling tower water is supplied to the condenser at a suitable temperature.
4. Steam is provided to heating coils in the reboiler.
5. Record all temperatures, pressures, and flow rates for all streams.
6. Determine the boil-up rate and supercritical vapor velocity from a heat balance over the condenser.
7. Evaluate HTU and HETP for the packing.

18 Multi-Phase Flow

Multi-phase flow is complex in fluid dynamics. It involves defining geometry, three-dimensional, different flow regimes, and different physical properties of the combined fluids transport and moving in unit operations. The simplest possible case of two-phase flow is a bubble rise in a stagnant flow. The boiling situation upward starts with a single phase and enters the nucleation bubbling regime, followed by bubble growth to form a slug, and then slug formation and regime, nearly covering the cross section of the pipe. Slug breaking during churn and slug-annular flow are described through the retention of liquid film on the tube wall. As the gas velocity increases, shearing stress on film generates drop-annular flow, which is moving upward in mist evaporation and then total vapor.

Horizontal flow is more asymmetric, starting with bubbly flow to pseudoplug flow and then stratified flow, whereby an increase in gas velocity generates an increase in shear on the liquid phase accompanied by disturbance at the interface, which in turn generates wavy flow, then plug flow, and then annular stratified flow.

The analytic aspect of multi-phase flow includes computations for total mass flow rate, volumetric rate, void fraction, and total volume of gas in the flow regime including variations in physical properties, e.g., density ratio of phases term, surface tension, viscosities, and equipment equivalent diameter of tube and area of transport phenomena. This complex modeling defines the quality of two-phase relative to specific surface conditions. The homogenous equilibrium flow model assumes equilibrium of momentum, heat, and mass transfer based on the equilibrium between gas and liquid velocities, i.e., relative velocity, drift velocity, and drift flux, whereby the volumetric flux of a component relative to a surface is correlated at the average velocity.

For homogenous two-phase flow (one-dimensional equilibrium), the pressure drop can be calculated:

- Input variable, diameter, density of liquid and gas, viscosity of liquid and gas, and flow rates
- Calculate effective viscosity and Re
- Find friction factor from single-phase charts
- Calculate effective density
- Calculate pressure drop

When the condition is separated flow, steady flow of the phases is considered together but with differing velocities. Simultaneous equations describe continuity, mass flux, momentum, and energy balances. Martinelli correlation method might be used to define friction factors for all liquid flow. The assumption is that we know how to

DOI: 10.1201/9781003384199-19

calculate the pressure gradient if either fluid alone were flowing in a pipe. For example, the laminar representation is a simplified model of separated flow that was able to correlate Lockhart–Martinelli results. This model is an analytical approach to the predictions of separated flow. The Martinelli charts enable us to calculate pressure drop and voidage. The Martinelli correlation can be quite easy to predict stratified, annular, slug, and separated flow regimes.

19 Bio-Chemical Engineering

CATALYTIC ENZYMES

Immobilized enzymes for industrial reactors are directly correlated to substrate concentration in catalyzing and enhancing biochemical reactions. Competitive inhibition and non-competitive inhibition are key factors that affect the rate at which enzymes breakdown. Typically, the inhibitor attaches to enzyme sites forming unreactive conditions. Both a combination of a reduction in the reaction rate and a decrease in the active concentration of enzyme exacerbate unreactive conditions.

Inhibitor effectiveness is the reciprocal of affinity, where reversible inhibition is characterized by an equilibrium between enzyme availability and inhibitor concentration. This equilibrium is a measure of the affinity of the enzyme for the inhibitor. Irreversible inhibition is characterized by a progressive increase with time.

FERMENTATION

Fermentation is a metabolic process that produces chemical changes in organic substrates through the action of enzymes. In biochemistry, it is narrowly defined as the extraction of energy from carbohydrates in the absence of oxygen. Fermentation is a process in which an organism converts a carbohydrate, such as starch or sugar, into an alcohol or an acid. For example, yeast performs fermentation to obtain energy by converting sugar into alcohol. Bacteria perform fermentation, converting carbohydrates into lactic acid—the process of sugars being broken down by enzymes of microorganisms in the absence of oxygen. Microorganisms such as bacteria and fungi have unique sets of metabolic genes, allowing them to produce enzymes to break down distinct types of sugar metabolites.

The main function of fermentation is to convert NADH, a chemical compound found in all living cells, back into the coenzyme NAD+ so that it can be used again. This process, known as glycolysis, breaks down glucose from enzymes, releasing energy. Based on the end product formed, fermentation can be classified into four types, namely, lactic acid fermentation, alcohol fermentation, acetic acid fermentation, and butyric acid fermentation, following glycolysis in the absence of oxygen. Alcoholic fermentation produces ethanol, carbon dioxide, and NAD+. Lactic acid fermentation produces lactic acid (lactate) and NAD+.

Fermentation helps break down nutrients in food, making them easier to digest than their unfermented counterparts. For example, lactose—the natural sugar in milk—is broken down during fermentation into simpler sugars—glucose and

DOI: 10.1201/9781003384199-20

galactose. Fermentation begins with glycolysis which breaks down glucose into two pyruvate molecules and produces two ATP (net) and two NADH. Fermentation allows glucose to be continuously broken down to make ATP due to the recycling of NADH to NAD+. Fermentation reactions occur in the cytoplasm of both prokaryotic and eukaryotic cells. In the absence of oxygen, pyruvate does not enter the mitochondria in eukaryotic cells.

20 Industrial Inks, Dyes, and Pigments

COMPOSITION OF PRINTING INKS

Printing inks are composed of pigments, extenders, synthetic and natural resins, waxes, liquid minerals, and drying oils and solvents. Many of these compounds are complex organic syntheses. Accurate formulation selections and balancing based on physicochemical properties are important in final printing quality.

Organic pigments are purposed for coloring (DIN 55944). Compared to inorganic pigments, they are more intensive, brighter, and more transparent. Nonetheless, they are more sensitive to sunlight, chemical degradation, and resistance to heat. In addition, they are more costly than inorganic pigments. The color index number refers to the chemical composition of a pigment group. These important differences affect manufacturing processes, coatings, particle size distribution, particle hardness, color intensity, and hue.

In pigment selections, considerations such as alkali and soap resistance to washing agents are important. Most yellow pigments are of the azo family. Most organic pigments are also from the azo group including red. The basis reaction for the synthesis of azo dyes and pigments is the reaction of primary aromatic amines with nitric acid forming a diazonium salt. Natrium salt of the azo dye is made from the solution of the diazonium salt coupled in an alkali medium. The main coupling components are aniline, naphthol, acetyl acetate anilide, and pyrazalone. Other pigments such as benzidine yellow and orange are azo dyes of pyrazalone coupling. Red pigment permanent carmine, a mono-azo pigment, is a result of naphthol coupling. Violet pigment is realized by using chin acridone and dioxazine pigments. Blue and green derivatives are generated from complex copper phthalocyanines. Green pigments are chlorinated copper phthalocyanines. Black inks are primarily composed of carbon black pigment.

A basic formulation of the offset lithographic vehicle consists of hard resin (40%), drying oils or alkyd resin (30%), and mineral oils (B.P. 260–310°C) (30%). The resin macromolecule allows for the penetration of mineral oil into the substrate to build the required printing film. Synthetic resins are made by polycondensation of organic esters. Another natural resin, rosin, is obtained from coniferous and is 90% abietic acid. Resin color, iodine number, acid number, and solution viscosity are all measures to characterize a resin material. In addition, waxes are used to improve the rub resistance and slipping effects of the final printing product.

DOI: 10.1201/9781003384199-21

Inks that dry by oxidative polymerization need driers to promote the oxidation process. Driers are composed of organic salts of Mn, Ca, and so on. Certain metals catalyze oxidation and are added in combination with driers.

Most printing inks are grinded on three roll mills. To produce inks with different rheological properties, mills will have different rolls. For large quantities of ink, Perl mills are used to account for time-consuming washing with greater heat generation and uncoated pigments (hard) dispersion.

In order to address suspected health, safety, and environmental concerns posed by the use of heavy metal-based pigments, companies have reformulated many standard color and concentrate products using synthetic organic materials. New materials are spurred, in part, by suspected toxicity problems with lead, chromium, and cadmium, in addition to potential waste disposal issues. Concerns have been raised that disposal of pigmented plastic products in landfills is a possible source of groundwater contamination, and that incineration of these products may release hazardous matter into the atmosphere or generate toxic residue. The obvious answer was to replace these potentially hazardous materials with organic substances considered to be less harmful, but that solution was more easily stated than accomplished. The heavy metal-based pigment, lead chromate, for example, is a standard pigment in green, yellow, and brown concentrates. But replacing lead chromate with synthetic organic pigments includes the following: Azo condensation or benzidine yellow may yield a loss of opacity, although an accompanying gain in tinting strength will occur. Red concentrates posed an even greater challenge. Lead molybdate, the heavy-metal pigment often used in red formulas, has a blue-shade undertone and a yellow-shade surface appearance. The difficulty in substituting synthetic organic pigments for this heavy metal lies in duplicating shades and maintaining opacity. Shade duplication is necessary for many blow-molded household chemicals/detergent and automotive oil bottle customers whose products have very exact brand-identifiable colors.

Generally, higher pigment loadings are necessary with non-heavy metal-based concentrates to achieve the same letdown ratio and opacity. Color matching remains a compromise between the customer's needs and the price of incorporating higher levels of a more expensive non-heavy metal system. Three factors must be considered when making these substitutions. The first consideration deals with how close the original color is to the new color. This color differentiation is made using a color spectrophotometer. Normal commercial matches will yield color differences of less than 2.0 Cie Lab units. A Cie Lab unit is the standard measurement of difference in brightness, yellow-blue and red-green colors. The second consideration deals with differences in opacity on a percentage basis due to the poor refractive indices of organic pigments compared to their heavy metal counterparts. The third consideration deals with the inherent higher costs of organic pigment compounds versus their heavy-metal counterparts which will influence end-product costs. There are many custom-formulated products on the market currently based on heavy-metal pigment systems. Synthetic organic pigments include the following (Table 20.1):

> Azo condensation or benzidine yellow may yield a loss of opacity, although an accompanying gain in tinting strength will occur. Red concentrates posed an even greater challenge. Lead molybdate, the heavy-metal pigment often used in red formulas, has a blue-shade undertone and a yellow-shade surface appearance. The difficulty in

TABLE 20.1
Synthetic Organic Pigments

Color	Formula	Non-Heavy Metal Equivalent	Difference Cost
Green	CM 45303	CM 45154	+50%
Yellow	CM 56046	CM 56483	+150%
Red	CM 77410	CM 74051	+25%
Red	CM 77685	CM 74052	+60%
Orange	CM 77408	CM 74042	+175%
Orange	CM 77306	CM 74068	+300%

substituting synthetic organic pigments for this heavy metal lies in duplicating shades and maintaining opacity. Shade duplication is necessary.

Many companies discontinued marketing color concentrates containing cadmium-, chromium-, and lead-based pigments used to color plastic products. To provide customers with adequate lead time to make the transition to alternate color concentrates or alternate suppliers, these heavy metal-based pigments were replaced with organic-based substitutes. This initiative is being taken to lessen the environmental impact that plastic products have on the incineration of municipal solid waste and on the land disposal of the resulting ash. The United States is moving toward an integrated waste management system. This process includes the reduction of the volume and toxicity of solid waste where feasible and practical, materials recycling, solid waste incineration with energy recovery, and landfilling.

Heavy metals in printing inks in the colors white, light blue, dark blue, and red are dependent on the determination of Pb, Se, Sb, Hg, Cd, As, and Ba.

PREPARATION OF THE SPECIMENS

DISSOLUTION

Printing ink specimens (about 2 g) are mixed with about 40 mL of nitric acid and left overnight at room temperature. Then they are heated for 1 hour at 40°C and then for further 7 hours at 60°C water bath temperature. Undissolved solids were separated by centrifugation, followed by filtration through a filter paper MN 640d (Macherey-Nagel), and the clean solution was diluted to 50 mL with bi-distilled water.

WATER EXTRACTION FOR THE BARIUM DETERMINATION

Printing ink specimens (about 1–2 g) were mixed with 15 mL of bi-distilled water and intensively shaken overnight at room temperature. The water phase was then separated and analyzed. Heavy metals in the printing ink:

1. The determination of heavy metals included X-ray spectrometry (Se)
2. Flame-AAS (Pb, Cd, Sb), cold vapor (Hg), and hydride-technic (As)

TABLE 20.2
Detection Limits of Heavy Metals

Metals	Detectability Limit (ppm)
Pb	10
Se	10
Sb	20
Hg	0.1
Cd	1
As	4

The detectability limit for the different metals, related to the mass of the specimens, is mentioned in Table 20.2.

Colorant producers, compounders, and their end-use customers withstand the worst of the difficulties involved in switching to alternative heavy-metal-free (HMF) colorants. Complex compromises are necessary to balance lightfastness, heat stability, color, and cost in HMF alternatives. For instance, colorant suppliers say processors of high-performance engineering thermoplastics must exercise greater control than ever before over key processing parameters. Some HMF alternatives are more sensitive to temperatures and pressures and, subsequently, are trickier to run. Some HMF colorants are more difficult to disperse, causing compounders difficulty in concentrate production. And some HMF alternatives lack the thermal conductivity of their heavy-metal predecessors and that means longer cycle times. Perhaps more importantly, processors of commodity resins-film producers in particular must be prepared to shell out the cost upcharge materials containing HMF colorants pose. In a sector of the processing community historically driven by cost-competitive pressures of every type, being forced to factor in the higher cost of some HMF colorants is a painful reality. Cadmium pigment substitutes have been sought for the last 40 years because of the high cost of cadmium. Household chemical consumer marketers: Procter & Gamble and Lever Brothers are requiring containers that were heavy-metal free.

Regulatory requirement mandates yearly inventory and emission reporting to document releases to the air, water, and land of substances considered toxic. Among the several hundred chemicals listed, are broad categories of compounds made from metals that include antimony, barium, chromium, cadmium, copper, lead, nickel, and zinc. Such metals are basic to core inorganic pigments used to color plastics. They are also basic to UV inhibitors, flame retardants, heat stabilizers, and other additives. Reporting and labeling requirements for an ever-lengthening list of materials considered carcinogenic and toxic are also spelled out in Safe Drinking Water and Toxic Enforcement Act, also known as Prop 65. Any vendor selling in California must identify products containing ingredients on the lists. This includes pigments such as cadmium compounds, lead chromate yellows and molybdenum oranges, and the organic yellow pigment family of diarylides if they contain benzidine or its salts.

Other states have established similar Right-to-Know legislation involving material ingredients.

Regulators wanted to ban colorants with more than 100 parts per million of four types of inorganic ingredients—cadmium compounds, lead, mercury, and hexavalent chromium—from any packaging that comes in contact with food. Cadmium pigments are virtually insoluble and are not readily absorbed into the body. Neither the Environmental Protection Agency (EPA) nor the International Agency for Research on Cancer has classified cadmium as a known human carcinogen. Because of cadmium's high heat stability, plastics colored with cadmium pigments or concentrates containing them can be recycled many times without loss of color. Cadmium pigments are primarily used in durable goods made from engineering thermoplastics, not in disposable packaging made from commodity thermoplastics. However, during incineration, bottom and fly ash disposal of cadmium pigments that may be converted to more soluble forms remains a concern (Table 20.3).

Molding re-sterilizable surgical handles in different colors for the medical market is used as a coding system. The different color handles quickly alert surgeons and their assistants as to which knife is for which purpose. The only way to dispose of many medical plastic parts, like the handles, is through incineration. Handles are molded at temperatures around 700°F in FDA-approved polysulfone. To find an HMF alternative, the cadmium-free orange alternate turned brown during processing, and to get green, a yellow cadmium pigment is added into a blue, so heavy metal Cd was used. Yellow organic pigments were touted as a safe alternative to inorganics, but it was found that when processed at high temperatures, diarylide pigments release a known carcinogen, dichloro-benzidine.

Alternative HMF should possess non-bleeding characteristics, be chemically inert, and be relatively inexpensive. There is no single group of HMF alternatives satisfying all of these characteristics. The alternatives are mixed-metal oxides, containing, for example, nickel, titanates and antimony, and organic pigments. Individual HMF alternatives can be preblended to suit specific applications. Albeit mixed-metals oxides might be viable replacements for duller colors; organics are often brighter.

TABLE 20.3
Colorants

Color	Pigment
Red	Perylene, quinacridone
Yellow/orange	Diarylides, disazo, pyrozolone, anthrones, hlansas, iron oxide
Green	Phthalic, chrome oxide
Blue	Indanthrone, iron, ultramarine
Violet	Carbazole, ultramarine, quinacridone
Brown	Iron oxide, ferrite
Block	Carbon black, iron oxide
White	Titanium dioxide, barium sulfate

Equipment enables the detections of materials in the parts-per-million and parts-per-billion range. Analysis of even HMF alternatives may show up trace amounts of lead or cadmium. Inorganic pigments provide better thermal conductivity compared with organic pigments with increased cycle times. Slower cycles mean higher product and manufacturing costs. Add to that the fact that the purchase price of many HMF alternatives is higher than their heavy-metal predecessors. Also, HMF organics are said by some to narrow the processing window for engineering thermoplastics. Some alternatives and alternative mixtures cannot be processed at temperatures exceeding 550°F. More critical attention has to be paid to key processing variables such as injection pressure and backpressure, shear, and processing temperatures. More sophisticated process controls and monitors may be required. Overall, sources agree that HMF colorants can influence mold design and machinery selection. For example, with organic pigments, processors should be prepared for color shifts, owing to the solubility and temperature sensitivity of the materials. If a large part is molded at one location on a 750-ton injection machine, and a small part destined for assembly on the same product is molded with the same resin and HMF color concentrate on a 50 tonner, the colors may not match.

Materials like polysulfone and polyphenylene sulfide run at high temperatures. These types of engineering thermoplastics operate under parameter monitoring and process control conditions as a matter of course. Steady-state conditions are important to run materials such as HMF colorants. Finely tuned fill, backpressure, and temperatures are key process parameters; nonetheless, mold gating or flow channels may have to be redesigned to eliminate shear heat. Warpage may also be a problem caused by these ingredients. For critical appearance applications, the end-product OEM and colorant supplier will have to be more directly involved with manufacturing. Organic alternatives require higher loadings. Some migrate at higher temperatures. Suppliers must have as much information on a particular run in a particular processing system as possible. Or else, imagine an improperly specified colorant migrating in a drying hopper or on film rollers. Heat stability will not be as important a consideration as cost and availability.

FLUORESCENT PIGMENTS

A wide variety of special-effects colorants are available for providing different visual effects in the finished product. Fluorescent pigments emit light to produce brilliant colors, and pearlescent pigments produce the deep luster of a pearl. Aluminum flakes produce a satiny silver luster, and bronze powders produce a whole range of gold colors. Daylight-fluorescent pigments are solid solutions of fluorescent dyes in synthesized polymeric resins. The carrier resins used in the manufacture of fluorescent pigments were originally condensation products of melamine, formaldehyde, and toluene sulfonamide. Depending on the molar ratios used, a glass-like thermoplastic or thermoset carrier resin would result. These resins had thermal stabilities to a maximum of 400°F (200°C), which was acceptable for LDPE or PVC plastisol applications but was obviously unacceptable for higher-temperature plastics or even in processes around 400°F where a high percentage of regrind was used. A carrier resin based on a complex thermoplastic polyamide system was invented that gave acceptable heat stability in molding conditions up to 600°F (305°C). This greatly increased the marketability of fluorescent pigments to the plastics trade. A carrier resin based on a thermoplastic polyester system, while not as stable as the polyamide system, was nevertheless stable to approximately 550°F (285°C). Fluorescent pigments for plastics must possess several definite properties: They must be transparent, colorless, friable, and brittle enough to be mechanically ground to a fine particle size, compatible with fluorescent dyes, and sufficiently heat stable and chemically stable to withstand the molding conditions of thermoplastic and thermoset resins. A high-melt thermoplastic pigment based on the polymerization of aliphatic di-isocyanates with melamine and sulfonamide was developed in Germany. It displayed some processing advantages but was withdrawn from the marketplace due to high cost. The final classification is a thermoset resin synthesized by the emulsion polymerization of benzoguanamine, melamine, and formaldehyde. Nonfluorescent colors are reflective or subtractive in nature. White light contains all the hues of the rainbow, from reds through oranges, yellows, and greens, to blues and violets. When this white light strikes a nonfluorescent orange surface, the orange light is reflected, and the other hues—the violets, greens, yellows, and reds—are absorbed or subtracted from the incident light. This absorbed light is usually dissipated as heat. When the same white light strikes a fluorescent orange surface, the same orange light is reflected as with the non-fluorescent pane. The shorter wavelengths of light—the violets, blues, and yellows—are still absorbed. However, instead of being dissipated as heat, the fluorescent dye converts these shorter wavelengths of light into orange hues. This emitted light, when combined with the normal reflected light, gives the increased intensity of color that is observed with a fluorescent orange. Fluorescent pigments are available in a full range of colors, from magenta to pink, red, orange, yellow, green, and blue. The red shades are gained by the use of Rhodamine dyes—red shade and blue shade. Yellows are obtained with dyes from the methine or amino naphthylamide families. Oranges employ blends of these reds and yellows. Since there are no truly fluorescent green or blue dyes, phthalocyanine green is combined with a fluorescent yellow methine dye to make fluorescent green pigments and optical brighteners (coumarin type) with the phthalocyanine blue to make fluorescent blues. It should be noted that while these greens and blues are cleaner and brighter than conventional pigments, they do not have the high luminosity obtainable in the reds, oranges, and yellows. The dye content of the fluorescent pigments used in plastics applications is generally in the range of 2.5–7.5%. There are two important factors that make it necessary to hold the dye content to such relatively low levels. First, the pigment carrier resin will dissolve a finite percentage of dye. Additional dye over this limit results in bleed and migration problems. Second, higher dye loading results in a phenomenon known as quenching. There is an optimum level of dye concentration at which the available white light is absorbed and emitted in an excited, fluorescent state. When the concentration is increased, the dye molecules internally absorb the available energy, and the result is an actual decrease in the fluorescent-light emission. This second factor explains why fluorescent pigments are bright luminescent powders, whereas fluorescent dyes appear as dull and drab powders at full concentration and exhibit the bright fluorescent effect only when in a dissolved and diluted form. Since the dye content of fluorescent pigments is relatively low, the effect of quenching is seldom a problem. Usually, the reality of economics will hold the pigment concentration to the minimum level required to give the desired aesthetic effect. It is well known that fluorescent colorants are less light stable than conventional pigments. This is due to the sensitivity of the fluorescent dye molecule to UV radiation. Thus, great care must be taken in using fluorescent colorants in outdoor applications. Lightfastness can be improved in two ways.

- Combination with nonfluorescent pigments of similar hue results in an apparent increase in lightfastness. However, there is a loss of brightness because the nonfluorescent pigment absorbs part of the white-light energy normally emitted by the fluorescent molecule. This brightness loss has to be balanced against the color retention desired by the molder.
- The addition of the UV absorber gives increased stability without affecting the daylight fluorescence of the color. Recommended types are benzophenones and benzotriazoles. However, the user again must weigh the benefit of this increased light stability against the additional cost.

APPLICATION AREAS

The major market for fluorescent colorants has been in thermoplastics. (The only thermosetting polymers using an appreciable quantity of fluorescents are polyesters and polyurethanes.) Polyethylene and PVC have the largest share, while the polypropylene market is growing steadily. The remaining important resins are general-purpose and impact polystyrene. Due to the transparent nature of the fluorescent colorants, the color achieved in any one particular plastic polymer will depend on the natural color and transparency of the plastic. The best transparent fluorescent color effects are obtained with general-purpose polystyrenes and acrylics, whereas some grades of cellulose acetates and propionates that have very slight blue or yellow hues have up to 50% lower reflection or "edge glow."

High-impact polystyrene and natural ABS give more pastel shades than polyolefin resin with equivalent loading. Therefore, to arrive at a color match for a 1% orange in HOPE, for example, a pigment loading of at least 1.5 % orange is necessary for high-impact polystyrene. It has been estimated that 90% of the direct sales of fluorescent colorants are to the custom color supplier, while the remainder is supplied to the molder who performs his own in plant resin coloring. A further breakdown of the custom color business suggests that 85–90% of fluorescent purchases are processed into color concentrates by single- or twin-screw extrusion, Banbury, or two-roll-mill methods. The remaining 10–15% goes into dry-color or unit-bag formulations.

The largest sales area for fluorescents in plastics is the toy and recreation market, followed by safety applications and packaging. The research and development effort on fluorescent pigments is a response to the requests for improved processing properties by molders and color-concentrate manufacturers. This has resulted in the development of heat-stable fluorescent orange pigments for Surlyn-coated golf balls and in ongoing programs to improve the blow molding processibilities for toy and bottle

TABLE 20.4
Solid Solution of Fluorescent Dyes in a Polyamide Resin

Specific Gravity	1.1938 @ 24°C
Average particle size	10–15 μm
Softening point	130°C (266°F)
Decomposition point	250°C (482°F)

applications, where the bright eye-catching effects of fluorescent pigments increase the salability of the molded item (Table 20.4).

DISPERSION

Fluorescent pigments differ from conventional organic pigments in that they represent a dye solution in a proprietary resin matrix. The resin matrix allows the dye to develop its fluorescent effect and be compatible with thermoplastic resins at the same time.

While conventional pigments disperse best at a high shear force, i.e., at a low resin temperature, fluorescent pigments disperse best at a resin temperature above 400°F, at which the matrix melts into the resin. The recommended optimum running temperatures to maintain excellent heat stability and obtain good dispersions are as follows:

400–450°F (205–232°C) in Polyethylene
425–450°F (218–232°C) in Polystyrene

HEAT STABILITY

The molder should test for heat stability under actual operating conditions to determine the suitability of the fluorescent pigments if temperatures are above 470°F (243°C), if some scrap is to be recycled, or if extended hold times are required.

ADDITIVES

It has been found that metal stearates have a negative effect on the color stability of fluorescent pigments. If stearate must be used, it is recommended to use calcium stearate. Polyethylene waxes and proprietary polymeric lubricants that do not contain stearate or silicone have been found to be satisfactory in polyolefins.

LIGHTFASTNESS

The lightfastness of the fluorescent pigments, while quite good, is slightly less than that of standard conventional pigments. The user should conduct tests to determine if the article colored with fluorescent pigments will meet specific outdoor lightfastness requirements.

PLATE-OUT

As with some other organic pigments, fluorescent pigments will exhibit some plate-out during molding. However, the high strength of the pigment enables lower loadings, which particularly allows suitable for products that typically have long production runs, such as blow molded bottles, closures, and some toys.

TOXICITY

Toxicity tests have been conducted on fluorescent pigments with the following results (Table 20.5).

Fluorescent pigments are compatible with the most common resins such as polyethylene, polypropylene, general-purpose polystyrene, high-impact polystyrene, and ABS. Bright fluorescent colors can be achieved with pigment loadings of 1% or less depending on the thickness of a part, type of resin, and light stability requirements. Higher loadings of pigment may be necessary for opaque resins such as natural ABS and for blown or extruded film where 4–5% pigment may be required.

UV CURING OFFSET INKS

UV curing offset inks have relatively good printing properties and instantaneous drying at the high running speeds of modern presses. These inks will only dry under the influence of UV energy, they have exceptional roller and duct stability and can be left on the press without the need for frequent wash-ups.

UV curing overprint varnishes are designed to give a high gloss protective coating that can be applied by a variety of on-press techniques. Wet litho, dry-offset and in-line coating versions are available, and they all benefit from the advantages of UV-instant cure, exceptional press stability, and quick turn-round of printed sheets.

Low-viscosity dry offset varnish is designed for general-purpose applications, where exceptionally high gloss and exceptionally good slip are important. These properties are generally maintained when applied over both UV and conventional

TABLE 20.5
Fluorescent Pigments Toxicity

Acute Oral Toxicity, LD50 g/kg	15.38
Acute dermal toxicity, LD50 g/kg	3.00
Acute dust inhalation, LC50 mg/L air 4 hours	2.88
Eye irritation	Mildly irritating

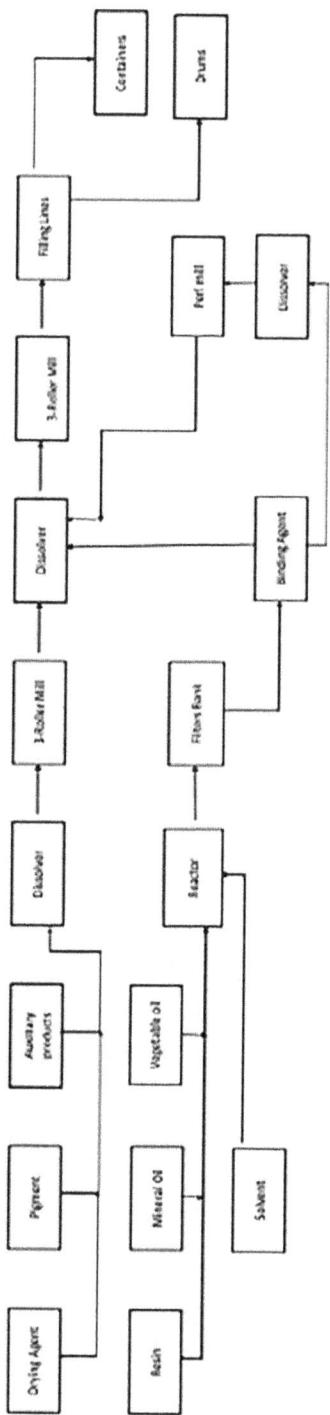

FIGURE 20.1 Printing inks—manufacturing process.

inks, but some loss of gloss may be expected when printing wet-on-wet over conventional inks. The low viscosity is necessary for rapid flow out to achieve maximum gloss.

INK MANUFACTURING

The role played by binding agents in ink manufacturing is of the utmost importance. For determining the properties of the ink, binding agents are thus produced under very stringent conditions. The raw materials—resins, vegetable oils, mineral oils, and additives—that have successfully completed quality control are processed using pre-defined production methods. The production is monitored by means of a temperature recorder and timer which are connected to the reactors. The rheological data are controlled closely. The entire reaction process can be followed on measuring instruments. After successfully completing quality control, the binding agents are pumped through filters into the storage tanks. Only reliable ink formulations are used, and these are constantly controlled throughout the production process. The constituents scheduled to be used in manufacturing the ink are weighed and checked against the computerized formulation to ensure that production requirements are respected. They are then pre-dispersed in a dissolver before undergoing final dispersion in a three-roller mil unit. The Perl mill represents the most modern method for producing large quantities of ink but requires optimal formulation and mechanical control. Each formulation includes exact production guidelines and the departments monitoring production must ensure that these guidelines are carefully followed, e.g., the rheological requirements. The term rheological data are used to designate the tack, viscosity, and surface tension of the ink, i.e., its consistency. Other properties are also controlled, such as fineness, shade, coloring strength, transparency, drying, and so on. Only when all these data have been successfully checked against the pre-defined threshold values, will the ink be cleared for filling. Offset and letterpress inks are filled automatically into 1- and 2.5-kg vacuum-sealed tins.

Unlike the oily offset and letterpress inks, flexographic and gravure inks offer considerable fluidity. Here, too, Perl mills are used and allow us to achieve an optimal degree of pigment dispersion. However, three-roller mills also come into filling into storage tanks or containers or whether they must be further adjusted.

The fineness of the ink is checked through a microscope. T shade, color strength, and transparency are checked against the standard sample. The tack level is determined with a tack-o-Scope, and viscosity and surface profile are measured by means of a drop-bar viscosimeter and telesurvey. Ink drying is monitored against the known characteristics of ink series (Figure 20.1).

21 Industrial Plastics

PLASTIC MATERIALS

Plastics are classified into thermosets and thermoplastics. Thermoplastics can be homopolymers, copolymers, or blends. For example, polyethylene is produced in a low-pressure process to provide a high-density PE (0.948–0.965). The high-pressure process provides for low-density PE (0.915–0.925). Stabilizers against heat and UV light exposure and antioxidants might be added to the master batch. Similarly, anti-static, anti-slip, antifog, impact modifiers, and colorants are all used for their designated properties. In addition, polypropylene has excellent stress crack resistance and water barrier properties. Polystyrene is adapted for impact using impact modifiers. Its impact resistance is determined by the impact modifier characteristics. It is easily foamed for insulation, cups and dishes, containers, and packaging. PVC, ABS, polycarbonates, and polyamides (nylon) have been universally used as well in several applications for residential and industrial use.

Processing of these granulated polymers to form into required shape, sheet, bottle, or custom molded is accomplished through

1. In-mold softening
   a. Compression molding
   b. Rotational molding
2. Extrusion molding
   a. Injection molding
      i. Parison injection into form
      ii. Injection blow molding
   b. Extrusion
      i. Profiles
      ii. Sheets—vacuum forming/thermoforming
      iii. Filament
      iv. Coating
      v. Film blowing
      vi. Blow molding
3. Hot melt
   a. Lamination
   b. Adhesion

DOI: 10.1201/9781003384199-22

22 Coating

High-quality printing of packs is taken for granted. Converting operations, such as folding box gluing or weighing and filling, and subsequent transport impose often extreme stresses. For this reason, most packs are varnished or coated after printing to add protection and increase sales effectiveness by higher gloss. Improved protection of a pack's contents is also provided by coating. Converters realized the advantage of inline coating and developed special equipment for sheetfed offset machines to enable one-pass printing and coating of sheets. The coating methods described here allow the application of a water-based dispersion coating to the freshly printed sheet in the same pass through the machine. These methods, however, do not produce high gloss surfaces as can be obtained using nitro varnishes. The thickness of the coatings has a direct effect on the gloss, which in turn also depends on the material being coated, the inks printed on the sheet, the type of coating varnish, and the drying properties. The weight of the pile being formed in the delivery and its effect on individual sheets limit the thickness of the coating.

The coating fountain roller unit has its own drive as standard. This enables stepless dosage control of coating volume. Coating fountain roller and coating metering roller continue to rotate during stoppages. The diagonal setting of rollers offers an additional means of dosage control and permits a differentiated coating varnish supply. Coating requires an additional doctor device for lateral dosage control and adjustment to the format. Coating application is direct. Coating may be applied across the whole of the sheet or to the pattern. The coating device replaces the last lower printing unit in multi-color five-cylinder system machines. The coating device has three rollers:

- Coating fountain roller. Centered and held firmly in the side walls. It carries coating liquid from the fountain to the coating device.
- Coating metering roller. With adjustable eccentric bearings, it controls the coating dosage or constancy of feed between the dip roller and the applicator roller.
- Coating form roller. With adjustable eccentric bearings, it transfers the coating to the stencil cylinder. The stencil cylinder takes the place of the blanket cylinder and is linked directly to the impression throw-on and throw-off control it carries a coating blanket.

CYLINDER AND ROLLER SETTING

The stencil cylinder is set to the impression cylinder and in the same manner as a blanket cylinder is set to suit the material being handled. Applicator setting to the stencil cylinder and the coating fountain roller may be made while the machine is running. Set values are maintained even when resetting the stencil cylinder. The

DOI: 10.1201/9781003384199-23

coating metering roller may be set in relation to the coating fountain roller while the machine is running. The coating form roller with pneumatic throw-off is coupled to the electric impression throw-on and throw-off control and may also be independently controlled. The coating device should conform to the latest safety regulations of the industrial safety inspectorate.

The separate coater is installed between the last printing unit and the delivery. It has a varnish applicator device and its own impression and form cylinders. Both cylinders are integrated into the tooth track of the machine. The form cylinder is coupled to the on-and-off control of the machine and is designed to take both a blanket to print solids or a relief plate for pattern varnishing. Makeready to the impression cylinder is governed by the requirements of the material being coated, as in printing unit makeready. The applicator device has a chromed dip roller, a dosage roller, and an applicator roller. The two latter rollers are rubber-covered. The dip roller and the dosage roller are linked by a toothed wheel gear and are driven by a separate variable-speed motor. The speed of the dip roller regulates the amount of varnish being applied. A further means of adjusting feed is given by setting the dosage or applicator roller harder or softer against the dip roller. The amount of varnish can therefore be controlled to a wide range to adapt to the most varied requirements. The dip roller runs in a trough with constant level control. A pump that may be placed directly into the varnish container conveys the coating varnish to the trough. The pump may be used to pump off varnish after the completion of work. It is easily reversed. The trough is removable to facilitate cleaning. In production, the applicator roller is driven by the blanket cylinder. During stoppages, all three rollers are driven by a separate motor to prevent caking. Throw-on and throw-off of the applicator roller are by pneumatic action coupled to the machine. It is also possible to control this action independently of the machine. Sheet feeding is via a transfer drum in a positive gripper bite. The coater should meet the requirement of the industrial safety inspectorate.

23 Industrial Adhesives

PACKAGING SYSTEMS

The general packaging market is a broad-based category including many adhesive applications. Adhesive types for this area include emulsion, dextrin (hydrolysis of starch), and hot melt adhesive. The emulsion products can be based on polyvinyl acetate, ethylene vinyl acetate, acrylic, and acrylic copolymer resins. The dextrine products are modified starches. Hot melts are based on ethylene vinyl acetate or polyethylene. There are two basic substrates involved. First, for the case seal and glued lap applications, is Kraft liner corrugated board, which may be coated or uncoated. Second, for carton seal, side seam and window mounting is chip board. This is often coated and/or printed in the glued area (Table 23.1).

Ethylene vinyl acetate (EVA) is typically used as a hot melt adhesive for carton and shipper containers applications—general purpose—very fast setting, good hot tack, refrigerator to high temperature, excellent stability; polyethylene (PE)—hot melt: General purpose—medium set speed, low hot tack, refrigerator to high temperature; high viscosity for extrusion grade; excellent stability; liquid adhesives: General purpose—high-solids resin product; fast speed of set; non-solvent for easier waste disposal; usable on extrusion or wheel applicators; it can also be formulated solvent added for penetration on tough board stock and speed in cold weather and extrusion viscosity; in addition—general purpose—fast setting product with high tack and high solvent for penetration into tough board stock.

For window applications, general purpose—for adhesion to a wide variety of window stock including mylar (polyester), tricite, polystyrene, cellophane, and polyethylene; good tack, high solids, and non-solvent for easier disposal; low temperature; non-solvent for easier waste disposal.

BASE CUP ADHESIVES

Base cup hot melts are usually blends of synthetic rubber, ethylene vinyl acetate, and tackifying resins. They are formulated for a good balance of adhesion (both to PET and HDPE), cohesion (shear strength), and high and low temperature resistance. Without exception, the base cup application involves adhering a high-density polyethylene (HDPE) base cup to a polyethylene terephthalate (PET) bottle. Depending on the machine, three, four, or six dots of hot melt are shot onto the base cup. The PET bottle is then lowered onto the cup, and the whole assembly is compressed so that the hot melt has a chance to set up. Like many other hot melt applications, potential failures can result if compression time is too short. Over-compression can also cause failures. Over-compression spreads out the glue dots to the point where only a thin film of adhesive remains. As a consequence, the shear strength of the assembly

DOI: 10.1201/9781003384199-24

TABLE 23.1
Trouble Shooting Problems—Adhesives

Adhesive Type	Problem	Cause	Solution
HOT MELT	Stringing	Application temperature is too low	Raise application temperature
		Viscosity of the adhesive is too high	Raise application temperature
		Hot melt is too tacky	Raise application temperature
		Substrate is too far from the application tip.	Position the substrate closer to the tip
		Substrate temperature is too low. Cold substrate causes viscosity to increase at the point of application	Pre-warm substrate to increase surface temperature
		Poor nozzle shutoff	Clean nozzle and repair
	Charring or gelling of adhesive in reservoir	Reservoir temperature is too high.	Lower reservoir temperature
		Surface oxidation	Maintain level of adhesive in the reservoir. Keep the reservoir covered. With some adhesives, a nitrogen blanket may be necessary
		Poor adhesive heat stability	Change adhesive; if not possible, hold reservoir temperature at minimum temperature and change reservoir contents often
	Dripping nozzle	Worn spool valve	Replace valve
		Nozzle dirty	Clean nozzle with recommended solvent
		Orifice worn	Replace part
	Air in adhesive at the applicator tip	Reservoir out of adhesive	Bleed lines and refill the reservoir
		Improperly seated valve	Reseat or replace the valve
		Moisture in or on adhesive	Switch to another lot number. Open boxes to air
	Improve adhesion to substrates	Adhesive bond shows that glue film mostly on one surface caused by improper penetration of the secondary surface	1) Apply heavier glue film or apply adhesive hotter. Sometimes both approaches are necessary 2) In the case of very conductive surfaces such as metal, heat to assure proper penetration of the secondary substrate

TABLE 23.1 (Continued)
Trouble Shooting Problems—Adhesives

Adhesive Type	Problem	Cause	Solution
		Bond failure shows adhesive on both surfaces with leggy strings. Usually caused by disturbing the bond during the setting stage	1) Apply cooler. 2) Check compression to assure no shifting during adhesive setup 3) Use faster setting product
		Poor adhesion on coated substrates caused by poor penetration of substrate coating	1) Apply adhesive hotter 2) Combine substrate quicker 3) Apply a larger volume of adhesive
		Glossy adhesive surface usually caused by poor contact with the secondary substrate	Adjust compression to assure intimate contact of substrates
		If bond failure occurs after trying the above solutions, it may be an adhesive problem	Suggest evaluating a new adhesive
	Substrates popping open out of compression unit	Hot melt not cooling rapidly enough	Lower application temperature
		Too much adhesive applied	Decrease the amount of adhesive applied
		Substrate too hot to achieve proper cooling of the adhesive	Cool the substrate to room temperature
	To increase open time		1) Increase application temperature 2) Increase the amount of adhesive
	Bubbles in hot melt glue line	Moisture level of the substrate is high, causing moisture vapor	Lower temperature. If this falls, refer to the section on "Air in Adhesive at Applicator Tip"
	Improve hot tack		Lower temperature of the adhesive. Select a higher viscosity hot melt
LIQUID Adhesives			
	Pop-open out of compression	Compression time too short or too much adhesive (adhesive wet)	Reduce the amount of adhesive. Increase compression time. Change to faster adhesive
		Open time is too long or too little adhesive (adhesive dry)	Increase the amount of adhesive. Move the applicator closer to compression, and change adhesive
		Uneven compression (or inadequate)	Adjust compression. Box should be snug, but without distortion

(Continued)

TABLE 23.1 (Continued)
Trouble Shooting Problems—Adhesives

Adhesive Type	Problem	Cause	Solution
		Improper scoring	If score position and/or depth causes excessive spring back suggest a change to different box lot. May also be solved by increasing compression time or adjusting the adhesive amount
	Poor adhesion (after drying)	Substrate contamination or coating	Check for coating change and make substrate or adhesive substitution
		Lack of surface penetration or "wet out"	Increase the amount of adhesive. Change adhesive or substrate

is much diminished. It is important to maintain an application temperature of at least 325°F. Any less and the hot melt does not completely wet out the two plastics and poor adhesion results. Low application temperature can also cause poor cut off and stringing, which makes for sloppy machining. Because bottles are subjected to high shear stress in storage (especially when ICS trays are used), bottle suppliers conduct a series of tests to assess the performance characteristics of base cup adhesives.

CIGARETTE ADHESIVES

The manufacturing and packaging of cigarettes can involve several specialized adhesive products. Increased production speeds and the abundance of specialty papers and filter materials have combined to require a special adhesive for almost every operation. Cigarette manufacture begins with the two-stage assembly of the filter plug. First, a polyvinyl acetate emulsion is used to anchor the cellulose acetate filter material to the plug wrap. Next, a hot melt is applied to seal the outer seam of the filter, surrounding the tow material. Plug wrap hot melts are applied with a simple gravity system and extruded through a Kaymich nozzle. The finished plugs are then conveyed to the maker/tipper area where they are combined with the freshly made cigarettes. At the maker, a continuous rod of tobacco is encircled by cigarette paper and sealed with a side seam adhesive. They can be applied with a simple gravity feed system like that employed for plug wrap hot melt. In line with the maker is the tipper, where precut cigarette rods are combined with the filter plugs and wrapped with tipping paper. Because of the high line speeds, tipping adhesives must handle shear well. Today machines regularly run at 7,500–10,000 cigarettes per minute.

FOAMED HOT MELT

Foamed hot melt adhesives are used in various industrial gasketing and sealing operations. Some of the advantages of foamed hot melt over traditional gasket and sealants are adhesion to a wide range of materials and improved filling and anti-sag characteristics, faster curing times, and lower thermal density. Applications are in the areas of automotive and product assembly, insulation laminating, and the appliance industry. For example, experience in product assembly centers around foamed-in-place gaskets for household appliances and sealing refrigerator and freezer cabinets. This automotive experience, while somewhat limited, is spearheaded by the replacement of traditional hand-applied gaskets with robotically applied foamed-in-place products. These applications include the lamination of automotive trim carpet to various substrates via the use of foamed sprayed pressure-sensitive hot melts and the replacement of mastic materials with foamed pressure-sensitive hot melts for attachment of weather-stripping materials for automotive trunks and windows. This competitive situation, while increasing, still is light when compared to usual niche markets. National Starch and Chemical, H.B. Fuller, and Dexter Hysol lead the competitive efforts. To date, however, specially formulated foam products coupled with unique foam melt/robotics still afford the title of leaders in the new area of foamed hot melt adhesives.

WOOD ADHESIVES

Wood adhesives are used to combine various species of soft and hard woods with themselves and other wood product substrates. Adhesives for this market include formaldehyde-linked resins from phenol, resorcinol or urea, powdered casein, polyvinyl acetate, and catalyzable polyvinyl acetate-based products. Types of bonds include laminating, finger jointing, edge gluing, and other assemblies. Types of bonding procedures include cold press, hot press, radio frequency press, and clamp carrier. Wood adhesives are divided into two basic classes. The exterior or wet use is catalyzed and becomes water resistant. The interior or dry-use products have to have high shear strength in non-water contact conditions. For example, wood adhesives consist of compounded polyvinyl acetate adhesives for interior and exterior finger jointing, edge gluing, laminating, and assembly. Major competitors are National Starch, National Casein, Adcon, Franklin, and Fuller.

BAG ADHESIVES

There are three major types of bags produced in the paper bag industry:

1. Grocery—This is an SOS-style bag made from a single ply of low-grade natural kraft. It is normally bonded with a starch adhesive cooked up in the bag plant.
2. Specialty—This includes the bleached, wax-coated, and poly-coated kraft small bags.
3. Multiwall—This is a larger bag composed of multiple plies of natural kraft paper. It often contains a polyolefin film or a kraft/poly-coated foil laminated sheet as one of the plies. This bag is used for shipping chemicals, fertilizer, cement, and so on.

Adhesives for bag making fall into two major categories: Bag forming and bag closure. Bag-forming applications include side seaming and bottoming on all types of bags and spot pasting and film laminating on multiwall bags. Bag closure or sealing may be direct or with a pre-applied adhesive which is heat reactivated after filling. Bag-forming adhesives consist of modified polyvinyl acetate and vinyl acetate ethylene for side seam and bottom on specialty bags, hot melt for film side seam, polyvinyl alcohol/clay/resin for seam and spot pasting on natural kraft, and tackified SBR latex for film laminating and spot pasting. Bag-sealing adhesives include hot melts for direct seal and pre-applied pinch bottom heat reactivation and modified polyvinyl acetate for delta seal. Competitors include Eastman, Fuller, National Starch, Port City, and Findley.

DISPOSABLE PRODUCTS

In categorizing "disposable products," including all diapers, both baby and adult incontinence; feminine care products; hospital gowns and drapes; and other miscellaneous items. The composition of these products is usually a double thin film substrate or a non-woven fabric or tissue adhered to the substrate. These products typically contain an absorbent layer. Hot melt adhesives have the majority share of these products, although cold glues are used in heavy volume primarily in the lamination of hospital disposables. The hot melts are a combination of EVA, elastomeric, and amorphous poly propylene (APP)-based products and can be either pressure sensitive or contain no residual tack at room temperature. These adhesives require high shear and peel resistance, especially at elevated temperatures, and are usually softer products with good hand. The method of application varies from standard nozzle extrusion to slot extrusion and spray. The disposable marketplace represents very high-volume usage of adhesives and is dominated by a handful of major suppliers. These suppliers include National Adhesives, HB Fuller and Century Adhesives, and Findley.

PRESSURE-SENSITIVE ADHESIVES

The pressure-sensitive adhesive market is a broad market that traditionally consisted of tapes, tags, and labels. Because of improvements in pressure-sensitive adhesive technology, many new assembly applications are being developed in automotive, packaging, bag manufacturing, floor covering, medical construction, and various other markets. Pressure-sensitive adhesives are permanently tacky materials that bond to a wide variety of substrates with light pressure. These adhesives can be hot melt, emulsion, or solvent type. Production includes the hot melt and water-based styrene butadiene rubber (SBR) varieties. There are three basic designations for tag, tape, and label adhesives. These are permanent, freezer, and removable grades. Product assembly products may or may not fall into one of these categories and are often based on other specific end-use requirements. Many adhesive companies compete in this broad market and have carved out specific niches. The major companies are National Adhesives, Fuller, Findley, and Malcolm-Nichol.

LAMINATING ADHESIVES

There are numerous industrial laminating operations that utilize adhesives. These major areas in which adhesives are utilized are paper and paperboard laminating, foil laminating, and specialty laminating. Paper and paperboard laminating is often called Layflat laminating. Combined structures must have minimal curl or warping. Paper stocks should lie smooth without blisters or wrinkles. Typical applications include printed papers to chipboard for gift boxes and printed litho labels to corrugated board stock for point-of-purchase packaging. In foil laminating, very thin aluminum foil is bonded to chipboard, kraft paper, paperboard, or unbleached kraft paper, primarily for insulation or food packaging applications. These laminates must exhibit heat, water, grease, and corrosion resistance to varying degrees, depending on the end use.

Panel laminating adhesives usually are compounded vinyl acetate ethylene products. Dextrine, polyvinyl alcohol/clay, and compounded polyvinyl acetate adhesives are used in paper laminating. Foil laminating adhesives utilize acrylic copolymers or casein/latex combinations. Major competitors include National Starch, Fuller, Universal, and Morton-Thiokol.

COHESIVE COATING WELDS

A cohesive is a water-based coating based on natural latex rubber that has been highly formulated to machine well on standard gravure and flexographic printing presses as well as a wide variety of other flexible packaging converting equipment. In addition to excellent machinability, these cohesives have been formulated to give a fiber-tear, destruct bond/seal on paper, board stock, and many plastic films—without the aid of heat or dwell time—only pressure. This is accomplished by applying the cohesive coating to a substrate, drying it, and then later combining coating to coating and applying pressure to instantaneously form an aggressive seal (Cold Seal). These products possess two major characteristics—cohesion and adhesion. Products with a greater cohesive nature tend to seal easier but have less versatility to adhere to a variety of substrates. Products with a greater adhesive nature have greater adhesion latitude but do not seal as well at low pressures. Major competitors include Croda, National Starch, Fuller, Technical Coatings, Valley, and Ajax.

LABELING ADHESIVES

Adhesives for labeling containers are made from a variety of chemical bases comprising both liquid and hot melt formulations. The containers being labeled are also of a wide variety of materials such as glass, plastics (of various kinds), and metal. This diversity of substrate and applicator machines has necessitated a variety of adhesive formulations. The contents of labeled containers span a wide variety of products from beer to motor oil to tuna and a large diversity of customers from large to small. The major liquid adhesive groups are Jelly Gum, Casein, Resin Ice-Proof (non-casein), and Resin (Emulsion). Hot melt may be rubber, polyvinyl acetate, rosin, or polyethylene-based. The two basic types of bottle labeling machines are the picker and rotary machines. The Meyer World Tandem and Super CM comprise the majority of picker machines. Krones and Jagenberg comprise the majority of rotary equipment. Can or wrap-around labeling has historically required a two-adhesive system. The first adhesive, the hot pick-up, pulls the label from the magazine onto the can. The second adhesive, the lap paste, seals the ends of the label together at a small overlap. A new type of can labeling was developed by Krones called the Canmatic. This process uses a hot melt adhesive to glue the entire label onto the can including the lap area.

Because of the large volume of adhesive used in this area, all major formulators compete heavily, including Fuller, National Starch, Reichold (Swift and Peter Cooper), United, and Adcon.

24 Electronic Materials Processing

There are four strategies to follow in the utilization of electronic materials:

1. Multilayer polymeric PCBs
   a. Use photographic techniques for vertical interconnect (smaller-diameter vias against punched or drilled interconnects)
   b. Apply thick film technology to create 3D systems (monolithic or multilayer structures)
2. Active connectors (incorporate circuitry)
   a. Filter connector—block electron leak and electromagnetic emission
   b. Include a thick film capacitor on a ceramic substrate
3. Optical memories—optical disk materials
4. Fiber optics—supply materials against an integrated system

The needed skills are in plastics, ceramics, glass, metals, materials sciences. The skilled techniques are in plating, substrate refinement, engineering new materials, stress against fracture, stiffness, and connectors. Specific experiences related to electronic materials processing include the following:

1. The use of Haake cone and plate viscometers, scanner graphic readers, and analytical software to analyze and calculate transport properties
2. The use of Dektak and Zeiss equipment to determine wet and fired film thickness and profile
3. Procedure on gold stripping, assay, and salvage
4. Procedure on cupellation of gold refining
5. Photo resolution of metal lines on insulating materials
6. Development of advanced materials and systems for IC packaging
7. Development of safety and precipitation procedures for Ni, Cu, and so on
8. Solutions to ensure acceptable waste treatment, effluent discharge treatments for COD and BOD, and biodegradable waste disposal management

THICK FILM SYSTEMS

Thick film compositions are dispersions of inorganic powders in organic fluid vehicles. The materials exhibit pseudoplastic or thixotropic rheology with viscosity an inverse function of shear rate. Electronic materials pattern is usually dried below 200°C and then fired in a conveyor belt oven adjusted to provide reproducible time-temperature profiles. The firing duration is in the range of 20–60 minutes, and the peak firing temperatures are in the range of 600–1,000°C. The key to producing compositions

DOI: 10.1201/9781003384199-25

reproducibly is control of intermediates and back integration to intermediates from readily available raw materials. Conductive oxides such as RuO_2 and $Bi_2Ru_2O_2$ are fire-able materials and are compatible materials that are sintered in ceramic materials for circuitry applications and for its durability. The binder interaction with the substrate influences the sintering kinetics of the functional phase. The understanding of the essential elements of these reactions is important in the development of these materials' combinations to the high-tolerance characteristics for reliable circuit performance. The resistivity is a function of the relative amounts of conductive alloy and semiconductor oxide. The conductivity phase is usually a good electrical conductor, which is thermodynamically stable under time-temperature conditions of firing. It must have small but finite solubility in the glass phase and must be completely by the glass. Two most desirable attributes of TFRs: (1) relatively small TCR and (2) the ability to reproducibly achieve sheet resistance values over many orders of magnitude by varying the volume fraction of the conducting oxide. The cubic structure of R1O6 atoms creates an octahedron. Each O atom is shared with one other octahedron to form a 3-D network of R2O6 stoichiometry tunnels within the framework, which are occupied by polarizable cations and anions. Therefore, the rutile structure of RuO_2 is where one O atom is shared with three octahedrons:

$$TCR (ppm/C) = (TRt - TRd) / TRd. \Delta T \times 10^6$$

A thick film model resistor of ruthenium in glass was used to study the effects of firing temperature on sheet resistance and temperature dependence of resistance (temperature coefficient of resistance). Printing techniques appear to play a role in the performance of the resistor. Control of drive pressure, screen mesh count, screen-nest clearance, and emulsion thickness appear to determine film thickness. Specific TR and TCR are directly proportional to firing temperature and profile, which are determining process parameters with respect to resistor stability and TCR shift. There is a need for a comprehensive model involving the effects of process conditions on microstructure development and the influence of this microstructure on charge transport. Selection of process parameters can yield relatively good quality resistors having predetermined properties and maybe without chemical additives. Proposed model

C= Co(T) F(Vc)   C = Conductance of a thick film resistor
Co = Conductance of a representative unit in the structure having temperature dependence of the whole
F = Function of the variation of isothermal conductance with volume fraction of the conducting oxide phase

THIN FILM SYSTEMS

PHOTO RESOLUTION OF METAL LINES ON INSULATING MATERIALS

Deposition on high-temperature materials with no evaporation may be developed with 1 mil or less resolution depending on the surface roughness of the substrate. Evaporation through masks presents shadowing problems. Blanket plating plus etching is limited to a large width-to-thickness ratio of etched lines. Printing metal suspension and firing is limited to > 2 mil lines and production of lower dimensions. A combination of photoresist, plating, and etching would produce fine-line metallization. After KMER removal, electroless gold is usually coated over copper patterns using the atomex-immersion gold solution. Coating only copper and not Mo-Mn is extremely thin (1,000–10,000 Å). Mo–Mn etchant is applied to utilize alkaline potassium ferric cyanide solution with ultrasonic vibration.

Seven mils lines pattern at 3 mil spacing on ceramic modules are produced. This can be tightened to a target of three mil lines at 2 mil spacings. In addition, 1 mil lines with 1 mil spacing require substantial improvement to surface roughness. The following steps are included in the process:

1. Ceramic glass substrate is cleaned through acid dipping, water wash, and acetone degreasing
2. Blanket metallization involves strong adhering metal film
   a. Dip 80% Mo – 20% Mn (organic liquid)
   b. Fire materials in H2-furnace at 1,500°C for 30 minutes
   c. Metal Xf = 1 mil
3. Photo Resist (PR)
   a. Pattern metal in covered PR
   b. Two layers 1-1 KMER
4. Metal plating
   a. Pattern is plated through PR mask
   b. Cu plating – Xf = 1 mil
5. Resist removal
   a. IRCL—PR strip (J-100 is used)
   b. Does not attack metallization or ceramic
6. Substractive etching
   a. Selective etch
   b. Does not attack metal pattern

CONDUCTIVE POLYMERS

Conductive unfilled polymers such as polyparaphenylene, phthalocyanine-doped Kevlar, polyacetylene, and several others work in charge-transfer salts such as tetra-thiafulvalene planar molecules stack up, permitting overlapping pi-electrons to conduct charge along the stack. Applications include electrical grounding, plastic batteries, semiconductors, wiring and shielding of equipment against electromagnetic and radio-frequency interference, and touch screens. EMI shielding via stainless steel fibers details a specific system for rendering polymers conductive by adding conductive fillers. Unlike conductive powders, stainless steel fibers can provide useful conductivity at loadings of 5–7 wt. %, to preserve the properties and processability of the unfilled resin. At these levels, the volume resistivity of the compound is reduced from about 1,016 to 1 ohm-cm. SS fibers come in three forms: Continuous filament (tow), sized and chopped fiber bundles, and air-laid non-woven web. The optimal fiber diameter is about 7 μm (0.28 mil) and chopped lengths range from 3 to 8 mm. EMI attenuation improves with decreasing fiber diameter and increasing length; in addition, it depends on dispersion during the manufacturing procedure.

THICK FILM—CASE STUDY

A model resistor consisting of ruthenium (Ru) in glass was used to study the effects of firing temperature on sheet resistance and temperature dependence of resistance. A comparison between the two sites manufacturing prints was made to be able to standardize site 1 BTU kiln model to site 2 Watkinson-Johnson (WJ) model. Firing temperature was shown to be the most important parameter in controlling resistor temperature coefficient of resistance (TCR).

EFFECT OF POSITIVE MATERIAL, KILN FIRING TEMPERATURE, AND DRYING ON TR AND TCR

A study of ink systems A (180) and B (185) was carried over the range of 10–100 K ohm/sq. It was observed that as the percent of positive material was increased the temperature coefficient of resistance (TCR) varied likewise. A kiln firing temperature of 870°C gave higher TCR than a standard of 855°C. Some resistive elements were dried first and then fired. It was observed that those gave higher TCRs.

EFFECT OF CLEARANCE AND SQUEEGEE DRIVE PRESSURE

The variation of total resistance with printing clearance was studied as a function of pressure. System A, which is less viscous than system B, showed a maximum variation. System B (high viscosity) showed similar behavior but would not give reproducible prints at higher pressure and higher clearance.

A screen 325 mesh was used for printing system A, while 250 and 200 mesh was used for system B.

EFFECT OF EMULSION THICKNESS

A study of the variation of the screen emulsion thickness and its effect on wet film thickness was conducted using emulsion thickness in the range of 0.002–0.008 inches. The deposited fired film resistor thickness pronouncedly increased as the emulsion thickness was increased and as the squeegee drive pressure was decreased. The temperature coefficient of resistance was increased as the total resistance increased for system B. Generally, the hot TCR was higher than the cold TCR. Data show a set of averaged results that reproduced these tests. The results show an ohmic slope that fits the general predictions of:

TR/sq = r 1/x where TR—Total resistance r—Resistivity x—Resistor film thickness

The results show that as the fired film thickness is increased, the temperature coefficient of resistance is decreased for System B.

CONCLUSIONS

Printing techniques have the utmost effect on thick film formation. The control of drive pressure, screen mesh count, screen—nest clearance, and emulsion thickness will provide a determined film thickness and hence a specific TR and TCR. The kiln firing temperature and profile are the most determining process parameters as far as resistor stability and TCR shift are concerned.

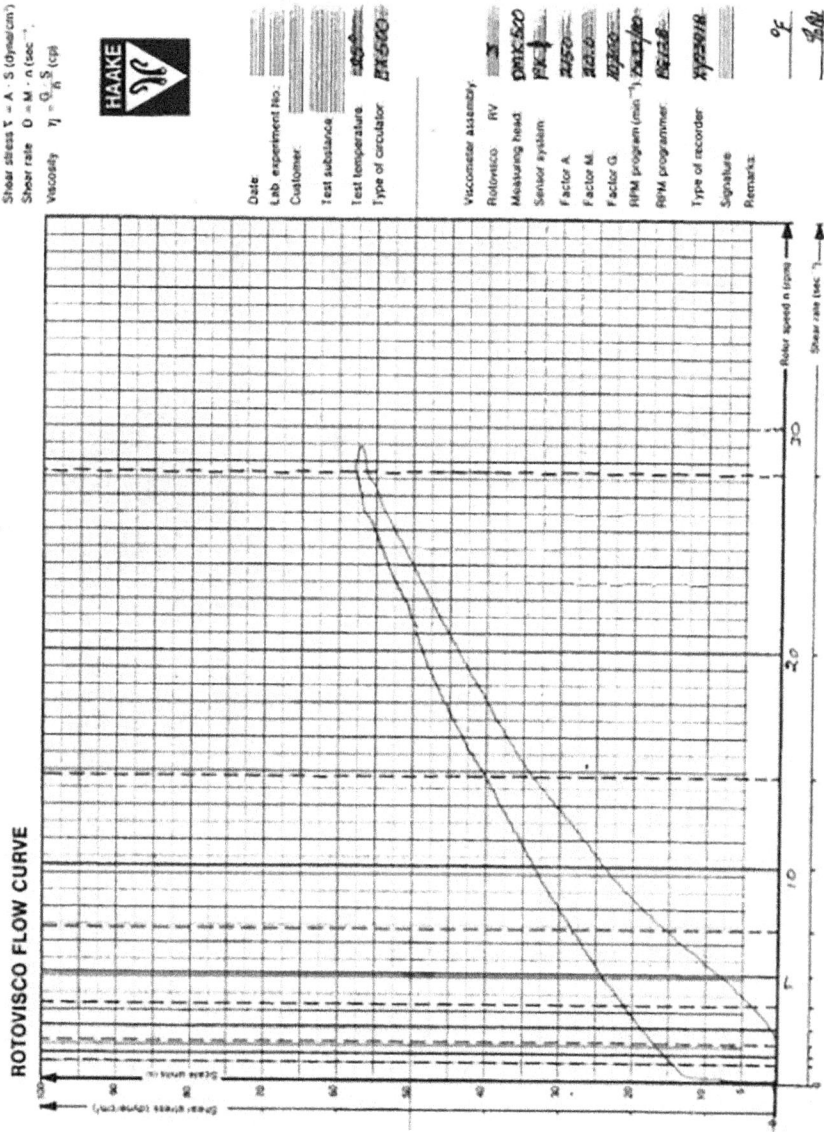

FIGURE 24.1   Rotovisco flow curve.

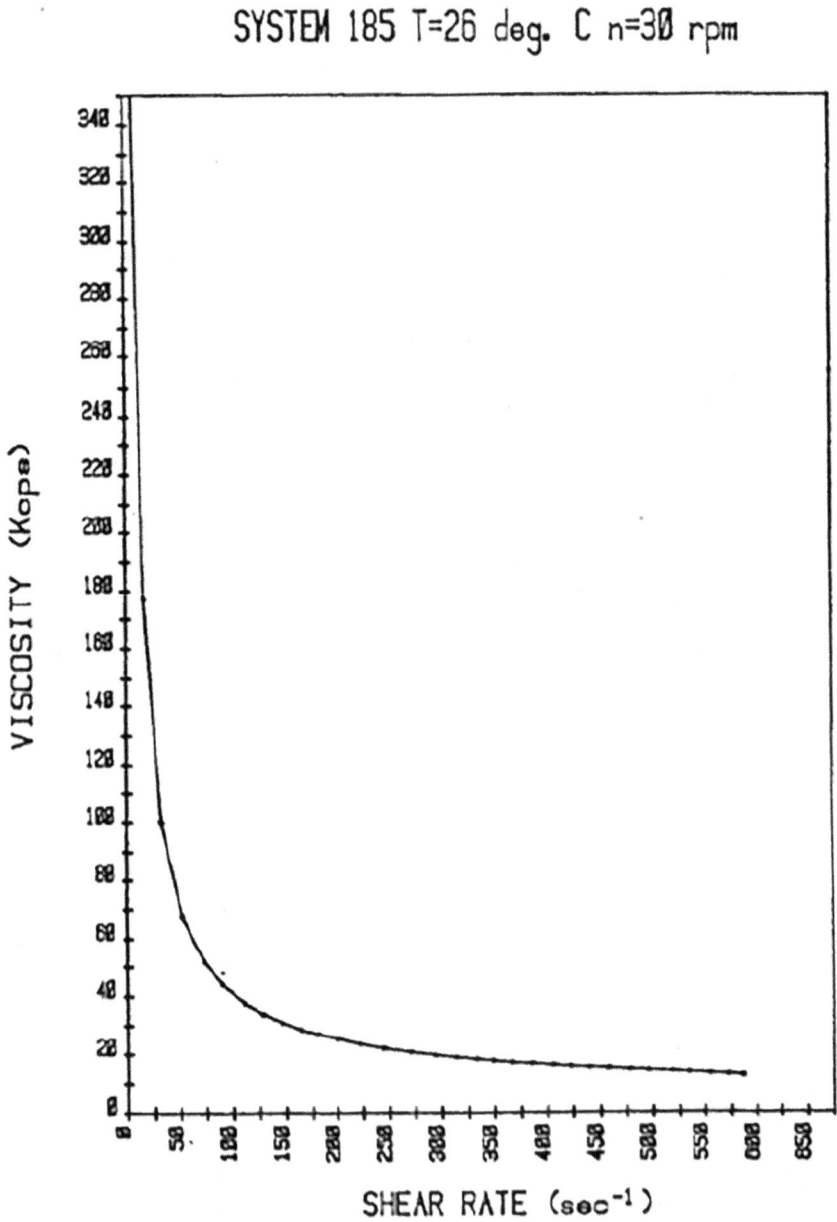

FIGURE 24.2 Electronic ink viscosity against shear rate.

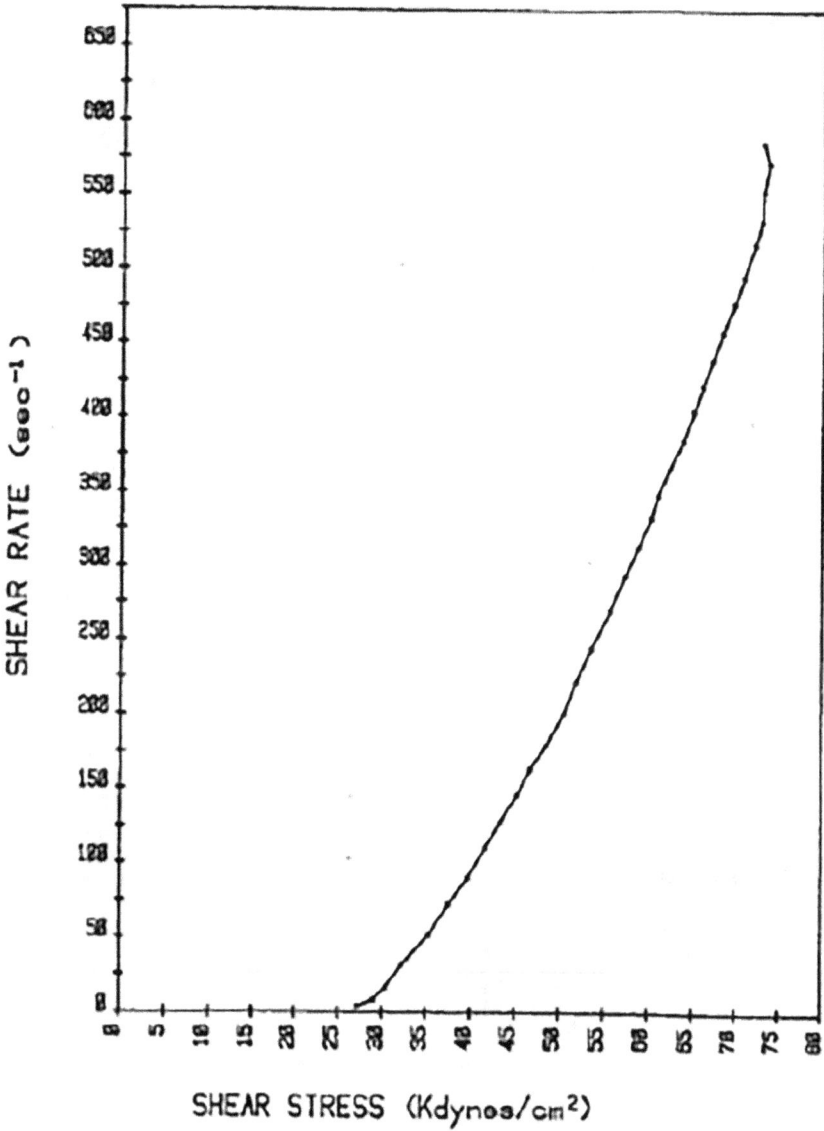

FIGURE 24.3 Conductive ink shear rate versus shear stress.

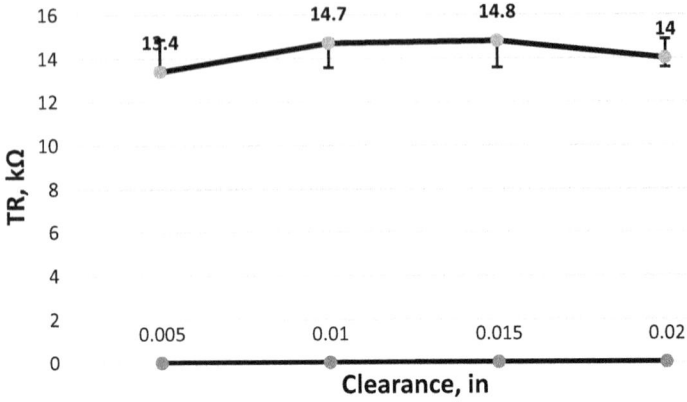

FIGURE 24.4   Total resistance against printing clearance.

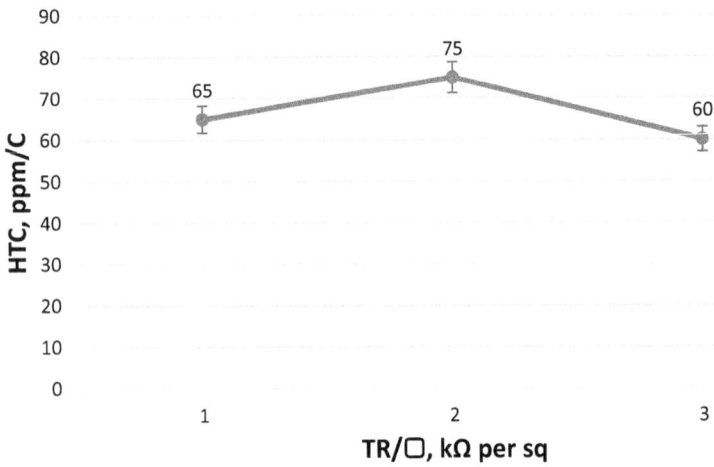

FIGURE 24.5   Temperature coefficient against TR.

| TR/□, kΩ per sq. | 9.5 | 47 | 88 |
| HTC, ppm/C | 65 | 75 | 60 |

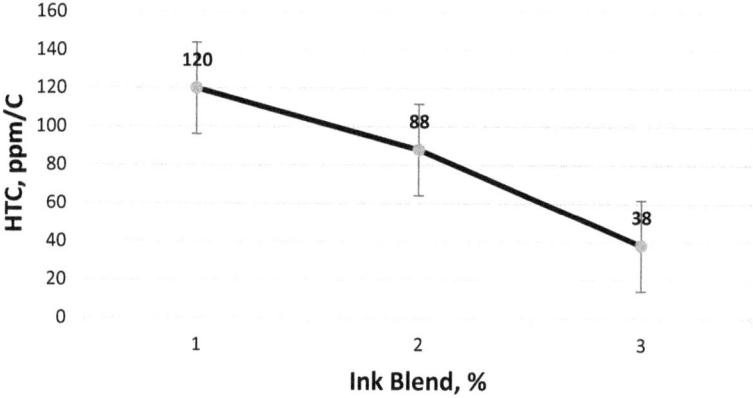

FIGURE 24.6   Hot temperature coefficient.

Ink Concentration, %	100	80	60
HTC, ppm/C	120	88	38

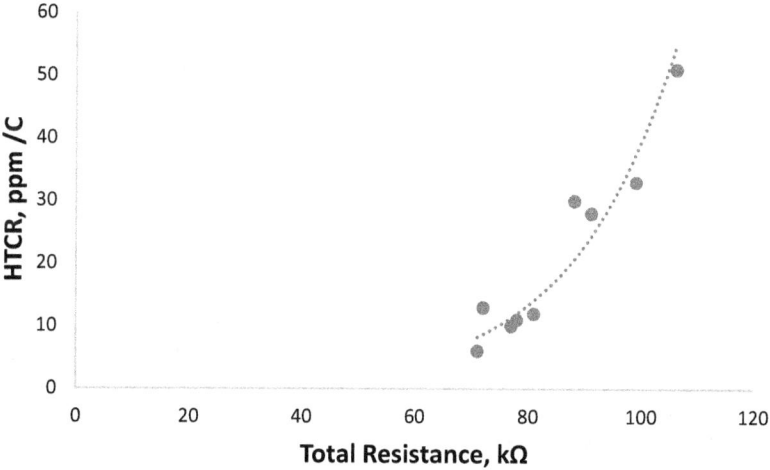

FIGURE 24.7   Temperature coefficient of resistance, ppm/C.

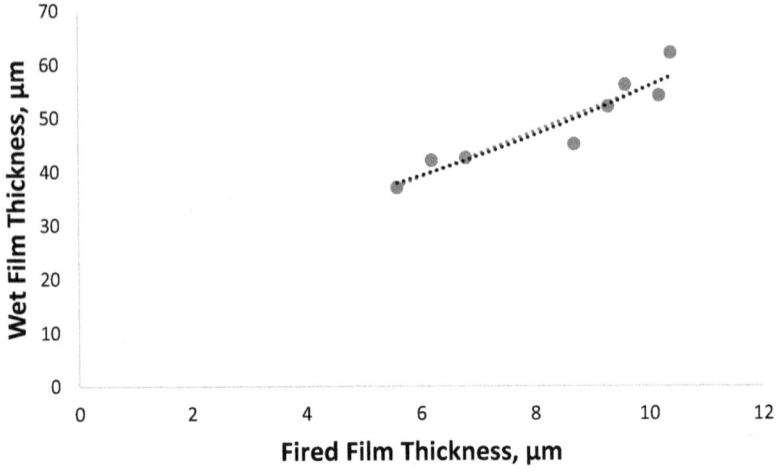

FIGURE 24.8   Wet/fired film thickness, μm.

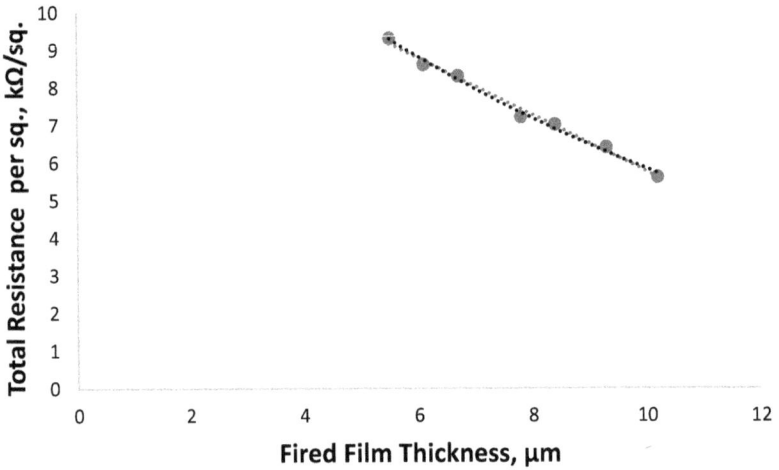

FIGURE 24.9   Total resistance per sq., kΩ/sq.

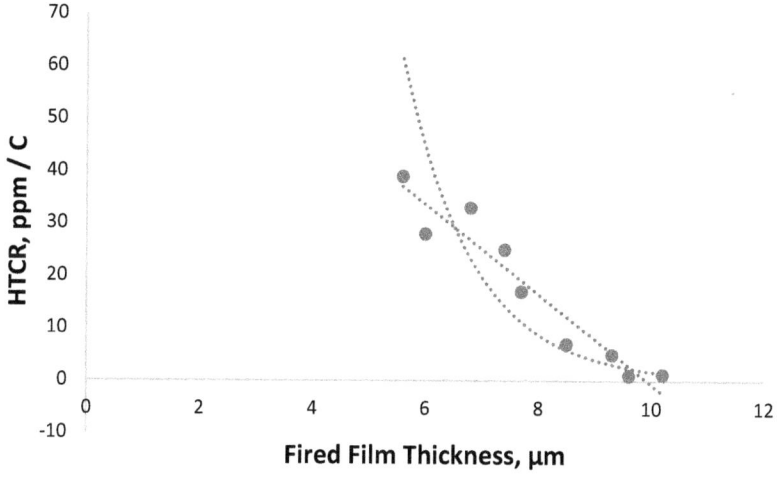

FIGURE 24.10   HTCR, ppm/°C.

25 Nuclear Power Plant— Reactor Heat Removal

The objective of this calculation is to perform a verification of one- and two-train residual heat removal cooldown rates. In this case, the nuclear plant uses cooling towers with finite basin storage capacity as the ultimate heat sink, and the heat loads during circulation can be limiting as far as the tower basin is concerned and hence the significance of RHR heat loads during recirculation.

Single-train operation RHR data are required, as the single-train operation is more severe due to 100% of fuel decay and spent fuel pit (SFP) heat being removed via one-component cooling water (CCW) train. This is consistent with the requirements of regulatory guide 1.139 to achieve cold shutdown within 36 hours after the reactor trip with offsite or onsite power (but not necessarily both) and assuming the most limiting single failure. This analysis will be based upon the latest heat load data for RHR, NSCW, CCW, and ACCW systems in the nuclear plant arrangement, as well as using the latest SFP data for auxiliary equipment load. This calculation aims to provide a review and verification of the methodology used by nuclear reactor design to compute RHR/CCW cooldown data presented in reference one (GP-5248). This is the first step in the development of an integrated program, which determines reactor cooldown, with and without offsite power available.

The reactor design presented two sets of data for determining RHR heat loads following reactor shutdown. In addition, RHR cooldown data are based on established background calculations. Background data were not verified and were based on design parameters and variable NSCW basin temperature. RHR heat loads were based on the time of the trip as a fraction of core-rated power. RHR heat load dissipated to CCW and beyond NSCW during shutdown with or without loss of off-site power (LOP) must be estimated from guidelines of reactor design assumptions. To ensure proper accounting of reactor coolant system (RCS) thermal mass and taking into consideration the auxiliary feedwater system operation, the heat rejection to each CCW/RHR train single- and two-train operation must be properly estimated. For simplified system analysis, the thermal capacity of RCS can be taken as 2.3603×10^6 Btu/°F.

Therefore, the RCS heat capacity is as follows:

$$984.2 \times 10^6 \sim \text{Btu from 557 °F to 140°F,}$$
$$495.9 \times 10^6 \sim \text{Btu from 350°F to 140°F}$$

Variations in the NSCWs tower performance and heat exchanged via ACCW and CCW systems will incur penalties on the RHR heat loads. This calculation is intended to cover only those cases where the RHR system is used for reactor cooldown such as reactor trip with or without loss of off-site power. Analysis of RHR and NSCW heat

DOI: 10.1201/9781003384199-26

TABLE 25.1

Uranium Effectiveness

	Calculated Results		Vendor Data	
Item	Overall U	Effectiveness, ε	Overall U	Effectiveness, ε
RHR HX	386	0.638	379	0.634
CCW HX	246.6	0.714	243.8	0.709

loads following a loss of coolant accident (LOCA), viz., safety injection/recirculation, is beyond the scope of this analysis.

Heat transfer calculations should have the capabilities of calculating heat exchanger overall heat transfer coefficient and thus heat exchanger effectiveness as a function of heat exchanger configuration, fluid flow rates, and flow arrangement. The calculated results should be compared to vendor data (Table 25.1)

The calculation computed the heat loads for one- and two-train operations of the RHR-CCW system using 350°F initial RCS temperature and NSCW from reactor design calculations. Reactor design heat loads were verified for existing RCS temperature, which was a proof of the program operability and validity.

The difference (Q_T-Q_D) of the heat load and the decay heat at an interval was considered and then divided by the RCS thermal mass (capacity, Btu/°F = 2.3603×10^6) to get ΔT_{RCS} and hence new RCS temperature. This procedure was repeated for each time step (1 hour) until $T_{RCS} \leq 212$ °F for one train and ≤ 140 °F for two trains. For two-train operations, the same flows (RCS, CCW, and NSCW) were used as for one-train operations. Also, the same heat exchanger parameters (RHR/CCW) were used.

The spent fuel pool heat load was divided by 2 for two-train operations, and the RCS thermal mass and decay heat load were halved for two-train operations. Simply, an imaginary isothermal barrier is assumed to exist across the reactor vessel and spent fuel pit to account for two-train calculations (Tables 25.2 and 25.4).

DESIGN CRITERIA

1. Unless an engineered safeguard feature (ESF) signal, e.g., safety injection (SI), is present, the auxiliary component cooling water (ACCW) system will operate during normal plant cooldown.
2. The NSCW temperature and ACCW heat loads will be greater with off-site power available, because of the reactor coolant pump operation.
3. The normal controlled valves modulate the RHR heat exchanger tube side; RCS flow to limit cooldown to 50°F/hour. These valves fail open, and thus for loss of off-site power (LOP) operation, the flow modulating feature is lost due to loss of instrument air.
4. For the spent fuel pit, use nominal design heat load for the high-density configuration of 17.38 10^6 Btu/hour.
5. The RHR system time of activation is 4 hours after shutdown.
6. For one-train cooldown, the estimates for NSCW supply temperatures to the CCW HX are higher than two-train values.

TABLE 25.2
Two-Train Cooldown

Time, hour	Decay heat, × 10⁶ Btu/ hour	RHR HX—Heat Load, × 10⁶ Btu/hour	RHR HX RCS, Temp, °F	CCW, Inlet T, °F	CCW, Outlet T, °F	CCW-HX NSCW, Inlet T, °F
4	63.9	209.4	350	127.1	211.5	97.5
5	57.8	107.9	226.7	111.9	155.4	96.1
6	55.4	73.4	184.2	106.0	135.7	94.9
7	54.2	61.4	168.9	103.5	128.3	94.0
8	51.7	57.0	162.8	102.1	125.1	93.2
9	49.9	53.9	158.3	101.0	122.7	92.5
10	48.1	51.5	154.9	100.1	120.8	91.9
11	46.9	49.6	152.0	99.2	119.2	91.3
12	45.6	48.0	149.7	98.6	118.0	90.9
13	44.4	46.9	147.7	98.0	116.8	90.5
14	43.8	45.4	145.8	97.5	115.8	90.2
15	42.6	44.4	144.4	97.1	115.0	89.9
16	41.9	43.4	142.9	96.8	114.2	89.7
17	40.7	42.4	141.6	96.5	113.6	89.6
18	40.1	41.3	140.2	96.3	112.9	89.5
19	38.9	40.4	139.2	96.2	112.5	89.5

Note—CRCS = 1.18 × 10⁶ Btu/hour   SFPload = 8.69 × 10⁶ Btu/hour (two-train(17.38 × 10⁶)/2 Btu/hour

TABLE 25.3
One-Train Cooldown

Time, hour	Decay Heat, × 10⁶ Btu/ hour	RHR HX—Heat Load, × 10⁶ Btu/hour	RHR—HX RCS, Temp, °F	CCW, Inlet T, °F	CCW, Outlet T, °F	CCW—HX NSCW, Inlet T, °F
4	125.9	208.5	350.0	128.1	212.2	97.5
5	118.0	180.5	315.0	122.9	185.7	96.1
6	112.5	159.4	288.5	118.9	183.1	94.9
7	103.4	143.5	268.6	115.8	173.7	94.0
8	102.3	131.8	253.7	113.4	166.6	93.2
9	98.5	122.4	241.7	111.5	160.8	92.5
10	95.5	114.5	231.6	109.8	155.9	91.9
11	93.7	108.2	223.5	108.3	152.0	91.3
12	91.2	103.5	217.4	107.3	149.0	90.9
13	89.4	99.5	212.2	106.3	146.4	90.5

(Continued)

TABLE 25.3 (Continued)
One-Train Cooldown

				RHR—HX		CCW—HX
Time, hour	Decay Heat, × 10⁶ Btu/ hour	RHR HX—Heat Load, × 10⁶ Btu/hour	RCS, Temp, °F	CCW, Inlet T, °F	CCW, Outlet T, °F	NSCW, Inlet T, °F
14	88.2	96.1	207.9	105.6	144.4	90.2
15	85.7	93.6	204.6	105.0	142.7	89.9
16	83.9	91.0	201.3	104.4	141.1	89.7
17	81.5	88.6	198.3	104.0	139.7	89.6
18	79.7	86.2	195.3	103.5	138.3	89.5
19	77.8	83.9	192.5	103.2	137.1	89.5
20	76.0	81.8	189.9	102.8	135.8	89.4
21	75.1	79.7	187.4	102.6	134.7	89.4
22	74.2	78.2	185.5	102.3	133.8	89.3
23	73.6	76.8	183.8	102.1	133.0	89.3
24	73.0	75.7	182.4	101.8	132.4	89.2
25	71.8	74.8	181.3	101.7	131.9	89.2
26	70.6	73.8	180.0	101.5	131.2	89.1
27	69.9	72.6	178.6	101.3	130.6	89.1
28	69.3	71.8	177.5	101.1	130.0	89.0
29	69.3	70.9	176.4	101.0	129.5	89.0
30	69.2	70.4	175.7	100.8	129.2	89.0

TABLE 25.4
RHR Temperature Profile

T, °F	V, ft3/lbm
105	0.016147
108.86	0.016162
113.78	0.016180
120	0.016204
128.78	0.016243

ASSUMPTIONS

1. The result of ACCW operation under normal cooldown conditions implies a nuclear service cooling water (NSCW) inlet temperature to the component cooling water (CCW) heat exchanger higher than the basin temperature by the NSCW temperature rise across the ACCVW heat exchanger.
2. A NSCW temperature of 95°F to the CCW heat exchanger is a good average temperature for cooldown analyses.

3. The SFP cooling system should be assumed operating, as it stops automatically only upon loss of off-site power, but be manually loaded upon the diesel generators.
4. The modeling assumes that the heat load on ACCW remains constant throughout the cooldown period. This load is evenly distributed between the two ACCWs heat exchangers.
5. At least one reactor coolant pump (RCP) will be operated if off-site power is available.
6. For two-train cooldown, all common heat loads, e.g., ACCW, SFP, and so on, are equally distributed between the two NSCW trains.
7. As a simplification of system transients, the RCS may be approximated by a uniform mass at the reactor coolant temperature.

SYSTEM DESCRIPTION

The primary function of the residual heat removal system is to remove heat energy from the core and RCS during plant cooldown and refueling operations. The RHR system consists of two parallel flow paths taken from the hot leg of loops 1 and 4. Each flow path contains

- RHR pump
- RHR heat exchanger
- Valves and instrumentation

Heat is transferred from the reactor coolant to the CCW which is circulation on the shell side of the RHR HX. The RCS cooldown rate is controlled by manually regulating the reactor coolant flow through the tube side of the RHR HX. A bypass line around the RHR HX contains a flow control valve, which automatically maintains a constant return flow to the RCS.

SYSTEM DESIGN

During the first phase of cooldown, the temperature of the RCS is reduced by transferring heat from the RCS to the steam system through the use of the steam generators, The RHR is placed in operation approximately 4 hours after reactor shutdown when the pressure is approximately 400 psig and the temperature is 350°F. The design pressure and temperature are 600 psig and 400°F, respectively. The estimated heat transfer rate at design conditions is 32.8×10^6 Btu/hour.

Generally, while in cold shutdown conditions, residual heat from the reactor core is removed by the RHR system. The number of pumps and heat exchangers in service depends upon the residual heat removal load at the time. Normal operation includes the power generation and hot standby operation phases, and when the reactor is at normal operating temperature and pressure. During normal operation, the RHR system is not in service but is aligned and ready for operation as part of the safety injection system (SIS). Plant shutdown is defined as the operation which takes the reactor plant from shutdown temperature and pressure to cold (ambient) conditions.

The cooldown rate is limited by the allowable equipment cooling rates based on stress limits and on the limits set by the operating temperature of the CCW, which services the RHR HX. As the reactor coolant temperature decreases, the reactor coolant flow to the RHR HX is increased.

Maximum RHR heat load occurs during single train operation 6 hours after reactor shutdown (Table 25.4).

CCW inlet temperature = 120°F
CCW flow rate through SFP HX = 1.98×10^6 lb./hour
SFP heat exchanger load = 17.38×10^6 btu/hour

Therefore, $T-120 = Q_{SFP}/(m. Cp)_{SFP} = 17.38 \times 10^6/1.98 \times 10^6 = 8.78°F$
Hence, T = 128.78°F

RHR HX

TOTAL HEAT TRANSFER AREA = 6,898.25 ft²
Maximum heat load on CCW in RHR HX = 168.2×10^6 Btu/hour

The volume of RHR available for CCW on the shell side is calculated:

Shell inner diameter = 42.75 inches
Shell length = 27.6927 ft.
Shell total volume = $\pi/4 (42.75/12)^2 (27.6927) = 276.035$ ft³
Total length of U-tubes = 35,132.476 ft.   O.D. = 0.75 inch
Total volume of U-tube = $\pi/4 (0.75/12)^2 (35,132.476) = 107.785$ ft³
Shell volume available to CCW = 168.25 ft³

SFP HX

Total heat transfer area = 5,479.37 ft²
Shell I.D. = 41 inches
Shell length = 23.343 ft.
Shell total volume = $\pi/4 (41/12)^2 (23.343) = 214.025$ ft³
Total length of U-tubes = 27,906.213 ft.   O.D. = 0.75 inches
Total volume of U-tube = $\pi/4 (0.75/12)^2 (27,906.213) = 85.615$ ft³
Shell volume available to CCW = 128.41 ft³

CCW HX

Total heat transfer area = 35,018.94 ft²
Shell I.D. = 6.4375 ft.
Shell length = 42.8 ft.
Number of tubes = 4,100   O.D. tubes = 0.75 inch   Length of a tube = 43.5 ft.
Shell total volume = $\pi/4 (6.4375)^2 (42.8) = 1,393.1$ ft³
Total volume of U-tube = $\pi/4 (0.75/12)^2 (43.5)(4,100) = 547.2$ ft³
Shell volume available to CCW = 845.9 ft³

The reactor decay heat is based on the assumption that the reactor is operating at the full engineered safety features design rating of 3,579 MWT (including R.C. pump power of 14 MWT) at the time of shutdown. This gives a value of reactor power (P_o) of 3,565 MWT. Instantaneous residual heat decay is plotted as a function of the power level at shutdown against time. The heat load is calculated as follows:

$$Q = 3{,}565 \ (\text{MWT}) \times 10^6 \times 3.4121 \ (\text{Btu/hour.w}) \times (P/P_o)$$
$$Q = 12{,}164 \times 10^6 \ (P/P_o)$$

where P/P_o is the decay heat as a fraction of output at power as recorded. The RHR system bypass control will be operated in normal shutdown and safe shutdown modes to hold the CCW temperature $\leq 120°F$ and the cooldown rate of the primary system ≤ 50 deg./hour.

Reactor design tabulates CCW heat loads from the RHR during cooldown for a 3,425 MWT plant and 105°F CCW temperature for 20 hours after shutdown. These data are per train as each RHR train is sized to provide one-half of the total RHR capacity necessary to meet the design cooldown requirements. The 20-hour cooldown period is based upon both CCW and RHR trains operating.

During the early period of the transient, <12 hours after shutdown, the heat rejection to the CCW is considerably in excess of that computed using the residual decay heat data. However, eventually, the two values essentially come together. The difference in these values is greatest when the rate of change in RCS temperature is large, but the heat load approach is when the RCS temperatures have stabilized (Table 25.5).

Using CCW design data

$$Q_T = 2{,}077.7 \times 10^6 \ \text{Btu}$$

Using residual heat data

$$Q_T = 1{,}510.8 \times 10^6 \ \text{Btu}$$

A difference of 566.9×10^6 Btu over 16 hours. This difference is due to the inclusion of thermal mass of the reactor, steam generators, pressurizer, piping, and so on, the core, and water inventory. The RCS was treated as a thermal uniform mass being cooled from 350 to 140°F. The total system quantity of heat was computed to be 432.1×10^6 Btu.

The RCS thermal capacity, $C_{RCS} = 566.9 \times 10^6 \ \text{Btu}/(350–140) \ °F$

Therefore, $C_{RCS} = 2.700 \times 10^6 \ \text{Btu}/°F$

This value represents the thermal energy absorbed or given as the system temperature changes (Table 25.6).

The residual heat decay data does not account for the effects of RCS thermal mass, which could be a significant effect during the early phases of the transient.

For cases where credit can be taken for auxiliary feedwater system operation during the first 4 hours after shutdown, failure to do so puts the NSCW system with in excess of $1,000 \times 10^6$ Btu, or approximately 25% of the total heat load for the first 24 hours after shutdown and therefore imposes a much more severe peak heat load on the NSCW cooling towers.

TABLE 25.5
Comparison of Heat Load Data

		CCW Data		Residual Heat Data	
Time from Shutdown, hours	RCS T, °F	Ht. Load/Train, × 10^6 Btu/hour	Total Ht. Load × 10^6 Btu/hour	P/P$_o$	Total Ht. Load × 10^6 Btu/hour
4	350.0	116.2	232.4	0.0102	124.1
5	309.1	115.3	230.6	0.0099	120.4
6	265.5	114.2	228.4	0.0094	114.3
7	219.8	112.5	225.0	0.0089	108.3
8	188.0	83.2	166.4	0.0085	103.4
9	172.6	69.1	138.2	0.0080	97.3
10	164.8	61.9	123.8	0.0078	94.9
11	160.6	58.0	116.0	0.0077	93.7
12	152.9	50.9	101.8	0.0075	91.2
13	148.0	46.4	90.8	0.0074	90.0
14	145.3	43.9	87.8	0.0073	88.8
15	143.6	42.4	84.8	0.0071	86.4
16	142.4	41.3	82.6	0.0069	83.9
17	141.4	40.4	80.8	0.0067	81.5
18	140.7	39.7	79.4	0.0065	79.1
19	140.0	39.1	78.2	0.0064	77.8
20	140.0	39.1	78.2	0.0062	75.4

TABLE 25.6
Thermal Capacity

Time Span, hours	RCS T, °F—Initial	RCS T, °F—Final	Heat Load, 10^6 Btu/hour
4–5	350.0	309.1	110.4
5–6	309.1	265.5	117.7
6–7	265.5	219.8	123.4
7–8	219.8	188.0	85.9
8–9	188.0	172.6	41.6
9–10	172.6	164.8	21.1

In calculating the spent fuel decay heat, the load may be assumed to come from 1/3 core just completed refueling plus 10/3 cores from previous years. Pit temperature would be 128.0°F with one train in service or 116.6°F with the two-train operation.

TABLE 25.7
Process Variables

Item	Inlet T, °F	Flow Rate, lb./hour	Flow Rate, gpm	ΔP, psi
Residual heat removal—HX	105	2,480,000	5,000	19.5
Spent fuel—HX	105	1,980,000	4,000	13.55
RHR pump seal-water cooler	105	2,480	5	2.5
CCW-HX (shell side)	146.9	4,502,000	9,165	10.0

Note:
1. Pressure drop should be used ±0.1 psi, as it is possible to size flow meters to such tolerances
2. Design inlet temperature
3. Design values

TABLE 25.8
Reactor Coolant Pumps (RCS)

Item	Flow Rate, gpm	ΔP, psi	ΔT, °F	Heat Load, Kw
Thermal barrier	40	30	70	
Upper oil cooler	170	3		135
Lower oil cooler	6	2		10
Motor air coolers (two per pump)	2 × 190	6		2 × 145

The initial temperature for the transient mode is 95°F. This is conservative because the highest basin temperature in steady-state operation is 89°F. The storage basins are sized to provide for 30 days of system operation following LOCA, or safe shutdown with loss of make-up water (Tables 25.7 and 25.8).

Total flow rate = 4 (40 + 170 + 6 + (2 × 190)) = 2,384 gpm, or 596 gpm per RC pump. All flow rates are specified in gpm using a reference temperature of 105°F.

Heat load from the thermal barrier,

$$Q = m. \, Cp. \, \Delta T$$
$$= (40 \times 496.7)(0.998)(70) = 1.388 \times 10^6 \text{ Btu/hour}$$
$$Cp = 0.998 \text{ @ } 105°F \text{ (100 psia)}, \qquad 1 \text{ kw} = 3,413 \text{ Btu/hour}$$

Therefore,

Total heat load = 4 (1.388 × 10⁶ + (135 + 10 + (2 × 145)) × 3,413) = 11.49 × 10⁶
Btu/hour
Or 2.873 × 10⁶ Btu/hour per pump

During plant start-up and normal operation, all four pumps are in service. During plant shutdown at 4 hours with off-site power available, only one pump is in service. There will be heat load on ACCW by the rest of the thermal barriers. Hence,

$$\text{Total heat load} = 7.037 \times 10^6 \text{ Btu/hour}$$

The flow is maintained for pumps, not in service, so the total flow supplied is 2,384 gpm. During plant shutdown at 20 hours, refueling, and LOP cooldown, the RCPs are not in service, but the ACCW coolant is provided to RCPs. In these modes, the heat load on ACCW is negligible. On the other hand, with LOP hot shutdown, although RCPs are not in service, there is a thermal heat load on ACCW from the four thermal barriers. In the safety injection and recirculating modes, the ACCW pumps are tripped and there is no ACCW flow in the system. Accordingly, the NSCW tower basin and supply temperature to the CCW heat exchanger decrease during the first 4 hours after shutdown due to total heat loads lower than those associated with power generation. During these 4 hours, the RHR pump system is activated, resulting in a large jump in system heat load. However, the thermal mass of the tower basin is affected such that the basin temperature rises slowly as more heat is added and then falls as the heat load decreases to the cold shutdown value.

SAMPLE CALCULATION FOR DETERMINATION OF RHR HX EFFECTIVENESS

Total CCW flow $m_{CCW} = 4.502 \times 10^6$ lb./hour
CCW flow through RHR HX $m_{RHR} = 2.48 \times 10^6$ lb./hour
SFP HX heat load on CCW $Q_{SFP} = 17.38 \times 10^6$ Btu/hour
NSCW temperature into CCW HX T = 95°F NSCW flow through CCW HX
     $m_{NSCW} = 3.95 \times 10^6$ lb./hour
Two-train cooldown for 20 hours
T= 6 hours $Q_{RHR} = 112.8$ Btu/hour $T_1 = 111.3$
Neglecting heat load from RHR pump seal cooler,
$T_4 - T_1 = (Q_{SFP} + Q_{RHR})/m_{CCW} Cp = 140.2°F$
CCW HX effectiveness,
$\varepsilon = (m_{CCW} Cp. \Delta T)/(m.Cp)_{min}. \Delta T_{max} = (4.502 \times 10^6). (140.2–111.3)/(3.95 \times 10^6)$
     $(140.2–95) = 0.7289$
$\underline{\Delta T_{mean}}$ for CCW HX
$Q_{NSCW} = (mCp)_{NSCW}. (T_{NSCW} - 95) = 127.96°F$
$\Delta T_m = (T_{h,o} - T_{c,i}) - (T_{h,i} - T_{c,o}) / \ln (T_{h,o} - T_{c,i} / T_{h,i} - T_{c,o})$ h = hot (CCW)
     c = cold (NSCW)
i = In   o = Out
$\Delta T_m = 14.18°F$
RHR HX data C_{min} is tube (RCS) side,
$\varepsilon = (\Delta T)_{RCS}/(\Delta T)_{in} = 140–117.8/140–105 = 0.6343$
CCW HX data C_{min} is tube (NSCW) side,
$\varepsilon = (\Delta T)_{NSCW}/(\Delta T)_{in} = 133.5–100.8/146.9–100.8 = 0.7093$ (Tables 25.9 and 25.10)

TABLE 25.9

Effectiveness of RHR HX for Two-Train Cooldown

Time from Shutdown, hours	R.C.—T, °F	CCW Flow, lb./hour	R.C. Flow, lb./hour	CCW Inlet T, °F	CCW, Outlet T, °F	Effectiveness
6	250	2486581	1.48×10^6	111.3	156.8	0.5511
8	182.3	2499722	1.48×10^6	104.5	134.0	0.6404
13	148.4	2486407	1.48×10^6	98.4	117.4	0.6384
15	142.3	2498724	1.48×10^6	97.1	114.2	0.6387
16	140.6	2492507	1.48×10^6	96.6	113.3	0.6392

TABLE 25.10

Effectiveness of RHR HX for Single-Train Cooldown

Time from Shutdown, hours	R.C.—T, °F	CCW Flow, lb./hour	R.C. Flow, lb./hour	CCW Inlet T, °F	CCW Outlet T, °F	Effectiveness
6	339.3	2486658	1.48×10^6	120.0	184.9	0.4972
8	277.0	2488644	1.48×10^6	116.2	177.3	0.6389
12	235.4	2486539	1.48×10^6	109.6	157.4	0.6383
15	222.1	2488013	1.48×10^6	107.1	150.8	0.6388
20	210.0	2488763	1.48×10^6	104.5	144.6	0.6391
25	201.8	2487568	1.48×10^6	102.6	140.3	0.6387

Using the two-train cooldown data at 20 hours after shutdown, the integrated heat loads are obtained by taking the sum of the Q's for hours 5–15, doubling it, adding in the data for hours 4 and 16, and then dividing the grand total by 2. This is arithmetically equivalent to taking the sum of the average Q's for each hour.

Total heat (RHR HX parameters), $Q_T = 1{,}681.9 \times 10^6$ Btu

Total residual heat decay, $Q_{TR} = 1{,}186 \times 10^6$ Btu

Or, a difference of 495.9×10^6 Btu

ESTIMATED RCS THERMAL CAPACITY

The RCS is treated as a uniform mass being cooled from 350 to 140°F. The system consists of steel masses (reactor vessel, pressurizer, hot leg piping, cold leg piping, reactor coolant pump, and steam generators) (four units). Also, the primary and secondary sides have water mass (Table 25.11).

For 350 to 140 cooldown,

$$Q = (350-140) \times 0.648$$
$$= 136.1 \ 108.0$$

TABLE 25.11

Steel Mass Heat Capacity

Item	Metal	Weight, lbs.	Cp	Heat Capacity 10^6/°F
Steel mass without steam generators	S.S.	2,220,000	0.15	0.333
steam generators (4)	C.S.	2,868,000	0.11	0.315
Total				0.648

For primary water inventory (volume = 11,109 ft³),

weight (@ average temperature of 250°F, V_f = 0.017 ft³/lb.) = 11,109/0.017 = 653,470.6 lbs.

Secondary side SG water mass = 981,470.6 lbs.

@350°F, h_f = 321.8 Btu/lb.;   @140°F, h_f = 108.0 Btu/lb.

Therefore, Q = (981,470.6)(321.8–108.0) = 209.8 × 10^6 Btu
Hence,
The thermal mass of water inventory over 350 to 140 range is

209.8 × 10^6/(350–140) = 0.999 × 10^6 Btu/°F

Therefore, the system total thermal capacity is 0.648 × 10^6 + 0.999 × 10^6 = 1.647 × 10^6 Btu/°F

Total Q = 209.8 × 10^6 + 136.1 × 10^6 = 345.9 × 10^6 Btu

The RCS sensible heat can be approximated to the thermal mass as

$Q_{RCS} = C_{RCS} (\Delta T_{RCS})$, Or
495.9 × 10^6 = 2.3603 × 10^6 ($T_{RCS, out} - T_{RCS, in}$),

The initial $T_{RCS, out}$ after 4 hours of shutdown of the reactor is 350°F (Figures 25.1 and 25.2).

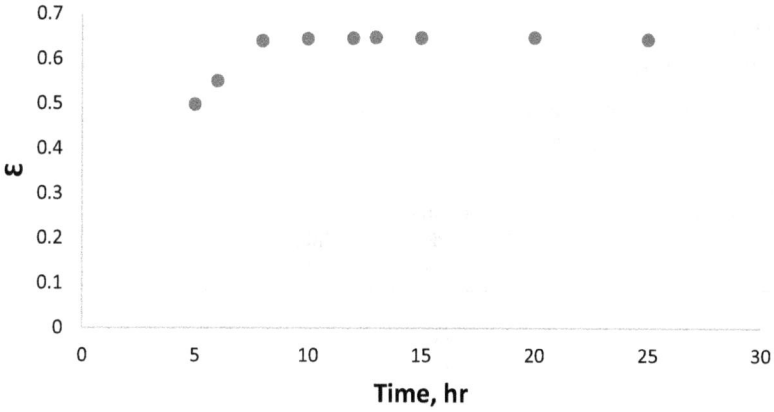

FIGURE 25.1 Effectiveness—RHR HX.

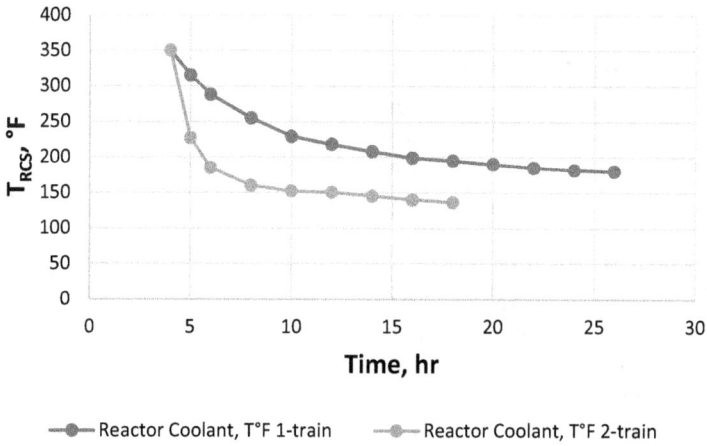

FIGURE 25.2 Reactor coolant temperature—cooldown.

TABLE 25.12
NSCW Temperature—Two-Train Cooldown

Time, hours	Loss of Offsite Power ACCW Operation		Offsite Power Available—Two-Train		Offsite Power Available—One-Train (Diesel)		Reactor Design
	Basin T, °F	CCW HX-Inlet T, °F	Basin T, °F	CCW HX-Inlet T, °F	Basin T, °F	CCW HX-Inlet T, °F	CCW HX-Inlet T, °F
0	90.0	91.6	90.0	92.4	90.0	92.4	
2	89.1	90.6	88.3	90.7	88.6	90.9	
4	88.3	89.7	87.2	89.4	87.7	89.9	97.5
6	90.0	91.4	89.0	91.1	89.5	91.6	94.9
7	90.5	91.7	89.3	91.4	89.9	91.9	94.0
8	90.6	91.8	89.4	91.4	89.9	91.9	93.2
9	90.6	91.7	89.4	91.3	89.9	91.8	92.5
10	90.5	91.6	89.3	91.2	89.9	91.7	91.9
12	90.3	91.3	89.1	90.9	89.7	91.4	90.9
14	90.0	90.9	88.7	90.3	89.3	90.9	90.2
16	89.7	90.5	88.3	89.9	88.9	90.5	89.7

TABLE 25.13
NSCW Temperature—One-Train Cooldown

Time, hours	Loss of Offsite Power ACCW Operation		Offsite Power Available ACCW Not Operating		Offsite Power Available—One-Train (Diesel)		Reactor Design
	Basin T, °F	CCW HX-Inlet T, °F	Basin T, °F	CCW HX-Inlet T, °F	Basin T, °F	CCW HX-Inlet T, °F	CCW HX-Inlet T, °F
0	90.0	93.2	90.0	90.0	90.0	94.9	
2	89.4	92.5	89.0	89.0	89.0	93.7	
4	89.0	92.0	88.3	88.3	88.3	92.9	97.5
6	91.7	94.6	91.0	91.0	91.0	95.5	94.9
8	93.1	96.0	92.5	92.5	92.5	96.9	93.2
9	93.5	96.2	92.9	92.9	92.8	97.2	92.5
10	93.7	96.4	93.1	93.1	93.0	97.3	91.9
11	93.7	96.4	93.2	93.2	93.1	97.3	91.3
12	93.8	96.4	93.2	93.2	93.1	97.3	90.9
15	93.6	96.1	93.1	93.1	93.0	97.0	89.9
20	93.3	95.5	92.7	92.7	92.6	96.3	89.4
25	92.9	94.9	92.4	92.4	92.2	95.6	
30	92.6	94.4	92.1	92.1	91.8	95.0	

TABLE 25.14
Heat Loads on CCW-HX

Item Identification	Heat Load, Btu/hour	Flow Rate, lb./hour
SFP HX	17.38×10^6	1.98×10^6
	7.6×10^4	2.48×10^2

TABLE 25.15
Heat Loads on Cooling Tower

Item Identification	Heat Load, Btu/hour $\times 10^6$	Flow Rate, lb./hour
CTB Aux. cooling coil	4.59	0.7×10^6
Reactor cavity	0.24	11.0×10^4
SI pump Lo cooler	0.1	2.0×10^4
SI pump MTR cooler	0.1	2.0×10^4
CNT charging pump Lo cooler	0.2	2.0×10^4
CNT charging pump MTR cooler	0.2	2.0×10^4
Containment spray pump & MTR	0.07	1.0×10^4
RHR pump motor	0.37	9.92×10^3
ESF Chiller	5.48	0.55×10^6
CCW pump MTR cooler	0.14	44.64×10^3
Pipe penetration area cooler	1.94	12.5×10^4
Diesel generator	21.08	0.75×10^6

TABLE 25.16
One- and Two-Train Residual Heat Decay

Time, hours	P/P_o	Residual Heat Decay, 10^6 Btu	Two-Train RC, T°F	Two-Train Total Heat Load, 10^6 Btu/Hour	One-Train RC, T°F	One-Train Total Heat Load, 10^6 Btu/Hour
4	0.0105	127.7	350.0	231.2	350.0	169.9
5	0.0095	115.6	300.0	225.6	339.3	168.2
6	0.0091	110.7	250.0	183.6	321.2	167.8
7	0.0089	108.3	205.0	146.6	297.6	151.6
8	0.0085	103.4	182.3	129.2	277.0	139.5
9	0.0082	99.7	171.2	120.0	261.9	130.5
10	0.0079	96.1	165.2	114.8	250.6	123.7
11	0.0077	93.7	161.5	105.2	242.0	118.5

(Continued)

TABLE 25.16 (Continued)
One- and Two-Train Residual Heat Decay

			Two-Train		One-Train	
Time, hours	P/P_o	Residual Heat Decay, 10^6 Btu	RC, T°F	Total Heat Load, 10^6 Btu/Hour	RC, T°F	Total Heat Load, 10^6 Btu/Hour
12	0.0075	91.2	155.3	94.2	235.4	114.4
13	0.0073	88.8	148.4	88.6	230.0	111.1
14	0.0072	87.6	144.6	85.2	225.7	108.4
15	0.0070	85.1	142.3	83.0	222.1	106.1
16	0.0069	83.9	140.6	82.1	219.0	104.1
17	0.0067	81.5	139.9	79.8	216.3	102.4
18	0.0066	80.3			214.0	100.9
19	0.0064	77.8			211.9	99.5
20	0.0063	76.6			210.0	98.2

26 Pharmaceutical Manufacturing

Controlled-release drug product solid dosage involves the movement of water through a semipermeable membrane. Rate-controlled delivery reduces the side effects, increases the efficacy, and decreases the dosing frequency, which improves the patient's condition. Conventional oral solid dosage (OSD) delivers drugs very quickly. Their efficacy is dependent on the amount of food in the stomach and the acidity of the stomach fluids. Controlled-release solid dosage slows drug delivery dissolution rates using various coatings. Controlled-release oral solid dosage systems are dependent on the osmolarity of the surrounding body fluids and temperature. These systems are permeable to water, and the release of active ingredients is dependent on the local conditions in the digestive tract due to coatings. The system continues to release its contents even after it has moved into the small intestine. The release process is dependent on the solute concentration within the OSD being greater than that of its surroundings.

The controlled-release system has a solid core containing therapeutic and osmotic agents. This core is coated with a semipermeable rate-controlling membrane that might have one or more laser-drilled orifices. This system mostly has a water-soluble drug and an osmotic agent in a monolithic core and delivers the drug in solution. In addition, multilayers have a separate osmotic characteristic and the drug delivers in solution or suspension. This system can deliver either water-soluble or insoluble drugs. In the moist environment of the GI tract, water enters the system through the semipermeable membrane and dissolves or suspends the drug core. The drug is released through the orifice(s) at a rate that is controlled by the osmotic and chemical properties of OSD structure and the membrane's permeability to water. This system provides a constant rate of drug delivery.

The controlled-release system is dependent on variables:

1) Ingredients
2) Permeability of the membrane coating
3) Solubility of the overcoating and a sub-coating
4) Size and number of the drilled orifices.

Controlled-release systems mechanisms are based on osmosis—the movement of water through a membrane by diffusion. Water moves toward areas of higher solute concentration until the solutions on either side of the membrane have equal concentrations. The osmotic potential is determined by the relative concentration of the solutions on either side of the membrane. The greater the difference in concentrations, the greater the osmotic potential.

DOI: 10.1201/9781003384199-27

The strength of osmotic pressure is controlled by the following:

1) Permeability of the membrane to water
2) Solubility of the components in the core
3) Concentration of solutes within the system
4) Consistency of the material being discharged from the system
5) The size/number of the discharge orifice(s)

The permeability of the membrane is based on its composition and thickness. The choice and optimization of the membrane material account for a considerable portion of the compound performance. The membrane not only controls the release of the active substance but must also ensure a constant volume for the system, as well as being resistant to mechanical stress and digestive fluids. Solute levels, which affect the osmotic potential, are adjusted using sodium chloride in the formulation of the OSD. The consistency of the drug component in the system is controlled by the solubility of the drug and the physical and chemical characteristics of the other excipients in the mix.

An elementary osmotic pump system (single-layer) and the expanding material that contains the active drug are both expelled from the system into the GI tract. In a multi-layer system (bi-layer), the osmotic drug layer gels and expands, while the osmotic layer expands. The combined force expels the drug-laden gel into the GI tract. The viscosity of each layer is based on the nominal molecular weight (size/length) of the polymer used in its manufacture. If the orifice were to become temporarily blocked, by a particle or undissolved core component, hydrostatic pressure within the drug reservoir would rise until it was enough to clear the blockage.

Excipients can be divided into two general categories:

- Compression characteristics variables—fillers, diluents, binders, adhesives, lubricants, anti-adherents, and glidants
- Biopharmaceutics stability—dis-integrant, polymers, resins, colors, flavors, buffers, and absorbers

The controlled-release manufacturing process involves the following steps:

1. Milling and charging, where components are received, milled into a fine powder, and sifted through screens into the initial production vessels
2. Granulation, the mixing of powdered components with moisturizing and binding agents and then blending with dry lubricants
3. Compression of the granulated mixture into solid cores
4. Sub-coating the cores, if a time delay is needed, by applying a water-soluble coating
5. Membrane coating by applying the organic solvent-based semi-permeable membrane
6. Laser drilling of the drug delivery orifice in the membrane
7. Drying the drilled cores to remove solvents used in membrane coating

8. Drug and/or color overcoating the drilled cores
9. Printing drug-specific information on the tablets
10. Sorting tablets to remove under- and over-sized tablets
11. Packing the tablets for shipment to clients

A controlled extended-release system is used for the management of hypertension and angina drug products such as calcium channel blockers (e.g., nifedipine). Drug delivery delay is a result of an aqueous sub-coating applied to the core before membrane coating. After delay, the drug is delivered using a membrane coating system over a 24-hour period. Other examples of extended-release OSD applications are for the treatment of hypertension using an antihypertensive that dilates blood vessels utilizing a bi-layer coating system.

There are two ways to manufacture sterile drug products. Aseptic Manufacturing and Sterile Fill-Finish: First, a process in which the drug product, container, and closure are first subjected to sterilization methods separately, as appropriate, and then brought together (aseptic manufacturing). Second is to fill and seal the drug product in a container and terminally sterilize it.

Many complex drug products are not amenable to terminal sterilization, leading to increased demand for aseptic manufacturing and sterile fill-finish capabilities. Aseptic processing is uniquely challenging because it requires careful planning, thoroughly trained personnel, and specialized facilities/equipment to properly execute. Cleanroom facilities and aseptic processes should be designed to minimize contamination risk from materials, equipment, and personnel. Sterile lyophilization requires investment in specialized equipment, facilities, and knowledge.

Drug products that are delivered via the parenteral, ophthalmic, inhaled, or otic route present an increased risk of infection or harm because they bypass many of the body's natural defenses. To ensure patient safety, the FDA requires that drug products delivered via these routes be supplied as sterile products. This designation includes many complex drug products, including ophthalmic suspensions, sterile injectables, lyophilized powders for injection, and aqueous-based aerosols for inhalation.

"Sterile products" refer to products that are going to be administered using an enteral route of administration. The "products" are going to be infused directly into the bloodstream or body tissue, it is extremely important they be "sterile." Consider that when a patient takes a tablet orally, their digestive system (acting as part of the immune system) could identify and kill any bacteria that may be present. With a parenteral route of administration, if any contaminants are present, they go right into the bloodstream, bypassing the digestive system.

Regulatory guidance is intended to help manufacturers meet the requirements in the FDA current good manufacturing practice (cGMP) regulations (21 CFR parts 210 and 211) when manufacturing sterile drug and biological products using aseptic processing.

Sterile injectables manufacturing involves the production of vaccines. A vaccine must be rigorously tested before the public gets it. We often think of vaccines as treatments for illness, rather they are preventative measures for specific targeted viruses such as the Flu vaccines. People who are well are given vaccines to keep them from getting sick. Therapeutic drugs are developed to cure diseases that already invaded the human body. They are tested and qualified for their efficacy in curing specific disease indications and proven safety through extensive clinical trials that define their side effects.

Researchers will study the virus and attempt to determine which type of vaccine may work best. There are several kinds of vaccines. Some have a weakened live virus, which triggers a protective immune response but does not cause illness. Some vaccines contain an inactive virus that creates an immune response in the body. Other types of vaccines utilize genetically engineered RNA or DNA, which triggers a protein that can prevent the virus from binding to our cells causing illness. The selection of which vaccine route to follow is dependent on experimental results that identify the best possible outcome. Following lab testing to ensure that a vaccine is working on the molecular cellular level, vaccines will undergo animal testing to predict safety for humans. This is followed by clinical trials targeting efficacy and safety for human use.

Phase 1 clinical trials involve a small number of healthy volunteers to test the vaccine for serious side effects. Phase 2 involves smaller studies evaluating efficacy. At this stage, dosage levels of the vaccine are evaluated for scheduling of dosages for multiple applications. Safety is constantly monitored by observing clinical trial subjects for any side effects and whether the immune response or antibody accumulation is effective to warrant moving forward to additional clinical studies. In Phase 3, larger field studies are conducted. This involves testing a population, vaccinating some while having a control group, and monitoring the effect over time to observe specific effects or difficulties. Common denominators among subjects are carefully evaluated for short-term side effects and dosage levels' impact on side effects that are statistically evaluated and documented. If Phase 3 shows that the vaccine is safe and effective, it is approved and registered for commercial use. Following a working vaccine approval by the FDA, there is considerable time to scale up mass production and distribution in commercial operations.

The main goal of vaccination is to inoculate huge numbers of people to develop immunity in the local community that would protect against larger outbreaks. Testing and monitoring for longer-term safety concerns continue even after the vaccine is generally available. Common side effects of vaccines include redness, swelling, associated pain at the location of injection, and low-temperature fever; severe side effects like allergic reactions are relatively uncommon. Vaccines are developed for their protective benefits, which outweigh the risks.

Vaccine newer biotechnology methods are based on cell culture that could allow for more rapid development. As recently experienced, with a pandemic circling the globe (COVID-19), there was an international effort to find a viable vaccine. Collaboration between the private sector and public research labs did accellerate vaccine development.

Recent approaches to expedite vaccine development utilize unconventional study designs such as for the coronavirus vaccine. Published reports on infectious diseases show that instead of traditional Phase 3 trials, volunteers at low risk of developing a severe form of COVID-19 (younger, healthy people without chronic conditions) are pursuing a human challenge study. These subjects could be exposed to the coronavirus, are monitored closely, and given care to prevent severe reactions. This type of study would involve fewer participants but could be done in less time. There are ethical concerns with this type of approach as typically clinicians' ethics rules generally forbid deliberately infecting human beings with a virus that might cause serious disease.

Besides vaccine trials that target preventing the disease, testing potentials of therapeutic treatments aim to make sick people well again continue to focus on drug development. For example, the recent development of a sterile injectable antiviral drug Remdesivir is intended for treating severe cases of coronavirus infection in a hospital setting. There are also drugs that could potentially address the immune system's response to a virus. Patients get sick as the immune system reacts against the virus, whereby they develop excessive inflammation. Therefore, anti-inflammatory drugs that might mitigate the immune system's effects might be considered for clinical trials.

All ingredients of vaccines play necessary roles either in making the vaccine, triggering the body to develop immunity, or ensuring that the final product is safe and effective. These include:

- Adjuvants that help boost the body's response to vaccine.
- Stabilizers that help keep vaccine effective after manufactured.
- Formaldehyde that is used to prevent contamination by bacteria during the vaccine manufacturing process.
- Thimerosal that is also used during the manufacturing process but is no longer an ingredient in any vaccine except multi-dose vials of the flu vaccine. Single-dose vials of the flu vaccine are available as an alternative.

The diseases vaccines prevent can be dangerous or even deadly. Statistically, the chances of a child getting diseases such as measles, pertussis, or another vaccine-preventable disease might be low, and the child might never need the protection vaccines offer. However, one chooses to vaccinate to ensure that the protection vaccines provide is in place if needed. Immunity is the body's way of preventing disease. At birth, the immune system is not fully developed, which can put a child at greater risk for infections. Vaccines reduce the risk of infection by working with body's natural defenses to help safely develop immunity to disease.

Vaccines use exceedingly small amounts of antigens (toxins) that help the immune system recognize and learn to fight serious diseases. Antigens are parts of germs that cause the body to activate the immune system. Getting a disease and getting a vaccine can both give future protection from that disease. With the disease humans get sick, but with the vaccine, they get that protection.

Sterile manufacturing engineering and process validation guides (US FDA, Guidance for Industry—Process Validation) are intended to manage risk, introduce quality by design (QbD), and ensure continued process reliability, consistency, and repeatability.

Sterile manufacturing guidelines are based on specific elements in the construction and installation of classified clean rooms, critical utility systems such as WFI, solution transfer systems, powder transfer systems, and monitoring and controls including automation of processes. Specialized technologies in the design of fill/finish lines of vials, cartridges, bottles for ophthalmic or otic purposes, pre-filled syringes (PFS), or ampoules, which might be filled and sealed aseptically or IV bags that are mostly terminally sterilized. In addition, considerations for personnel and material flow to prevent cross contamination are usually very well defined to separate in and out movements of sterile suites and cascading clean rooms through separate paths and including the use of path-through transfer systems for material flow.

Overall, paying attention to the design, construction, and installation should focus on process flow and employ standardization of equipment and methodologies as much as possible. Building sustainable premises, equipment qualifications, cleaning validation, and sanitization requirements should be all in consideration as part of the design of pharma/biopharma processing. Similarly, process validation and periodic revalidation of manufacturing systems are all important stages and steps that need to be on well-defined schedules to ensure the reliability of drug production.

The main emphasis of clean room classifications is to quantify particle limits when rooms are in operation (dynamic conditions) for non-viable particles. In addition, microbiological limits are emphasized when classified rooms are in operation for viable particles. Media fills are conducted to certify rooms for aseptic conditions of drug products (Tables 26.1 and 26.2).

TABLE 26.1
ISO Clean Room Classifications

Class ISO 146144-1 (Federal Standard 209E)	Average Airflow Velocity m/s (ft/min)	Air Changes Per Hour	Ceiling Coverage
ISO 8 (Class 100,000)	0.005–0.041 (1–8)	5–48	5–15%
ISO 7 (Class 10,000)	0.051–0.076 (10–15)	60–90	15–20%
ISO 6 (Class 1,000)	0.127–0.203 (25–40)	150–240	25–40%
ISO 5 (Class 100)	0.203–0.406 (40–80)	240–480	35–70%
ISO 4 (Class 10)	0.254–0.457 (50–90)	300–540	50–90%
ISO 3 (Class 1)	0.305–0.457 (60–90)	360–540	60–100%
ISO 1–2	0.305–0.508 (60–100)	360–600	80–100%

EN ISO 14644 Methodology

TABLE 26.2
ISO Clean Rooms Microbial Limits

Grade	Recommended Limits for Microbial Contamination			
	Air Sample cfu/m³	Settle Plate, Ø 90 mm, cfu/4 hours	Contact Plates, Ø 55 mm, cfu/Plate	Glove Print, Five Fingers, cfu/Glove
A	<1	<1	<1	<1
B	10	5	5	5
C	100	50	25	-
D	200	100	50	-

MICROBIOLOGICAL LIMITS

Monitoring of classified areas is continuous whether the room is in operation or not. Particles are measured through a continuous particle monitoring computerized system under measured conditions of temperature, pressure, and relative humidity. In addition, EMS automated systems are fitted with alarms to control alert and action levels in classified rooms limits of operation and regulatory requirements. Microbial counts are sampled and read using settling plates at both the aseptic core and equipment surroundings inside the classified clean room. The locations of these plates and frequency of sampling are based on formal risk analysis, sampling requirements, and validation protocols.

Various technologies are employed in sterile manufacturing such as Isolator technologies, Blow/Fill/Seal, aseptic customized tubing set as needed for product specifications, and terminally sterilized drug products, including sterile injectables (parenteral medicines). Protective personnel garments are specified for each clean room classification to cover hygiene, method of gowning, and required training and certification. General aspects of wall materials of construction, sinks, drains, changing rooms, airlocks (MAL) (ante rooms), and air supply and return circulation are major considerations in the design criteria and specifications used in clean room classifications. The use of sanitizers, disinfectants, detergents, and fumigation is tightly controlled and specified for room and equipment cleaning.

Moist heat (saturated steam, not super-heated steam), radiation, or ethylene oxide gasification might be employed in the sterilization of components, active ingredients, excipients, or drug products. Biological and chemical indicators are utilized to confirm sterility during routine production. In aseptic filling, liquid materials are filtered through a 0.22-µm sterile filter. Filter integrity is tested before and after use. Drug aseptic conditions require all filling operations to occur under class A (ISO 5 zone) conditions in the clean aseptic core; similarly, all capping will be under class A conditions. Aseptically filled vials, cartridges, or prefilled syringes must pass 100% integrity testing.

To verify sterile products, a sterility assurance level of 10^{-6} is achieved by over-kill methods and sterility tests must confirm process capability. PIC/S conventions highlighted the details for validation of aseptic processes and recommendations for sterility testing. In addition, it covered isolators for aseptic and sterility testing. Other guidelines governing these areas are EN ISO 14644-1, -2, ISPE guidelines, and US FDA aseptic processing guide.

QbD dictates component preparations and compounding of drug formulations are key steps in cascading the bulk product from beginning to end. These processes all require extreme care and technical know-how in transferring across clean rooms. Personnel flow and materials' flow separations and consideration for entering and leaving have rigorous requirements to ensure cross contamination and product safety. There are many guidelines that govern all these processes defining specific requirements.

Classified areas are functional in sterile manufacturing facilities. Personnel behaviors, personal protective equipment (PPE), and process flow should be governed by SOPs. Quality approach and the significance of systems validation to prevent cross contamination and mixing should be emphasized. Advances in disposable technologies and aseptic processing are important developments that need to be investigated on a case-by-case basis. Critical utilities to generate water for injection and solution transfer systems should be a core part of the integrated facilities design. Advances in control automation to facilitate process reliability and faster production in a compliant manner should be constantly upgraded and validated. Reference to barrier and isolator technologies and RABS to handle both API transfers and high-potency drugs are key developments in aseptic sterile facilities. The systematic management of risk to ensure efficacious and safe pharmaceuticals should be explained in terms of equipment qualification (EQ) and process validation. The importance of change control and analytical testing of raw material quality is systemic in drug manufacture.

The main compliance message is that FDA inspections focus—specific concentration on two areas: cGMP compliance and PAI for NDA. Documentation and procedures to cover WFI trending, pest control trending, and overall facilities adherence to SOP and organizational structure, including appropriate protocols and reports covering all aspects of product submissions for pre-approval, namely, exhibit batches data, stability batches data, microbiology, and chemistry data—all presented in an orderly manner with full documentation including all CAPAs and deviations (Table 26.3).

TABLE 26.3
Clean Rooms GMP Classifications

FDA Maximum Number of Particles Permitted/m³		COFEPRIS Maximum Number of Particles Permitted/m³					EMA and WHO Maximum Number of Particles Permitted/m³					ANVISA Maximum Number of Particles Permitted/m³				
	In operation		At rest		In operation			At rest		In operation			At rest		In operation	
Class	.5 µm	Class	.5 µm	5 µm	.5 µm	5 µm	GR	.5 µm	5 µm	.5 µm	5 µm	GR	.5 µm	5 µm	.5 µm	5 µm
ISO 5	3,520	ISO 5	3,520	29	3,520	29	A	3,520	20	3,520	20	A	3,520	20	3,520	20
ISO 6	35,200	ISO 6*	35,200	293	3,520,000	293	NA	NA	NA	NA	NA	NA	NA	NA	NA	NA
ISO 7	352,000	NA	NA	NA	NA	NA	B	3,520	29	352,000	2,900	B	3,520	29	352,000	2,900
ISO 8	3,520,000	ISO 7	352,000	2,930	3,520,000	29,300	C	352,000	2,900	3,520,000	29,000	C	352,000	2,900	3,520,000	29,000
NA	NA	ISO 8	3,520,000	29,300	Not defined	defined	D	3,520,000	29,000	Not defined	defined	D	3,520,000	29,000	Not defined	defined

Appendix I: Process Flow Diagram—Component Preparation, Compounding, Fill/Finish, and Packaging of Liquid Aseptic Operation

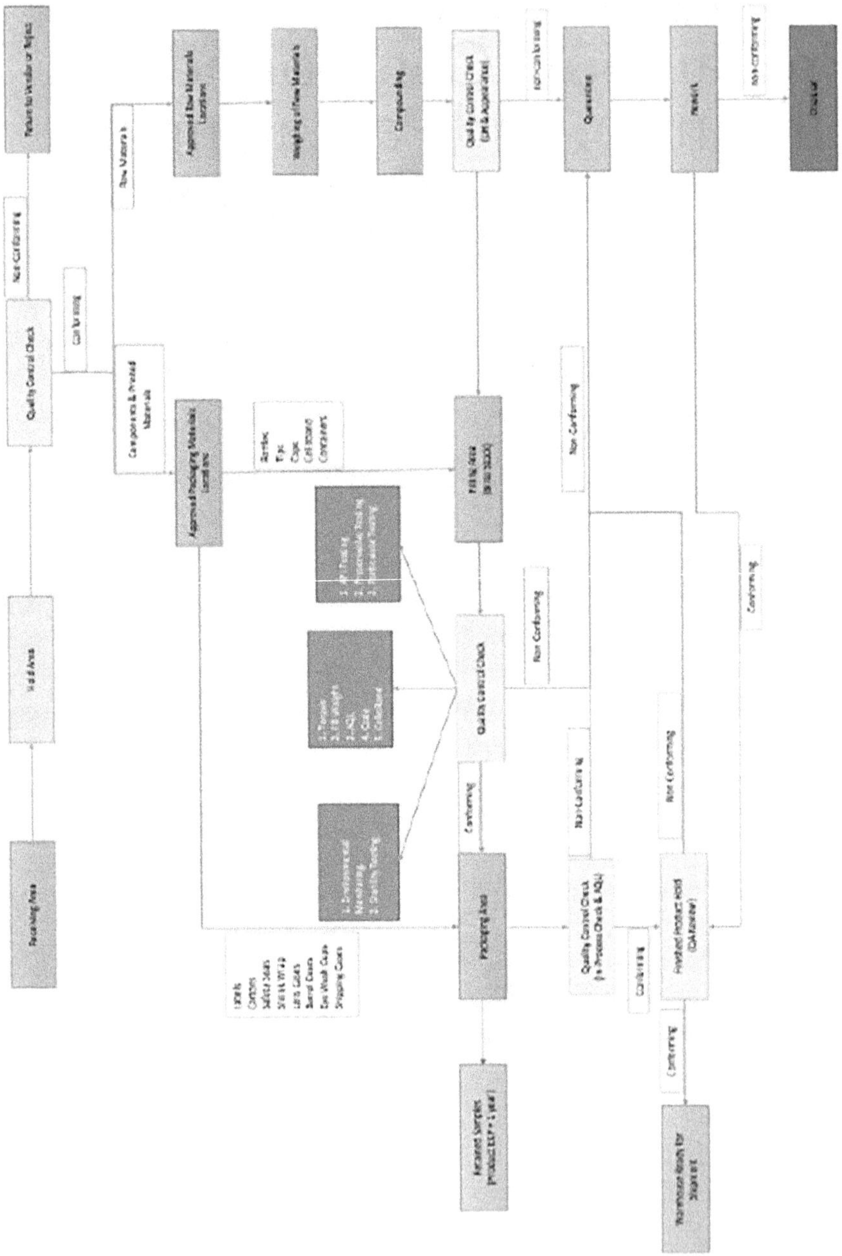

Appendix II: Master Validation Plan

1. PURPOSE

1.1 This procedure is intended to establish a uniform approach to process validation, including revalidation, for manufacturing drug products and medical devices at KC Pharmaceuticals, Pomona.

1.2 The purpose of this procedure is to define a validation plan that provides guidance to the various groups to assure that manufacturing systems, critical utility systems, support systems, equipment, and processes can operate as specified in a repeatable fashion.

2. SCOPE

2.1 This procedure applies to all manufacturing processes of drug products and medical devices at A.

2.2 This plan will provide basic validation requirements and guidance for validating existing facilities, equipment, and processes, as well as new equipment, facilities, and processes. This Validation Plan also provides guidance in the revalidation of existing facilities, equipment, and processes. It is an integral part of the A policy to ensure compliance with the appropriate FDA guidelines, ISO and cGMP requirements, and any appropriate customer requirements.

2.3 This plan also applies to software, hardware, and firmware validations for manufacturing processes as well as other processes that relate to QSR/ISO/cGMP requirements.

3. RESPONSIBILITIES

3.1 Validation is the responsibility of all departments associated with the development and manufacture of a product, including production, laboratory services, quality assurance, engineering, and business development.

3.2 The project leader shall plan the validation activities, ensure that validation activities are conducted, review and approve protocols and reports, and perform many of the validation activities related to products, processes, equipment, and facilities. The project leader shall coordinate validation activities with outside contractors.

3.3 The originator preparing a validation/revalidation must provide all documentation/information (as applicable). This includes the scope of the validation/re-validation (IQ, OQ, and PQ), equipment/room number, and the type of validation/re-validation (cleaning, product, software, stability, sterility, etc.). In addition, the validation/re-validation will not be closed out until an electronic copy has been submitted (and verified) to document control.

3.4 The individual who acquires production or laboratory equipment will be responsible for ensuring that all validation activities are completed. Engineering will provide validation support as requested.

3.5 The originator shall write the validation protocol, perform, or coordinate the tests, compile, and analyze the data, draft the report (using an applicable template), recommend the re-validation interval, and complete the approval process. If data are transcribed in tables in the final report, it will be subject to peer review to ensure accurate transcription.

3.6 The originator shall ensure that any documentation, new or revised, whether caused by or affected by the validation, is complete and current. The originator will implement, or administer the implementation of, any modifications or enhancements encountered during the validation.

3.7 The originator will complete a gap analysis report for validation studies in conformance to regulatory standards or requirements (i.e., software validation in conformance to 21 CFR 211, sterilization validation in accordance with ISO/ANSI/AAMI guidelines, etc.).

3.8 Any validation protocol that is open for more than 1 year will be expired. It is the responsibility of the originator to notify the coordinator of all validation protocols that will exceed 1 year.

3.9 Some validation processes are routine in nature and are well defined. To facilitate the control process, these validation processes may be written into procedures to describe in detail the validation process requirement. The procedures are approved and maintained under the A DCR process. The validation procedure specifies the validation parameters as well as required approval signatures in a specified (approved) form. Since the validation protocol requirements are described in the procedure and summarized and approved in the form, the approved form can be attached and used for archival purposes. The final report will need to go through the approval process accordingly.

3.10 Production is responsible to review and approve master validation plans (MVPs) and validation protocols and reports. Production also supports the execution of the protocol.

3.11 Engineering, process transfer, and QA are responsible to prepare, review, and approve validation protocols and reports. Engineering also performs quality risk analysis, provides data, and defines critical process parameters/steps where appropriate to ensure that process parameters used for validation are appropriate to produce product that meets pre-defined specifications.

3.12 Quality assurance is responsible to review and approve MVPs and validation protocols and reports.

3.13 QA or designee shall assign and electronically maintain numbers to validation protocols, as well as maintain the validation database and files. This individual shall ensure that protocols and reports meet the requirements of these procedure(s).

3.14 QA shall also be responsible for the database system (where validations are scheduled). Once the report has been signed off (i.e., revalidation interval), the originator will input recognition of the validation event into the system so that a report can be generated at the time of the next revalidation.

4. REFERENCES

4.1 Quality Manual
4.2 Document Control—Change Control System
4.3 Data Control and Data Recording Rules
4.4 Equipment and Process Change Control
4.5 Computerized System Validation
4.6 Non-Conformance
4.7 Documentation and Approval for Unscheduled Calibration, Maintenance, Repairs and Changes to Room Classification
4.8 Change Control Request
4.9 Equipment Qualification Installation Qualification (IQ) Checklist
4.10 Equipment Qualification Operational Qualification (OQ) Checklist
4.11 Equipment Information Input Sheet
4.12 Tooling Log
4.13 Equipment/Process Design Verification
4.14 Document Change Request (DCR) Flow
4.15 Equipment Change Control
4.16 Process Change Control
4.17 Product Change Control
4.18 Test Method Validation
4.19 Purchasing and Acquiring Inspection, Measuring, and Test Equipment for Manufacturing
4.20 Annual Product Review
4.21 Asset Inventory
4.22 Equipment Qualification—Manufacturing Systems
4.23 Guideline General Principles of Process Validation—FDA document
4.24 21 CFR—Part 820—FDA Quality System Regulation for Medical Devices
4.25 21 CFR—Part 211—FDA Drug cGMP

5. ATTACHMENTS

5.1 Attachment 1—Protocol Template
5.2 Attachment 2—Final Report Template
5.3 Attachment 3—Formatting Template

6. DEFINITIONS

6.1 Active Pharmaceutical Ingredient (API): Any substance or mixture of substances, intended for use in the manufacture of a drug product and that, when used in the production of a drug, becomes an active ingredient of the drug product.

6.2 AQL (Acceptance Quality Level) is defined as the maximum percent defective (or a maximum number of defects per hundred units of product based on sample inspection) that can be considered satisfactory as a process average.

6.3 Critical Process Parameters: A parameter whose value has a direct and measurable impact on the quality of the product.

6.4 Critical Quality Attribute: All attributes of the product that ensure the product is safe and effective. These are ensured by the API specifications and GMP controls.

6.5 Critical System: A system (including software and firmware), process, or piece of equipment used in the manufacturing of medical devices or pharmaceutical product which may affect the form, fit, function, or sterility of the manufactured product.

6.6 Concurrent Validation—Concurrent validation is acceptable. It is used rarely when data from three replicate production runs cannot be generated because only a limited number of batches can be produced. Prior to the completion of concurrent validation, validation batches can be released and used for commercial distribution provided they meet all validation acceptance criteria. Pre-validation batches produced prior to the concurrent validation batch(es) can be released and used for commercial distribution provided they were produced under GMP to meet all validation sampling and acceptance criteria, and there is no significant process change between the pre-validation batches and validation batches. If a concurrent validation batch fails to meet the acceptance criteria, batches that were already released for commercial distribution may be subject to recall.

6.7 Design Qualification (DQ): The documented verification that the proposed design of the facilities, systems, and equipment is suitable for the intended purpose.

6.8 Drug Product: The dosage form in the final immediate packaging intended for marketing.

6.9 Excipients: Substances which are introduced in the manufacturing process and become part of the drug product.

6.10 Factory Acceptance Test (FAT): Establishes documented evidence through vendor certification that the process equipment operates at specified ranges and conditions. Testing required typically includes utility verification, process parameter DOE, and method (feasibility) studies.

6.11 Gap Analysis: Gap analysis is an assessment to compare the actual performance with its expected standard performance.

6.12 Installation Qualification (IQ): A documented review of a system, facility modification, or process equipment and ancillary systems after installation to verify that key aspects of installation have been correctly performed. This includes verifying that calibration is complete, maintenance requirements and preventive maintenance have been scheduled, and operational procedures have been approved.

6.13 "Key" Parameters: A parameter in manufacturing which has an influence on the properties of the material in the appropriate process step. "Key" parameters will be treated as critical parameters as specified in this procedure.

6.14 MVP: A high-level document that establishes an umbrella validation plan for the entire process and is used to guide the project team in resource and planning.

6.15 Operation Qualification (OQ): A sound scientific study(s) that ensures that a system, facility modification, or process equipment operates as specified by the manufacturer or contractor and A requirements. An operational qualification may take the form of a design of experiments (DOE) in which the operational capability of the process is assured, and the process is optimized.

6.16 Process Capability Index: The capability of a process to meet the customer requirements determined by calculating the short-term process stability by determining the variations between a fixed number of samples. CpK is defined, as a minimum, by the CPU, CPL. The CpK must meet a minimum of 1.0.

$$CPU = \frac{UCL - \mu}{3\sigma} \quad CPL = \mu - LCL$$

where
UCL is the upper critical limit
LCL is the lower critical limit
μ is the average
σ is the standard deviation

6.17 Process Validation (PV): Process validation is defined as the collection and evaluation of data, from the process design stage throughout production, which establishes scientific evidence that a process is capable of consistently delivering quality products. Process validation involves a series of activities taking place over the lifecycle of the product and process. A process validation approach and documentation procedures are systematically consistent with the approach presented in this section and cover these requirements. This guidance describes the process validation activities in three stages:

6.17.1 Stage 1—Process Design: Process design is the activity of defining the commercial manufacturing process that will be reflected in the master production and control records. The goal of this stage is to design a process suitable for routine commercial manufacturing that can consistently deliver a product that meets its critical quality attributes. The commercial process is defined during this stage based on knowledge gained through development and scale-up activities.

6.17.2 Stage 2—Process Qualification (PQ): During this stage, the process design is confirmed as being capable of reproducible commercial manufacturing. This stage has two elements: (1) design of the facility and qualification of the equipment and utilities and (2) performance qualification (PPQ). During this stage, cGMP-compliant procedures must be followed, and successful completion of this stage is necessary before commercial distribution. Products manufactured during this stage, if acceptable, can be released.

6.17.3 <u>Stage 3—Continued Process Verification:</u> The goal of this third validation stage is to continually assure that the process remains in a state of control (the validated state) during commercial manufacture. A system for detecting unplanned departures from the process as designed is essential to accomplish this goal. Adherence to the cGMP requirements, specifically including the collection and evaluation of information and data about the performance of the process, will allow the detection of process drift. The evaluation should determine whether action must be taken to prevent the process from drifting out of control (§211.180 (e)).

6.18 Process Validation Addendum: An addition to an existing signed validation document to provide additional information or data to the original validation.

6.19 Process Revalidation: A validation performed on a part of a facility, a piece of equipment, or a process to assure that it remains in control (a state of statistical control). Revalidation is performed because of equipment and/ or maintenance changes to a validated part of the facility. Revalidation is typically not performed on a routine basis, unless required by a guideline.

6.20 Prospective Validation—Prospective validation is conducted prior to the distribution of either a new product or a product that has been significantly changed, i.e., made under a revised manufacturing process where the revision may affect product characteristics. It must be completed before the commercial distribution of the final drug product or device.

6.21 Site Acceptance Testing (SAT): Documented testing similar in nature to that of the Factory Acceptance Test, however, testing is conducted in the facility (site) on a much scaled-down version. SAT is conducted to ensure that no damage has occurred to the equipment in transit and that the unit functions properly. Determination of units for product OQ is specified by product specifications only.

6.22 Retrospective Process Validation: A validation conducted utilizing historic data generated from the routine operation of a part of a facility, a piece of equipment, or a process. Retrospective validation may be conducted for well-established processes that have been used without significant changes in raw materials, equipment, systems, facilities, or the production process. This validation approach may be used where:

6.22.1 Critical quality attributes and critical process parameters have been identified and documented. Appropriate in-process acceptance criteria and controls have been established.

6.22.2 No significant process/product failures attributable to causes other than operator error or equipment failures unrelated to equipment suitability have occurred.

6.22.3 Impurity profiles have been established for the existing API.

6.22.4 Batches selected for retrospective validation should be representative of all batches produced during the review period, including any batches that failed to meet specifications, and should be sufficient in number (three to nine batches) to demonstrate process consistency.

Retained samples can be evaluated to obtain data to retrospectively validate a process.

6.23 Worst Case: A set of conditions or process parameter limits which pose the greatest chance of product defects or failure but are within the acceptable range of process operation. The worst-case conditions or limits are not expected to produce product failure.

7. GUIDANCE

7.1 This procedure incorporates guidance from the US Food and Drug Administration (FDA), the Global Harmonization Task Force (GTHF), the International Organization for Standardization (ISO), and the International Society of Pharmaceutical Engineering (ISPE).

7.2 The Global Harmonization Taskforce (GHTF) and US FDA process validation guidance have substantial overlap in both pharmaceutical and medical device standards. Manufacturing products whose predicate regulations require process validation such as drugs, medical devices, API, and biologics should incorporate FDA standards plus assure the highest quality standards as being developed in process science and validation. FDA Process Validation Guide: General Principles and Practice was finalized in January 2011.

7.3 GHTF process validation standard, SG3/N99–10:2004, Quality Management Systems—Process Validation Guidance remains applicable and FDA 2011 standard explicitly stated that device firms were to follow GHTF guidelines ISO 13485:2003—Medical devices—QMS updated to harmonize with ISO 9001:2000. This is of particular interest to A, as a combination manufacturer that produces both drugs and devices, whereby there is a need to comply with both GHTF and 2011 FDA guidance standards.

7.4 Firms that produce drugs or medical devices are required to perform studies to manage risk and employ statistical tools. Besides the reference to ISO 14971, Application of Risk MANAGEMENT TO Medical Devices, GHTF standard reference the use of fault tree analysis (FTA) and Failure Modes and Effects Analysis (FMEA) to determine which aspects of a process pose the greatest risk to product quality. FDA guidance reference Design of Experiments to identify relationships between control and component inputs and process output characteristics in terms of significant interactions in the process. Proper documentation of statistical strategies in process validation is important to establish confidence in the results to ensure "high degree of assurance" requirements (820.75) such that this predicate is not violated, which will result in an inspection to declare the entire validation effort invalid. Therefore, risk assessment and statistical requirements should be employed to ensure compliance of a validated process.

7.5 Good Automated Manufacturing Process (GAMP) 5 as issued by ISPE define determinations whether software is, or is not, an integral part of equipment design. Validation plans or risk assessment documents would reference software 21 CFR part 11 (electronic records) impact. Although there are no specific requirements to separate validation efforts because of electronic record implications, company standards with respect to FDA part 11, ISO 13485, and EU Annex 11 tie validation of systems that process electronic records and electronic signatures (ERES) to the standard based on a separate computer system validation (CSV), which is audited separately.

7.6 Overall, the classic acceptance of three production runs to support a performance qualification has been established. FDA accepts that three production runs during process validation are a practical standard, and FDA recognizes that all processes may not be defined in terms of lots or batches. Three is scant for accepting arbitrarily a successful continuous trend. Ongoing monitoring of process variability and trending is a regulatory expectation. Trends in the process should be monitored to ensure the process remains within established parameters. Data collected should include relevant process trends. Data should verify that the critical quality attributes are being controlled throughout the process

8. PROCESS VALIDATION

8.1 Process Validation Approaches/Techniques—The preferred approach to process validation is prospective validation. There are, however, some circumstances when either concurrent validation or retrospective validation is an acceptable practice.

8.2 Building management systems and off-the-shelf programs that store labeling artwork and print and reconcile labels, have internal software processes that function independently of the equipment being monitored and operated; as such, these type systems might warrant their own validation as part of the facilities, in contrast, a PLC that was coded to operate the specific machine, for example, a heat sealer is arguably an integral part of that equipment validation. Software validation of the overall system, challenges of ladder logic as part of equipment qualifications combined with code documentation, and change control meet requirements under FDA guidance regulations.

8.3 Matrix Evaluations

8.3.1 Validation of different strengths may be done with reduced validation effort based on worst-case considerations and on scientific rational (matrixing and bracketing). If two or more drug substance (API) suppliers are to be used in the first validation of a finished drug product manufacturing process, a bracketing or matrixing approach could be chosen with respect to the number of validation batches (e.g., three batches with respect to the API of the first supplier and one batch with respect to the API of additional suppliers). At least one batch per API supplier should be required for the validation.

8.3.2 Packaging validations should be conducted for the finished drug product. For each packaging configuration, at least three packaging validation runs should be conducted on no less than 10% of each bulk finished drug product batch. If multiple packaging configurations exist (i.e., different bottle sizes and counts, etc.) a matrix or bracketing approach can be designed to validate the packaging process.

9. PROCESS VALIDATION REQUIREMENTS

9.1 Pharmaceutical Development—Process validation of drug products for use in early clinical trials is normally not required, where a single batch is produced or where process changes during development make batch replication impractical. The combination of process controls, calibration, and, where appropriate, equipment qualification controls drug product quality during development. Process validation as described in this standard should be performed when batches in pharmaceutical development are produced for commercial use, even when such batches are produced on a pilot or small scale.

9.2 Pharmaceutical Production—All critical manufacturing steps in pharmaceutical production necessary to produce the drug product beginning from the first formulation operation to the bulk drug product will be validated according to validated procedures.

9.3 Process Validation Procedure—Once it has been determined that a manufacturing process requires validation, the project leader, or designee, gathers all information needed to prepare the validation protocol and perform the validation. Manufacturing process validations involving more than one step will begin with an MVP prior to the approval of the validation protocols for each individual step. This MVP describes all validation activities necessary for the validation of the manufacturing process. In a multi-step production, for example, component preparation, compounding, and filling operations, each step is considered an independent process, i.e., each step is validated independently from the others. Validation of individual steps in general will proceed as follows:

9.3.1 Approve validation protocol with pre-determined acceptance criteria

9.3.2 Monitor critical process parameters

9.3.3 Evaluate analytical results

9.3.4 Approve validation report including all data necessary

9.3.5 Pre-validation batches may be produced prior to the validation batches and commercial batches afterward.

9.4 Critical Process Parameter/Steps—Critical processing steps, parameters, and ranges of an API or drug product should be determined during the design phase and included in the development report. Within the scope of a quality risk analysis, these parameters/steps must be challenged. With the use of risk analysis, conclusions should be drawn regarding critical impurities (e.g., genotoxic impurities). These conclusions are also to be part of the development report.

9.5 Prerequisites

9.5.1 Equipment Qualification (EQ)—Equipment associated with the process must be qualified and in current calibration at the time of validation.

9.5.2 Environmental Conditions—If the product is produced in a controlled environment, the HVAC equipment must be qualified or undergoing qualification concurrently (for new installations) and an ongoing environmental monitoring program implemented.

9.5.3 Regulatory Dossier, Drug Master File (DMF), and other regulatory filings: Changes to the process and analytical methods that affect the regulatory filings because of process changes mandating validation will be included in an update to the appropriate filing after validation but prior to the release of the product to the end customer. Ensuring the appropriate disposition of the product is managed in the change control procedure.

9.5.4 Development Report—For new products, the development report will identify the critical process parameters that are required for validation.

9.5.5 Analytical Methods Validation—Analytical methods must be validated in accordance with the standard on analytical validation.

9.5.6 Raw Materials—Raw materials must be qualified. Qualification includes written specifications and test methods and confirmation that the vendor has been entered into the vendor qualification program.

9.5.7 Deviations—All open deviations must be closed prior to the approval of the validation report.

9.5.8 Training—All required training for production log records, SOPs, and validation procedures must be conducted and documented prior to the start of the validation.

9.6 Process Validation Documentation—This section describes the main documents established during process validation.

9.7 MVP—This document is only applicable to the validation of a multi-step drug product/medical device manufacturing process. When an MVP is required, the following topics need to be addressed at a minimum, if applicable:

9.7.1 Scope

9.7.2 Responsibilities

9.7.3 Documentation

9.7.4 Resources/training

9.7.5 Process description

9.7.6 Critical process parameters/acceptance criteria in general

9.7.7 Equipment/maintenance/calibration

9.7.8 Cleaning

9.7.9 Schedule

9.7.10 Other items may be added as deemed necessary by the validation team.

9.8 Validation Protocol—For each validation step, a protocol must be written that details the production process, its critical parameters, and acceptance criteria. The protocol should contain the following at a minimum, if applicable:

9.8.1 General overview

9.8.2 Validation team

9.8.3 Responsibilities

9.8.4 System description

9.8.5 Raw materials

9.8.6 Release testing

9.8.7 Equipment qualification

9.8.8 Critical process parameters

    9.8.9   Acceptance criteria

    9.8.10 Composition of the formulation

    9.8.11 Description of the production process

  9.9 Validation—The validation report should address the following at a minimum, if applicable:

    9.9.1   Summary

    9.9.2   Purpose/scope

    9.9.3   Summary of analytical results

    9.9.4   Comments/deviations

    9.9.5   Conclusions

    9.9.6   Measures/recommendations

  9.10 Managing Changes to Approved Validation Documents—If an approved version of a validation document (MVP, validation protocol, or validation report) undergoes any changes to its content, then the latest version of such document must be edited and approved by all approving parties of the previous version. An addendum or amendment might be appropriate for this purpose.

  9.11 Deviations During Validation—Deviations will be documented as required by the standard on deviations, and site-specific SOPs. This should be addressed in the validation report. A validation study fails if one or more of the validation batches do not meet the predefined acceptance criteria. Corrective actions to resolve the issues causing the failure must be identified and implemented before another validation study may be conducted. Some issues, unrelated to the process, may result in aborted validation attempts and will not be considered validation failures. In such cases, the batch will be identified "No Validation Result" and a replacement validation batch will be performed within the same campaign if possible. Deviations should be closed prior to the approval of the final report. Specific issues which are not considered validation failures include, but are not limited to:

    9.11.1 Mechanical failure of equipment

    9.11.2 Failure of a utility system

    9.11.3 Confirmed operator error such as failure to take a validation sample

10. CONTROLS/RELEASE OF MATERIAL—APIS AND DRUG PRODUCTS MAY BE RELEASED FOR FURTHER PROCESSING OR COMMERCIAL DISTRIBUTION BEFORE THE VALIDATION REPORT IS COMPLETE AS LONG AS THE VALIDATION RESULTS ARE EVALUATED AND APPROVED BY QUALITY

11. REVALIDATION/CHANGE CONTROL

  11.1 Equipment Change—If a process is transferred to different manufacturing equipment within the same manufacturing site, the change must be evaluated regarding the degree of comparability to the validated process and the extent of revalidation required if any. For example,

11.1.1 The new manufacturing equipment is the same as the originally qualified equipment (i.e., like-for-like changes). In this case, the manufacturing process is considered validated in the new manufacturing equipment. This must be thoroughly documented and approved by the quality department.

11.1.2 The new manufacturing equipment is different from the originally qualified, but the difference is not relevant to the process step under discussion. For example, product transfer to a line of different materials yet compatible with the product(s). In this case, the manufacturing process is considered validated in the new manufacturing equipment. This must be thoroughly documented and approved by the quality department. Documentation should include a comparison of impurity profiles of at least three batches.

11.1.3 The new manufacturing equipment is different from the originally qualified equipment and the difference is relevant to the process step under discussion. This requires validation, the extent of which will depend on the specific equipment affected.

11.2 Change Control

11.2.1 Revalidation is required in the following cases:

11.2.1.1 Any major change is implemented. As a result of several minor changes that together are determined to potentially affect product quality. If more than 10% of the batches in a campaign fail with no assignable cause, the process must be reviewed. If appropriate, the process will be revalidated. The customer may also require revalidation. In addition to the annual product review, a review of the validation status will be performed after 6 years. The following attributes should be part of the review of the validation status. The review should be documented:

11.2.1.1.1 Review of current Master Batch Record

11.2.1.1.2 Review of the last Process Validation (protocol, report)

11.2.1.1.3 Review of all changes related to the manufacturing process and production equipment

11.2.1.1.4 Verification if the last process validation complies with the current requirements of this Standard

11.2.1.1.5 Review of analytical method changes and assessment of their influence on the previous purity profile comparisons

11.2.1.1.6 Review of deviations and complaints related to product and manufacturing process

11.2.1.1.7 Review of changes related to the specifications (critical raw materials, reviewed process step)

11.3 The following changes are considered major changes:

11.3.1 Equipment

11.3.2 Relaxing or deleting a raw material specification that is likely to affect product quality

11.3.3 Product specifications change not based on process performance

11.3.4 For pharmaceutical manufacturing, a change of the API supplier

11.3.5 Process changes

11.3.6 Batch size increase/decrease

11.3.7 Critical parameters

11.3.8 Addition/deletion of critical processing steps

11.4 The following changes are considered minor changes:

11.4.1 Editorial changes to batch records

11.4.2 Waste treatment/air emissions changes

11.4.3 Change to "equivalent" equipment (like-for-like)

11.4.4 Changes to non-critical parameters

11.4.5 Tightening of critical parameters

11.5 Influence on process validation/qualification of processes when changing vendors—There is no influence on the process validation status when the vendor for a raw material or intermediate changes if the specifications and quality are not changed. Considering the following criteria or individual process-specific requirements qualification measures could be necessary:

11.5.1 Intended use (reagent, solvent, process aid, etc.)

11.5.2 Step in which the raw material is used

11.5.3 Critical/non-critical raw materials

11.5.4 The following measures and requirements must be executed and documented under change control. The execution of lab experiments must be considered in every case.

11.6 Analytical testing of raw material quality—raw material, process aid, cleaning agent, and API, all require testing. Samples/batches of raw material from the new vendor are compared with three representative samples/batches from the existing vendor. If there are only two batches available, the comparison could be done using these two samples as an exception.

11.7 Assurance that manufacturing systems are fit for intended use should not rely solely upon verification after installation but be achieved by a planned and structured verification approach throughout the system lifecycle.

11.7.1 Product Life Cycle—quality risk management (QRM) and quality by design (QbD) concepts are incorporated into facility and system verification efforts. At the beginning of 2011, the US FDA published an update on pharmaceutical process validation titled: "Guidance for Industry—Process Validation: General Principles and Practice." Based on the principles of IGH Q8, Q9, and Q10 (Pharmaceutical Development, Quality Risk Management, and Pharmaceutical Quality Systems), the FDA guide (PVG) targeted risk-based GMP initiative. The PVG is structured on the lifecycle concept: The objective of process validation is a state of ongoing control across the entire product development and manufacturing lifetime. The specific architecture that the agency applies to the life cycle model is a three-stage model that begins with process design and ends only with the discontinuation of manufacture. This contains commissioning and qualification, which is referred to as stage 2a of the FDA process validation cycle. ASTM E2500–07, the standard guide for specification, design, and verification of pharmaceutical and biopharmaceutical manufacturing systems and equipment, is cited by FDA PVG as guidance for activities that verify that facilities, systems, and equipment are fit for their intended use as verified by qualification.

11.7.1.1 QbD—The fundamental principles of PVG are that quality must be designed into a process from the beginning. It cannot be ensured merely by inspection or sampling and testing. The key to defining facility and equipment quality requirements is based on the process knowledge gained during stage 1—process design. PVG states "This knowledge and understanding is the basis for establishing an approach to control of the manufacturing process those results in products with the desired quality attributes." FDA Expects manufacturers to:

11.7.1.1.1 Understand the source of variation

11.7.1.1.2 Detect the presence and degree of variation

11.7.1.1.3 Understand the impact of variation on the process and on product attributes

11.7.1.1.4 Control the variation in a manner commensurate with the risk it represents to the process and product

11.7.1.2 Design-Based Control Strategies—Critical aspects of manufacturing systems are typically functions, abilities, performance, or characteristics necessary for the manufacturing process and systems to ensure consistent product quality and patient safety. Failure modes and effects criticality analysis (FMECA) is an example of a type of risk assessment method that is practical in identifying specific risks and suggests controls, which is applicable to critical aspects identification. Other scientific approaches such as the design of experiments might be applicable.

11.7.2 To effectively apply QbD and QRM to the design and delivery of manufacturing facilities, equipment, and systems, a functional level of process knowledge regarding the intended use of assets must be available. Based on this knowledge, Subject Matter Experts (SMEs) can define the appropriate quality requirements according to regulations (guides), and specific requirements affecting product quality and patient safety with respect to product knowledge, process knowledge, regulatory requirements, and company quality standards. Product and process knowledge comes from earlier stages (FDA—Stage 1) of process design and development along with relevant manufacturing experience linking the engineering designs of the facilities and systems that support those requirements. Namely, the critical utility systems including water for injection (WFI) or purified water (PW), clean steam (CSM), clean compressed gases (CCG), Heating ventilation and air conditioning (HVAC), reverse osmosis water (RO), and classified clean rooms should all be governed by establishing a trail of documentation to cover: user requirement specifications (URS), machine design specifications (MDS), software design specifications (SDS), hardware design specifications (HDS), functional design specifications (FDS), design review (DR), installation, operation, and performance qualifications (IQ, OQ, and PQ). Similarly, manufacturing systems should follow all the aforementioned requirements in addition to the factory acceptance test (FAT), commissioning and site acceptance test (SAT), and following PQ a full equipment criticality analysis (ECA) to define gaps and equipment life cycle for sustainable compliance in producing safe and efficacious medicinal and pharmaceutical products.

11.7.3 ASTM E2500–07 provides the necessary high-level strategy for science and risk-based verification that facilities and systems are fit for use. In addition, commissioning and qualifications methodologies should be embedded in quality systems across the design, manufacture, and distribution of life sciences-regulated products. To facilitate this effort, ISPE published two guides in 2011—Science and Risk-based Approach for the Delivery of Facilities, Systems, and Equipment (FSE Guide) and Applied Risk Management for Commissioning and Qualification (ARM Guide). FSE guide emphasis is on new or flexible quality systems without significant legacy. The ARM guide focuses on embedded terminologies within established quality systems, whereby the culture is non-risk-based, and the organization quality system's maturity needs development.

11.7.4 FDA process validation guidance establishes a three-stage lifecycle: Stage 1—process design and development, stage 2 is divided into 2a for C&Q of equipment, systems, and facilities and 2b to focus on risk management. The goal is to support verification approaches to those systems that are fit for their intended use.

12. SITE VALIDATION REQUIREMENTS

12.1 Design Qualification (DQ)

12.1.1 The first element of the validation of new facilities, systems, or equipment could be design qualification (DQ).

12.1.2 The compliance of the design with GMP should be demonstrated and documented.

12.1.3 The verification that the proposed design of the facilities, systems, and equipment is suitable for the intended purpose is documented.

12.2 Installation Qualification (IQ)

12.2.1 The Installation Qualification should be performed on all new equipment, and on existing equipment that has had its installation modified.

12.2.2   The Installation Qualification verifies that proper installation was performed, appropriate manufacturing procedures are in place, required calibrations are complete and maintenance (if applicable) procedures are in place. Equipment Qualification/Installation Qualification (IQ) Checklist may be used in lieu of a formal protocol and final report.

12.2.3   The Installation Qualification provides documented evidence that all utilities for the process equipment are installed per the manufacturer's guidelines.

12.2.4   For all equipment using AC Power, utilities should be verified for the correct power supplied to each piece of equipment. Equipment should have a nameplate or a specification sheet attached and be verified with a voltmeter.

12.2.5   Grounding sources should be verified and documented, where applicable.

12.2.6   Environmental conditions (temperature and humidity) of the clean room or designated area should be verified and documented against the manufacturer's specifications as required.

12.2.7   Safety Features:

    12.2.7.1   When operating under normal conditions, verify that all **E-stops** (if any) operate as designed.

    12.2.7.2   When operating under normal conditions, verify that all **Safety Doors** (if any) operate as designed.

    12.2.7.3   Alarms: At each instant when an alarm condition is verified, the equipment must return to the initial operation condition.

12.2.8   In addition, the system as a "whole" is qualified: "Worst-case" conditions of main utilities are challenged, and all measuring equipment is calibrated.

12.2.9   Worse-Case Conditions: Utilities—Power, water, and/or compressed air (or other gases). For each set of conditions, turn off the listed utility, document how the equipment responds to the condition (equipment, where indicated, should respond as designed), and record the final response after the condition has been returned to normal operation conditions (equipment should regain stability).

12.2.10 All devices (i.e., pressure gauges, temperature gauges, and/or conveyor speed) that are critical to the proper operation of process equipment should be calibrated and clearly labeled to indicate the calibration status.

12.2.11 All auxiliary equipment (i.e., weight check for balance, chart recording or SCADA monitoring/alarm for storage chamber, oven, etc.) should also be identified and clearly listed in the IQ report.

12.2.12 All worst-case testing/verification of PLC software (I/O Logic) should be documented by the Engineer qualifying the equipment and verified by QA (signature/date).

12.2.13 Tooling specific to a given customer must be documented. The history for all tooling will include the product, device ID number, acquisition date, location, description, material, and passivation date (if required).

12.2.14 A formal documented summary of the Installation Qualification (IQ) is required following completion of the IQ (the Validation, Final Report shall be used as a template). This document shall serve as a history of the activities taken place as well as to indicate the completion of the IQ portion of the validation. The Installation Qualification must be closed (passing results with required signatories for all appropriate documentation) prior to beginning the Operational Qualification.

12.2.15 Any discrepancy or deviation to the protocol found during the execution of Installation Qualification (IQ) must be viewed as a deviation to the protocol and shall be documented in the "Exception to Protocol" section. This section must clearly state the findings and investigation, evaluate the root cause and how it would impact the system performance, and provide detailed corrective action or justification of why the system at its current state is acceptable.

12.2.15.1 Observation: Clearly state the findings and how it is different from the original specification.

12.2.15.2 Investigation: Describe the investigation performed to find the facts and the evaluation of the root cause of the deviation, the corrective action, and how it will affect the equipment or system performance.

12.2.15.3 Conclusion: Provide the details of the corrective action to repair the discrepancy and the testing performed of the corrective action. Or provide justification of why the system as in its status with its deficiencies is adequate for the intended service.

12.3 Operational Qualification (OQ)

12.3.1 The Operational Qualification is required to evaluate the limits of the capability of the manufacturing process. The method by which the extremes of the process limits impact the form, fit, function, or sterility of a product(s) varies among validations. However, the method(s) (test method) that will be used are required to be described in sufficient detail in the Rationale section of the Validation Protocol. The method(s) used to assess the capability of the process being validated and the acceptance limits are also required. Equipment Qualification/Operational Qualification (OQ) Checklist should also be used to document the Operational Qualification for equipment of limited functional parameters.

12.3.2 Equipment may be purchased where the factory acceptance test may be completed. With such comprehensive data, it is not necessary to complete an extensive OQ, but a reduced OQ, a site acceptance test, where data are captured primarily to ensure there is no damage to the equipment in transit and sampling for the product run(s) is determined by product specifications.

12.3.3 In such instances, where product builds are needed to satisfy OQ requirements, the number of units should be based on the engineering (project) team or design product specifications.

12.3.4 The closure of the Operational Qualification (OQ) will be signified by a formal report (the Validation, Final Report shall be used as a template) detailing the summary of all activities and the status of all items being qualified. The Operational Qualification must be completed and approved by Engineering (Technical Services) and QA (passing results with required signatories for all appropriate documentation) prior to beginning the Performance Qualification.

12.3.5 Any discrepancy or deviation to the protocol found during the execution of Operational Qualification (OQ) must be viewed as a deviation to the protocol and shall be documented in the "Exception to Protocol" section. This section must clearly state the findings and investigation, evaluate the root cause and how it would impact the system performance, and provide detailed corrective action or justification of why the system at its current state is acceptable.

12.3.5.1 Observation: Clearly state the findings and how it is different from the original specification.

12.3.5.2 Investigation: Describe the investigation performed to find the facts and the evaluation of the root cause of the deviation and the corrective action and how it will affect the equipment or system performance.

12.3.5.3 Conclusion: Provide the details of the corrective action to repair the discrepancy and the testing performed of the corrective action or provide justification of why the system in its current state with the identified deficiencies is adequate for the intended service.

12.3.6 Development of Procedures:

12.3.6.1 Operational procedures, maintenance procedures, and other relevant procedures shall be developed prior to Installation/Operational Qualification (IOQ) and Operational Qualification (OQ) protocol pre-approval submission.

12.3.6.2 The operational and maintenance procedures shall be finalized during the IOQ or OQ protocol execution.

12.3.6.3 The operational procedures and maintenance procedures shall be approved and issued prior to the approval of IQ or OQ protocol final report.

12.3.6.4 All planned Preventive Maintenance shall be made active in the maintenance management system, prior to the approval of the IOQ or OQ protocol final report.

12.4 Process-Development Activities (Engineering Run)

12.4.1 Engineering run may include, mixing study, dosing study, placebo batch runs, active batch runs, or any run that helps to evaluate processes or process parameters.

12.4.2 Depending on the complexity of a project, one or more engineering runs may be required to evaluate the equipment, process, and evaluate the operating parameters.

12.4.3 It is the project leader's responsibility to conduct engineering runs under a protocol with a clear definition of objectives and a summary report to conclude and document the findings.

12.4.4 The Engineering Run protocol is subject to pre-approval and final approval like any other protocol, and once the protocol is complete the protocol and all supporting documents must be submitted to QA for archiving.

12.4.5 Documentation for Engineering Runs

12.4.5.1 Upon initiating an Engineering Run protocol, the Project Lead is responsible for ensuring that all instructions required for proper execution of the Engineering study or run are captured and detailed in an appropriate manner. The set of instructions may be created as an attachment to the protocol and reviewed and approved as part of the DCR submission package for the Engineering Run protocol or as a separate controlled work order or batch record.

NOTE: Upon execution of the protocol for the Engineering Run, any exception to the protocol or actions that were taken which were not identified in the approved protocol may be identified in a separate section of the protocol titled, "Exceptions to the Protocol."

12.4.5.2 The following detail examples of acceptable instructional documentation which may be utilized as a part of the Engineering Run:

12.4.5.2.1 Example 1: The Project Lead may create a set of instructions based on the existing format of an approved batch record for that product line. The header of each page will only reference the corresponding Engineering Run protocol number and all pages will be numbered accordingly. These pages become part of the Engineering Run protocol as an attachment. Implementation and usage of the set of instructions may only be issued to production and used upon approval of the protocol. If an Engineering run involves Production personnel, training shall be performed and documented per A's Training procedure.

12.4.5.2.2 Example 2: As an alternative method to capture all instructions required as per the protocol, the Project Lead may create a new work order or batch record through A's Document Control, Change Control System The work order will be issued to the Production floor as a controlled document stamped with "Engineering Run."

If an Engineering run involves Production personnel, training shall be performed and documented.

Note: At no time may an "unapproved," "unofficial," or "draft" document be issued to the floor and used to capture study data collected as part of the Engineering Run. All engineering runs will be executed in accordance with the approved protocol and/or applicable regulations (i.e., cGMP) as detailed in the protocol.

12.5 Exhibition/Submission Lot

12.5.1 Instructions for exhibition or submission lots will be documented in a newly created work order or batch record through A's Document Control, Change Control System. The work order will be issued to the Production floor as a controlled document.

12.5.2 Training of all personnel involved will be performed and documented per A's Training procedure.

12.5.3 Deviations from the protocol and/or work order will be documented and approved.

12.6 Performance Qualification (PQ)

12.6.1 Three lots of statistical significance (determined by the Project Team) may be produced at the nominal setting (defined during the Operational Qualification). Each lot should be subject to predetermined visual and functional acceptance criteria, including content uniformity, if necessary, specified by the Project Team and/or Customer.

12.6.2 Deviation from the Protocol will be documented and approved.

12.6.3 A validation study fails if one or more of the validation batches do not meet the predefined acceptance criteria. Corrective actions to resolve the issues causing the failure must be identified and implemented before another validation study may be conducted.

12.6.4 Some issues, unrelated to the process may result in aborted validation attempts and will not be considered validation failures. In such cases, the batch will be identified "No Validation Result" and a replacement validation batch will be performed within the same campaign if possible.

12.6.5 It is the responsibility of the project leader to have all newly purchased equipment/or systems that require revalidation be tagged and subsequently logged with an Engineering (System ID) number.

12.6.5.1 This number shall serve as the system traceability number for any unit to be validated. Equipment Information Input Sheet shall be filled out to include equipment name, manufacturer, model number, and re-validation interval.

12.6.6 Any equipment that has completed all runs/cycles for the Performance Qualification must be summarized/analyzed in the Final Report for any equipment to be released to production for routine use. The closure of the PQ is defined by a formalized report signed by all required signatories like those of the original protocol. Prior to the release of the final report, all relevant documentation (procedures—SOPs, training, etc.) shall be released through Document Control for proper operation and training of the specified equipment.

12.6.7 Deviation from the Protocol will be documented and approved.

12.7 Future Process Validations and Replacement Equipment

12.7.1 Future Process Validations (New Product)

12.7.1.1 Following the verification of all acceptance criteria of the Equipment Qualification, the Installation Qualification and Operation Qualification should not be repeated for subsequent process validations. However, Performance Qualifications should be repeated to meet individual product specifications if no rationale for exclusion has been provided by Engineering or Quality Assurance.

12.7.1.2 To link new process parameter data to the same piece of equipment, a new template will be designed for each new product parameter as required.

12.7.2 Replacement Equipment

12.7.2.1 For all replacement or modified equipment, an Installation and Operational Qualification should be repeated to ensure new equipment is installed correctly and to define new parameters.

12.7.2.2 The Qualification should consist of one lot at the nominal setting subject to predetermined visual and functional acceptance criteria specified by the Project team.

12.7.2.3 New template data shall be entered through the same process described earlier for newly purchased equipment. All previous equipment will be archived in the database.

12.8 Revalidation

12.8.1 Revalidation may be required when any aspect of a validated system has been changed and these changes have the potential of affecting the product produced by that system. Revalidation also applies to validated systems such as water systems, HVAC systems, or other systems where such a system is integral to the manufacturing process.

12.8.2 If any part of a validated system is changed in a manner that will affect the validated parameters of that system, the critical system may require revalidation. If full revalidation is not required, additional validation or Quality Assurance testing may be required. If such testing is determined (through review as part of the change control system) to be required, the results of the testing will be documented as an addendum to the original validation.

12.8.2.1 If complete revalidation is required, the QA function will quarantine (or take out of service) the critical system until the validation is completed. If this is not feasible, the Validation function and Quality Assurance will develop a rationale for the non-validated system remaining in use during the period of the validation. This rationale will identify any additional sampling, additional monitoring, or data review that is required during the quarantine period.

12.8.2.2 Revalidation may be required for specific predefined critical systems at predetermined intervals (i.e., autoclaves).

12.8.2.3 If a system fails the revalidation, the QA function will issue an out-of-validation notice. This document invalidates the critical system, and the system cannot be used in manufacturing until it is released by validation via the appropriate documentation. The critical system will be identified (tagged) as out of validation.

12.8.2.4 Revalidation is necessary only when engineering changes have been performed on the validated system which impacts the validated parameters of the system. If a change to the validated system is made that does not impact the validated equipment parameters as justified by engineering rationale, revalidation is not necessary.

12.8.2.5 If a hot fix is required for a piece of software to maintain the integrity of the data within the program, the change is made. Testing of the hot fix will be performed but a full validation will occur during the next full update of the software.

12.8.2.6 Feature updates in software will require a revalidation.

12.8.3 If the supplier of the active pharmaceutical ingredient changes during the commercial lifetime, a new validation is necessary. The extent of the validation should be established in a risk assessment. Reduced sampling may be possible if the impact of the API on the individual process steps is not critical.

12.9 Revalidation Schedule

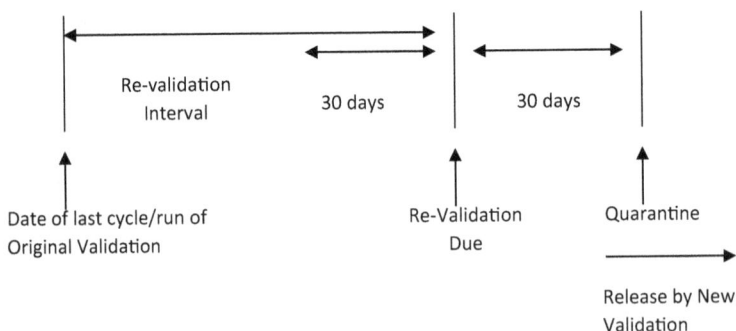

12.9.1   The date of revalidation is defined by the last manufacturing cycle/ run of the original validation plus the validation interval.

    12.9.1.1   For example, if the validation interval is 1 year and the date of the last run of the initial validation is April 26, 2022, then the revalidation will be due on April 26, 2023.

12.9.2   Completion of a revalidation should occur within 30 days of the scheduled revalidation date.

    12.9.2.1   The closure of a revalidation includes all preparatory work, the conclusion of all required runs, laboratory data, and the approved (by all signatories) final report.

12.9.3   **Equipment:** For any piece of equipment that fails to meet the **30-day** period for completion of a revalidation shall have all products manufactured (after validation expiration) on this piece of equipment quarantined until satisfactory results have been achieved.

12.9.4   **Gamma Sterilization:** For any cycle that cannot be revalidated during the scheduled time period, a deviation will be issued. Quarterly dose audit can be performed at any time within the time period of the specified calendar quarter. If samples are not available for the quarter to perform the dose audit, it should be documented in a memo. If there is a meaningful change in the production process, revalidation to evaluate the sub-process dose may be considered.

    12.9.4.1   The revalidation shall commence at the next lot(s) manufactured of the specified product.

12.10 Change Control System

  12.10.1 The QA function is responsible for assuring that a change control system is implemented and is operational at the plant level.

    12.10.1.1 The change control system will assure that any change to a validated system (equipment, process, or product), which impacts the validated parameters of that system, is reviewed by at least the quality function.

    12.10.1.2 If a change to a validated system involves a change in the controlled documentation, it will be controlled by the master change control procedure.

    12.10.1.3 If a change to a validated system impacts its validated parameters for which there is no engineering rationale, the QA function will take the appropriate action(s) to assure that the equipment is not used in the manufacturing process until a revalidation is complete.

    12.10.1.4 If an unscheduled change is made to a validated system/ process that impacts its validated parameters, the necessary actions will be taken to revalidate the system/process as quickly as possible. If a product manufactured during validation is to be released prior to the completion of the validation, the requirements for release will be defined in the Validation Protocol. Otherwise, the product will remain on hold until the Process Validation Technical Report is approved.

12.10.1.5 No modification to a validated part of a facility, acquisition of a new piece of equipment which requires validation, addition of a new process requiring validation, or alteration of an existing validated process may be performed without the involvement of the QA function. The QA function will assure that non-validated systems and equipment are identified and scheduled for validation if validation is appropriate.

12.10.1.6 The originator is responsible, upon receipt of applicable approvals to enter the appropriate information of the specified equipment in the approved record as well as the recommended revalidation interval.

12.10.1.7 The QA function is required to establish and maintain a record-keeping system that will identify and control all executed process validation documentation. The requirements for this system are:

12.10.1.7.1 Each Process Validation will be issued a unique number.

12.10.1.7.2 The QA function will assure that approved documentation is sufficiently controlled to prevent the removal of original documents without the knowledge of validation, loss of original documentation, or misfiling of original documentation.

12.10.1.7.3 The QA function will assure that the information in the Process Validation files is sufficiently protected or backed up so that, in the event of a catastrophe, the documents are protected or can be regenerated in a usable form.

12.10.1.7.4 The QA function will maintain a listing of all Process Validation documents and their status. This listing will be maintained current.

12.10.2 Refer to Quality Manual

12.10.3 Document Change Control

12.10.4 Equipment and Process Change Control

12.10.4.1 Equipment and Process Change Control

12.10.4.1.1 New Equipment

12.10.4.1.2 Existing Equipment

12.10.4.2 Equipment Change Control

12.10.4.3 Process Change Control

12.10.5 Product Change Control

12.10.5.1 Master Project Plan Template

12.10.5.2   Product Realization Process

12.10.5.3   Product Change Control

12.11 Document Archival

12.11.1   QA is responsible for document archival (both electronic and hard copy), upon receipt of any Protocol/Final Report.

12.11.2   Verify Originator has prepared the electronic package (scanned PDF copy of Protocol/Final Report and any applicable attachments), if not return the package to the originator and request electronic copies to be prepared.

12.11.3   Update Protocol Database with approval date for Protocol/Final Report.

12.11.4   Archive electronic copies.

Archive original documents using Asset Inventory

NOTE: Electronic/Hard copy archive should be completed within 30 days of receipt of the complete, approved package.

12.12 Revalidation Program Evaluation

12.12.1   For any process which requires periodic revalidation, MVP Reference # is assigned.

12.12.2   This MVP Reference # is to provide scheduling and tracking of the validation schedule. The Validation Schedule Crystal Reports will be generated on a minimally monthly basis to determine the status of the revalidation for the equipment (or equipment related/item/software) to be qualified by Manufacturing or Quality personnel.

12.12.3   The Engineering, Production, Laboratories Services, and QA function will, on a quarterly basis, evaluate the validation program for manufacturing for the upcoming quarter.

12.12.4   QA function will perform, on an annual basis for commercial products revalidation review of the requirement of all active equipment, evaluation of production process in conformance with new or revised regulatory standards, and requirements for materials.

12.13 Documentation

12.13.1   Good documentation is essential to a successful validation. Data-recording rules must be followed.

12.13.2   All entries must be made at the time the operation is performed.

12.13.3   Protocols must be designed to allow real-time or contemporaneous documentation of actual sample collection times during the execution of the protocol.

12.13.4   QA to verify the data collection methods (batch record, forms, etc.) referenced in the protocol to ensure the protocol is designed to capture the required data.

ATTACHMENT 1—PROTOCOL TEMPLATE

ATTACHMENT 2—FINAL REPORT TEMPLATE

ATTACHMENT 3—FORMATTING TEMPLATE

ATTACHMENT 4—EXAMPLE OF A MIXING STUDY

ATTACHMENT 1—PROTOCOL TEMPLATE

1. Introduction
   1.1 Identify equipment, process, or facility being validated. Use serial numbers, model numbers, and engineering identification numbers wherever possible
2. Purpose
   2.1 Explain the reason for the validation and what requirements are sought
3. Scope
   3.1 Identify the boundaries of the validation
4. Reference
   4.1 The originator will complete a gap analysis report for validation studies in conformance to regulatory standards or requirements (i.e., software validation in conformance to 21CFR211, sterilization validation or sealing validation study in accordance with ISO/ANSI/AAMI guidance, etc.)
5. Equipment
   5.1 List all equipment that will be used specific to this protocol. It is acceptable to reference work instructions or batch records to prevent repeating the list.
6. Materials
   6.1 List all materials that will be used specific to this protocol. It is acceptable to reference work instructions or batch records to prevent repeating the list.
7. Responsibility
   7.1 Identify what department (s) will perform each function of the validation. Indicate what work will be performed by an outside laboratory or contractor.

Activity	Validation	Engineering	Manufacturing	Laboratory	QA	Customer
Develop IOQ	x	x			x	
Approve IOQ	x	x	x	x	x	x
Execute IOQ	x	x	x			
Develop Summary Report	x					
Approve Final Report	x	x	x	x	x	x
Develop SOP	x	x	x			x
Perform Testing	x		x	x		
Perform Calibration	x	x				

8. Rationale
   8.1 Provide justification for the validation approach that is being taken.
   8.2 Identify the key input variables and the methods of controlling them.
   8.3 Identify the output variables and the methods of controlling them.
   8.4 Define and justify acceptable tolerances, including controls.

9. Procedure

    9.1 Design the experiment. Describe the procedure as it will be performed. Design necessary forms for results so they may be documented as the test is performed.

    9.2 Provide step-by-step instructions for the execution of the protocol.

10. Acceptance Criteria

    10.1   Isolate variables that could influence the outcome. Indicate conditions that will invalidate the procedure and require retesting

    10.2   The criteria shall provide the objective evidence, scientifically and statistically, that indicates the objectives of the protocol have been successfully fulfilled and will be supported by the collected data.

ATTACHMENT 2—FINAL REPORT TEMPLATE

Approval/Date
Validation
Engineering
Laboratory
Manufacturing
Other

Customer
QA/RA

---

1. Summary

    1.1   Provide an overview of the validation execution activities. State clearly the process or facility has/has not met acceptance criteria and is/is not validated.

2. Results

    2.1   x

3. Statistical Interpretation of Validation Data

    3.1   x

4. Conclusion

    4.1   x

5. Exceptions to the Protocol

    5.1   In the course of executing the protocol, any exceptions or deviations must be explained and justified. The exceptions should be detailed as shown in the following:

    5.2   Exception 1

        5.2.1   Observation: Describe what happened and did not happen

        5.2.2   Investigation: Detail the investigation

        5.2.3   Justification: Provide justification for why the exception did not impact the execution or results in the protocol

6. Data
    6.1   Summarize data. Data should be included or referenced. Write the protocol number on all data sheets. Referenced data (e.g., laboratory notebooks) should be verified.

7. Recommendation
    7.1   Indicate recommendations for acceptance or rejection, and process or equipment changes. Include revalidation criteria or schedule if applicable.

8. References
    8.1   Cite standard production methods, laboratory test methods, or other operating documents. Reference the revision level of the document used. Reference any industry, national or international standards that are applicable.

Note: Title and number of the report should be the same as the protocol.

ATTACHMENT 3—FORMATTING TEMPLATE

1. Numbering System
    1.1   The numbering system format for validation protocols is as follows:
        1.1.1   YYNNN   YY = Year
    NNN = Sequential number (Starting each year with 001)
        1.1.2   General page formats are as follows:
        1.1.3   First page format (centered at the top of the page)
            1.1.3.1   Site
            1.1.3.2   Protocol [or Final Report] #####
            1.1.3.3   Title
        1.1.4   Second and following page format header
            1.1.4.1   Page x of y
            1.1.4.2   Site
        1.1.5   All attachments should have the same header if possible so that they are identified as belonging to that document. Raw data that are attached should have the protocol number written on it.

2. Archiving of Protocol and Summary Report
3. Return the approved documents and any raw test data to the QA Coordinator. The protocol, final report, and attachments will become one document. Transfer the computer files to the network server as specified by Documentation.
    3.1   Filename extension conventions are as follows:
        3.1.1   Protocol YYNNN.P.docx (doc. or xls., etc.)
        3.1.2   Final Report   YYNNN.FR.docx
        3.1.3   Amended Final Report YYNNN.FRA.ext (or FRA2, FRA3, etc.)
        3.1.4   Interim (Status) Report YYNNN.INT.ext
        3.1.5   Other Report YYNNN.RPT.ext
        3.1.6   Attachment YYNNN.ATCH.ext (or ATCH2, ATCH3, etc.)
        3.1.7   Do not overwrite the program file extension
    3.2   ATTACHMENT 4

Example of a mixing study documentation of actual sample collection times
In-Process Homogeneity Test at the 45 Minutes of the Final Mix

Procedure	3.3 Expected Result	Actual Result	3.4 Acceptable Y/N	3.5 Performed by / Date
3.6 Homogeneity samples at the end of 45 minutes mix: lot # _____				
- Upon completion of 45 minutes mix, use the sampling fixture to take 50 grams of sample each from top, middle, and bottom of the bulk product in separately labeled container for, ○ BAK assay, ○ pH ○ Osmolality. ○ Appearance - Submit sample to lab for testing.	- 50 grams of sample is taken from the top portion of the bulk product - 50 grams of sample is taken from the middle portion of the bulk product - - 50 grams of sample is taken from the bottom portion of the bulk product -	☐ Sample quantity from the top: _____ grams. - Sample Time: _____ - Sample labeled as: _____ ☐ Sample quantity from the middle: _____ grams. - Sample Time: _____ - Sample labeled as: _____ ☐ Sample quantity from the bottom: _____ grams. - Sample Time: _____ - Sample labeled as: _____	3.9	3.10
3.7				
3.8				

REVISION HISTORY

DCR	REV	DESCRIPTION
140329	-	New QSP
150324	a	Updated to current document format including company name change to Site throughout the document where applicable. Section 4—References, SOP-xxxxxx—Equipment Qualification—Manufacutring Systems. Attachments—Updated to current template formats.
160027	B	Added to Section x.x—"If data is transcribed in tables in the final report, it will be subject to peer review to ensure accurate transcription."

Appendix III: Hand Calculations—System Architecture and Computations for Nuclear Reactor Heat Removal Systems Computer Program

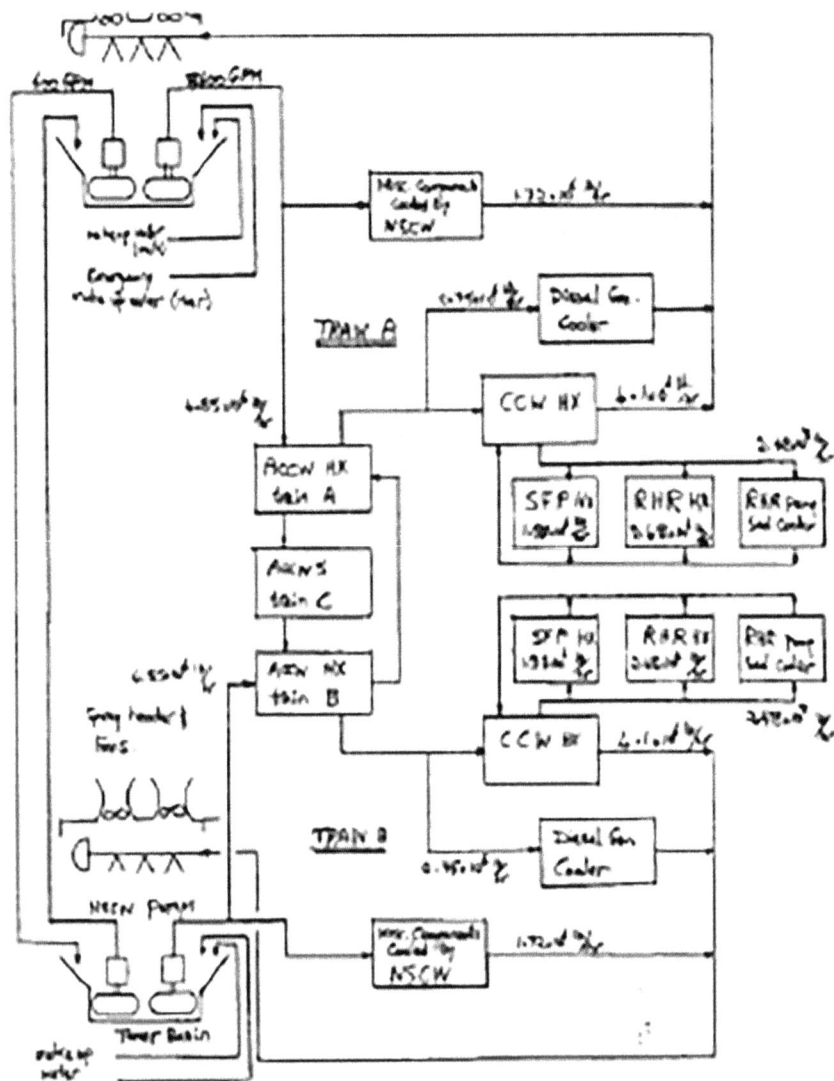

Block flow diagram of NSCW, CCW and ACCW systems

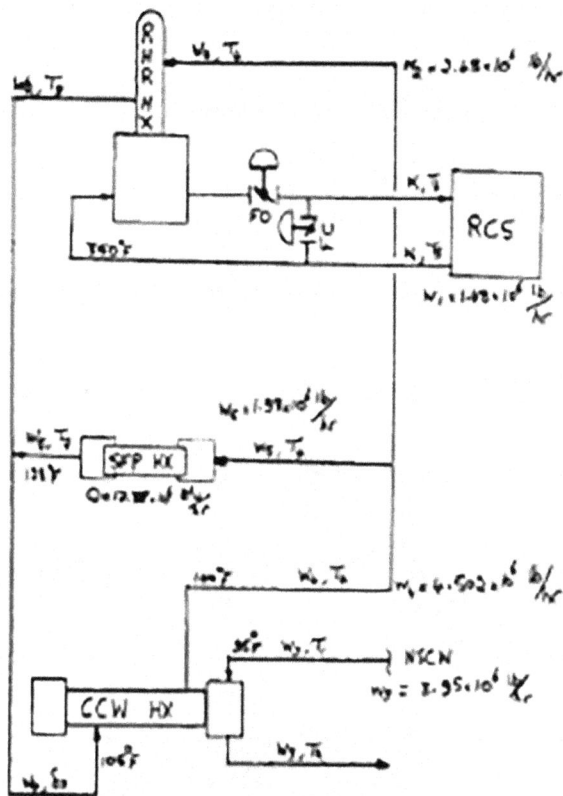

Schematic diagram of the Residual Heat Removal System.

FIG—CCW-HX-LOOP

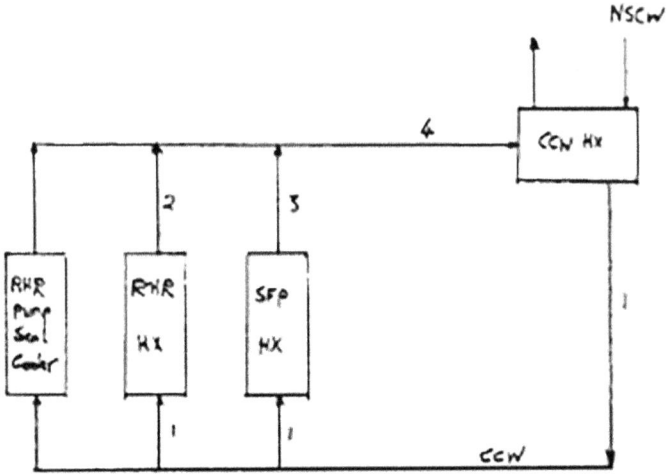

Numerical methods

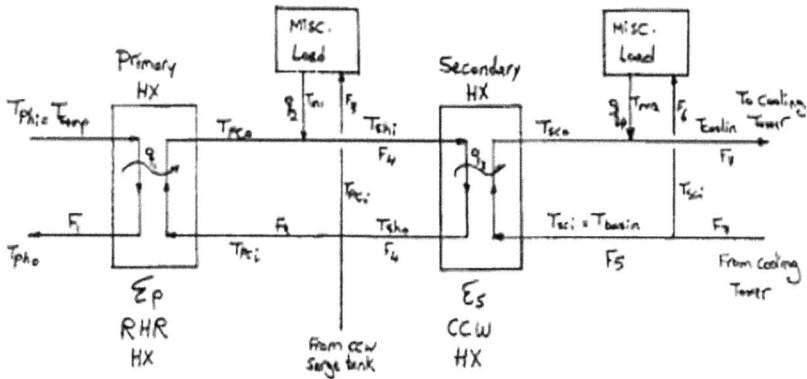

Know variables

$$T_{phi}, T_{sci}, q_2, q_4, \varepsilon_p, \varepsilon_s, \text{ All flowrates}$$

I. Iterative Calculation

1. starting from primary HX

guess T_{pci}

$$q_1 = C_{min} \varepsilon_p (T_{phi} - T_{pci})$$

$$T_{pho} = T_{phi} - q_1 / C_{pht} \cdot F_1$$

$$T_{pco} = T_{pci} + q_1 / C_{pcld} \cdot F_2$$

2. Find secondary HX hot side inlet temp., T_{shi}

$$T_{shi} = T_{pci} + \frac{q_1 + q_2}{C_p F_4}$$

3. Calculate heat load of secondary HX

$$q_s = C_{min} \, \mathcal{E}_s \left(T_{shi} - T_{sci} \right)$$

$$T_{sho} = T_{shi} - q_s / C_p \text{ hot} \cdot F_4$$

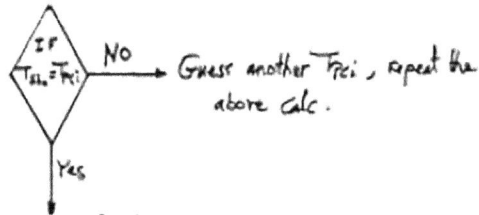

IF $T_{sh_o} = T_{ri}$? — NO → Guess another T_{ri}, repeat the above calc.

Yes

$$T_{sco} = T_{sci} + q_s / C_p \text{ cold} \cdot F_6$$

4. Find cooling tower inlet temp T_{coolin}

$$T_{coolin} = T_{sci} + \frac{q_s + q_u}{F_7}$$

II Straight forward Calculation

1. Start from primary HX

$$q_1 = C_{min_p} \mathcal{E}_p \left(T_{Phi} - T_{Pci} \right) \tag{1}$$

$$T_{Pho} = T_{Phi} - q_1/F_1 \tag{2}$$

$$T_{Pco} = T_{Pci} + q_1/F_2 \tag{3}$$

$$T_{m_1} = T_{Pci} + q_2/F_3 \tag{4}$$

$$T_{shi} = \frac{F_2 \cdot T_{Pco} + F_3 \cdot T_{m_1}}{F_2 + F_3} \tag{5}$$

$$F = F_2 + F_3$$

$$\Rightarrow T_{shi} = T_{cpi} + \frac{q_1 + q_2}{F_4} \tag{5'}$$

2. Secondary HX

$$q_3 = C_{min_s} \mathcal{E}_s \left(T_{shi} - T_{sci} \right) \tag{6}$$

$$T_{sco} = T_{sci} + q_3/F_5 \tag{7}$$

$$T_{m_2} = T_{sci} + q_4/F_6 \tag{8}$$

$$T_{sho} = T_{Pci} = T_{shi} - q_3/F_4 \tag{5'}$$

because $q_3 = q_1 + q_2$

3. Assume no heat loss

$$q_3 = q_1 + q_2 \qquad\qquad \text{(9)}$$

<u>Solution</u>

From equations (9), (1), & (6)

$$C_{min_p}\,\mathcal{E}_p\left(T_{Phi} - T_{Pci}\right) + q_2 = C_{min_s}\,\mathcal{E}_s\left(T_{shi} - T_{sci}\right) \qquad \text{(10)}$$

From (5')

$$T_{shi} = T_{Pci} + \frac{q_3}{F_4}$$

$$= T_{Pci} + \frac{1}{F_6}\,C_{min_s}\,\mathcal{E}_s\left(T_{shi} - T_{sci}\right) \qquad \text{(11)}$$

or,

From (5)

$$T_{shi} = \frac{1}{F_4}\left[F_2\cdot\left(T_{Pci} + \frac{q_1}{F_2}\right) + F_3\cdot\left(T_{Pci} + \frac{q_2}{F_3}\right)\right]$$

$$= \frac{1}{F_6}\left(F_1\,T_{Pci} + q_1 + F_3\,T_{Pci} + q_2\right)$$

$$= \frac{1}{F_4}\left(F_4\,T_{Pci} + q_3\right)$$

$$= T_{Pci} + \frac{q_3}{F_4}$$

Therefore from equations (10) and (11),

T_{shi} and T_{Pci} can be solved.

Rearrange equation ⑩

$$T_{shi} - T_{sci} = \frac{C_{minp}\,\mathcal{E}_p}{C_{mins}\,\mathcal{E}_s}\left(T_{phi} - T_{pci}\right) + \frac{q_z}{C_{min_s}\,\mathcal{E}_s}$$

$$\therefore \; T_{shi} = T_{sci} + A_1\left(T_{phi} - T_{pci}\right) + A_2$$

$$= A_1\left(T_{phi} - T_{pci}\right) + A_3$$

$$= A_1 T_{phi} - A_1 T_{pci} + A_3$$

$$= A_4 T_{pci} + A_5 \qquad\qquad ⑫$$

Rearrange equation ⑪

$$T_{shi} = T_{pci} + \frac{C_{mins}\,\mathcal{E}_s}{F_4}\left(T_{shi} - T_{sci}\right)$$

$$= T_{pci} + A_6\left(T_{shi} - T_{sci}\right)$$

$$= A_6 T_{shi} + T_{pci} - A_6 T_{sci}$$

$$= \frac{T_{pci}}{(1-A_6)} - \frac{A_6 T_{sci}}{(1-A_6)}$$

$$= A_7 T_{pci} - A_8 \qquad\qquad ⑬$$

From ⑫ & ⑬

$$A_4 T_{pci} + A_5 = A_7 T_{pci} - A_8 \;\Rightarrow\; T_{pci} = \frac{A_5 + A_8}{A_7 + A_4} \qquad ⑭$$

Then $T_{shi}, q_1, T_{pho}, T_{m1}, q_3, T_{sco}, T_{m2}$ can all be found

and, $\; T_{coolin} = \dfrac{F_5 \cdot T_{sco} + F_6 \cdot T_{m3}}{F_5 + F_6}$

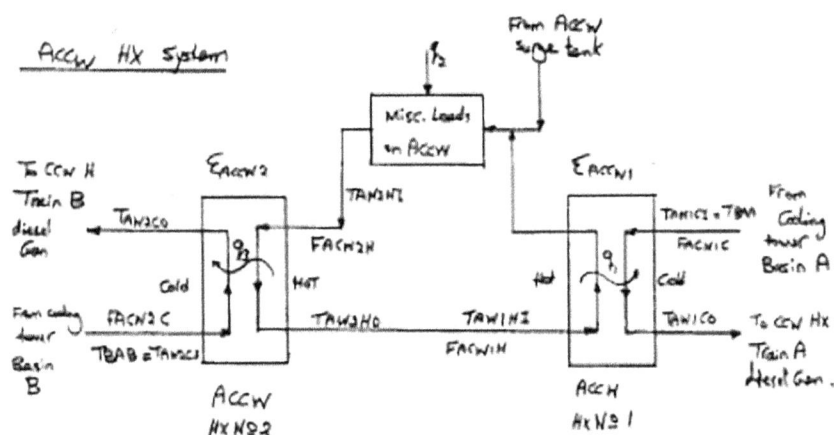

ACCW HX System

Known variables

T_{BAA} ($TAW1CI$), T_{BAB} ($TAN2CI$), misc. heat load q_2, All flow rates, $FACW2C$, $FACW1C$, $FACW2H = FACW1H$, and HX effectiveness ε_{ACCW1}, ε_{ACCW2}.

I. Straightforward Calculation :—

1. ACCW HX No. 1

$$q_1 = C_{min_1} \, \varepsilon_{ACCW_1} \left(TAW1HI - TAW1CI \right) \qquad ①$$

$$TAW1CO = TAW1CI + q_1 / FACW1C \qquad ②$$

$$TAW1HO = TAW1HI - q_1 / FACW1H \qquad ③$$

$$TAW2HI = TAW1HO + q_2 / FACW1H \qquad ④$$

2. $Accw$ Hx N° 2

$$q_3 = C_{min\,2}\,\varepsilon_{ACC\,N2}\left(TAN2HI - TAN2CI\right) \qquad (5)$$

$$TAN2CO = TAN2CI + q_3/FACW2C \qquad (6)$$

$$TAW2HO = TAW2HI - q_3/FAW2H \qquad (7)$$

3. Assume no heat loss

$$q_2 = q_1 + q_3 \qquad (8)$$

and

$$q_2 = C_{min\,1}\,\varepsilon_{ACC\,N1}\left(TAW1HI - TAW1CI\right) + C_{min\,2}\,\varepsilon_{ACC\,N2}\left(TAW2HI - TAW2CI\right) \qquad (9)$$

<u>Solution</u>

From equations (4)(3), (1)

$$TAW2HI = \left(TAW1HI - q_1/FACW1H\right) + q_2/FACW1H$$

$$= TAW1HI - \frac{C_{min\,1}\,\varepsilon_{ACC\,W1}\left(TAW1HI - TAW1CI\right)}{FACW1H} + q_2/FACW1H \qquad (10)$$

From eqn (9) + (10), $TAW1HI$ and $TAW2HI$ can be solved.

The effectiveness of the RHR heat exchanger may be calculated from the following relationship

$$\text{effectiveness} = \frac{(mc_p)_c \, (T_{c,out} - T_{c,in})}{(mc_p)_{min} \, (T_{H,in} - T_{c,in})}$$

in which

$$C = CCW$$

$$min = \text{flow capacity of reactor coolant}$$

$$H = \text{Reactor coolant temperature}$$

The effectiveness of a heat exchanger is a function of flowrates and the geometry. It may be affected by ambient conditions which influence the heat capacity. Σ is also a function of N_{tu}.

$$N_{tu} = UA/(mc_p)_{min}$$

From eqⁿ ③

$$TAW2HI = TAW2CI + \frac{q_2}{C_{rin}\cdot\Sigma_{ACM2}} - \frac{C_{rin}\cdot\Sigma_{ACCN1}}{C_{rin}\cdot\Sigma_{ACCN2}}\left(TAWIHI - TAWICI\right)$$

$$= TAW2CJ + A_1 - A_2\left(TAWIHI - TAWICI\right)$$

$$= A3 - A2\cdot TAWIHI + A2\times TAWICI$$

$$= A4 - A2\cdot TAWIHI \qquad\qquad (11)$$

From eqⁿ ④O

$$TAW3HI = TAWIHI - \frac{C_{rin}\cdot\Sigma_{ACN1}\left(TAWIHI - TANICI\right)}{FACWIH} + \frac{q_2}{FACWIH}$$

$$= TAWIHI - A5\left(TAWIHI - TANICI\right) + A6$$

$$= \left(1 - A5\right)TAWIHI + A5\cdot TAWICI + A6$$

$$= \left(1 - A5\right)TAWIHI + A7 \qquad\qquad (12)$$

From eqⁿ ⑪ & ⑫

$$A4 - A2\cdot TAWIHI = \left(1 - A5\right)TAWIHI + A7$$

$$\left(1 - A5\right)TAWIHI + A2\cdot TAWIHI = A4 - A7$$

$$\Rightarrow \qquad TAWIHI = \frac{A4 - A7}{1 - A5 + A2}$$

Then, $TAW2HI$, q_1, q_3, $TAWICO$, $TANIHO$, $TAW2CO$, $TAW2HO$
Can all be found.

Flow chart

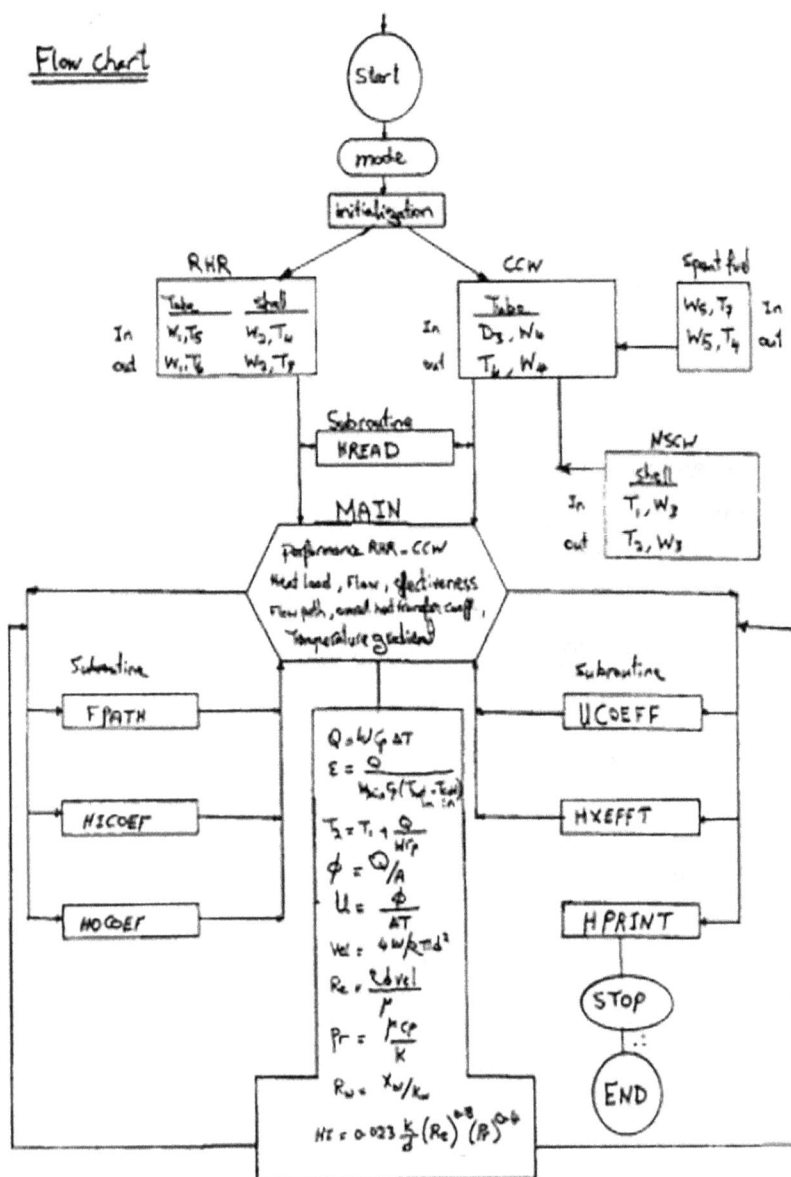

Start

mode

initialization

RHR

	Tube		Shell	
In	W_1, T_5		W_2, T_u	
out	W_1, T_6		W_3, T_7	

CCW

	Tube	
In	D_3, N_u	
out	T_4, W_4	

Spent fuel

W_6, T_7	In
W_5, T_4	out

Subroutine
KREAD

NSCW

	shell	
In	T_1, W_2	
out	T_2, W_3	

MAIN

Performance RHR - CCW
Heat load, Flow, effectiveness
Flow path, overall heat transfer coeff,
Temperature gradient

Subroutine
FPATH

HICOEF

HOCOEF

$Q = W C_p \Delta T$

$\varepsilon = \dfrac{Q}{W_{c_0} C_p (T_{c_i} - T_{c_0})_{\ln \Delta A}}$

$T_2 = T_1 + \dfrac{Q}{W C_p}$

$\phi = Q/A$

$U = \dfrac{\phi}{\Delta T}$

$Vel = \dfrac{4W}{\rho \pi d^2}$

$Re = \dfrac{d \cdot vel}{\mu}$

$Pr = \dfrac{\mu c_p}{K}$

$R_w = X_w / k_w$

$Hi = 0.023 \dfrac{k}{d} (Re)^{0.8} (Pr)^{0.4}$

Subroutine
UCOEFF

HXEFFT

HPRINT

STOP

END

```
I N D E X                                                      PAGE   1

    1                 SUBROUTINE FPATH(N,FH,FC,FT,FS)
                 C
                 C    THIS SUBROUTINE DETERMINES THE TUBE & SHELL FLOWS
                 C
    2                 IF(N .EQ. 1) GO TO 10
    3                 FT=FC
    4                 FS=FH
    5                 GO TO 20
    6            10   CONTINUE
    7                 FT=FH
    8                 FS=FC
                 C
    9            20   CONTINUE
                 C
   10                 RETURN
   11                 END


I N D E X                                                      PAGE   3

    1                 SUBROUTINE HICOEF(TAVGT,FMASS,DI,NTUBE,NTUPAS,HI)
                 C
                 C    THIS SUBROUTINE CALCULATES TUBE SIDE FILM COEFFICIENT, HI
                 C
    2                 REAL NTUBE,NTUPAS
                 C
    3                 P=PSL(TAVGT)
    4                 VISCO=VISL(P,TAVGT)*3600.
    5                 THCOND=CONDL(P,TAVGT)
    6                 SPVOL=VCL(P,TAVGT)
    7                 DEN=1./SPVOL
    8                 PRANDL=VISCO/THCOND
    9                 AREA=3.1416*(DI**2/4.0)/144.
   10                 FVOL=FMASS/DEN
   11                 VEL=(FVOL/AREA)/(NTUBE/NTUPAS)/3600.
   12                 RENOLD=3600.*(DI/12.)*DEN*VEL/VISCO
   13                 HI=0.023*(RENOLD**0.8)*(PRANDL**0.4)*(THCOND/(DI/12.))
                 C
   14                 WRITE(6,11) P,VISCO,THCOND,SPVOL,DEN,PRANDL
   15            11   FORMAT(2X,'P,VIS,THCON,SPVOL,DEN,PRAN =',2X,6E14.6,/)
   16                 WRITE(6,12) AREA,FVOL,VEL,RENOLD,HI,TAVGT
   17            12   FORMAT(2X,'AREA,FVOL,VEL,REND,HI,T =',6E14.6,/)
                 C
   18                 RETURN
   19                 END
```

```
            C
            C       THIS PROGRAM CALCULATES THE PERFORMANCE OF RHR-CCW HEAT
            C       EXCHANGER LOOP
            C
      1             REAL NTUBEP,NTPASP,NSPASP,NTUBES,NTPASS,NSPASS
            C
      2             COMMON /MISCQ/ QMIS1(2),FMIS1(2),QMIS2(12),FMIS2(12)
      3             COMMON /TMISC/ TPCII(2),TMIS1(2),TSCII(12),TMIS2(12)
            C
      4             QCOLER=0.
      5             TCOLER=0.
      6             FCOLER=0.
            C
            C       READ INPUT DATA
            C
      7             CALL HREAD(ITYPEP,NFLWP,TSUMP,ARHR,FRHRH,FRHRC,OORHR,DIRHR,DISHLP,
            &       IPTCHP,DPTCHP,NTUBEP,NTPASP,NSPASP,IBAFP,DBAFP,FQUIP,FOUOP,
            &       TKRHR,
            &       ITYPES,NFLWS,TBA,ACCW,FCCWC,DOCCW,DICCW,DISHLS,IPTCHS,DPTCHS,
            &       NTUBES,NTPASS,NSPASS,IBAFS,DBAFS,FQUIS,FQUOS,TKCCW)
      8             TPHI=TSUMP
      9             TSCI=TBA
            C
            C
            C       FIND THE TOTAL HEAT LOAD  & FLOW FROM MISC. LOADS
            C
     10             QTOTM1=0.
     11             FTOTM1=0.
     12             QTOTM2=0.
     13             FTOTM2=0.
     14             DO 10 I=1,2
     15             QTOTM1=QTOTM1 + QMIS1(I)
     16             FTOTM1=FTOTM1 + FMIS1(I)
     17        10   CONTINUE
     18             FCCWH=FRHRC+FTOTM1
     19             DO 20 I=1,12
     20             QTOTM2=QTOTM2 + QMIS2(I)
     21             FTOTM2=FTOTM2 + FMIS2(I)
     22        20   CONTINUE
            C
            C       FIND HEAT EXCHANGER EFFECTIVENESS
            C
            C       FIND TUBE & SHELL SIDE FLOW
            C
     23             CALL FPATH(NFLWP,FRHRH,FRHRC,FTUBEP,FSHELP)
     24             CALL FPATH(NFLWS,FCCWH,FCCWC,FTUBES,FSHELS)
            C
     25             TAVGTP=129.
     26             TAVGSP=112.
     27             TAVGTS=117.
```

```
I N D E X
                                      MAIN PROGRAM                              PAGE   6

    28            TAVGSS=130.
             C
    29            CALL UCOEFF(FTUBEP,FSHELP,TAVGTP,TAVGSP,DORHR,DIRHR,DISHLP,IPTCHP,
             &      DPTCHP,NTUBEP,NTPASP,NSPASP,IBAFP,DBAFP,FOUIP,FOUOP,TKRHR,UO1,
             &      HIRHR,RIRHR,HORHR,RORHR,RTRHR,RFIRHR,RFORHR)
    30            URHR=UO1
    31            CALL UCOEFF(FTUBES,FSHELS,TAVGTS,TAVGSS,DOCCW,DICCW,DISHLS,IPTCHS,
             &      DPTCHS,NTUBES,NTPASS,NSPASS,IBAFS,DBAFS,FOUIS,FOUOS,TKCCW,UO2,
             &      HICCW,RICCW,HOCCW,ROCCW,RTCCW,RFICCW,RFOCCW)
    32            UCCW=UO2
             C
    33            CALL HXEFFT(ITYPEP,URHR,ARHR,FRHRH,FRHRC,EFFT)
    34            EFTRHR=EFFT
    35            CALL HXEFFT(ITYPES,UCCW,ACCW,FCCWH,FCCWC,EFFT)
    36            EFTCCW=EFFT
             C
    37            WRITE(6,56) HIRHR,RIRHR,HORHR,RORHR,RTRHR
    38     56     FORMAT(2X,*HIRHR,RIRHR,HORHR,RORHR,RTRHR =*,2X,5E14.7,/)
    39            WRITE(6,57) RFIRHR,RFORHR,URHR,EFTRHR
    40     57     FORMAT(2X,*RFIRHR,RFORHR,URHR,EFTRHR =*,2X,4E14.7,/)
    41            WRITE(6,58) HICCW,RICCW,HOCCW,ROCCW,RTCCW
    42     58     FORMAT(2X,*HICCW,RICCW,HOCCW,ROCCW,RTCCW =*,2X,5E14.7,/)
    43            WRITE(6,59) RFICCW,RFOCCW,UCCW,EFTCCW
    44     59     FORMAT(2X,*RFICCW,RFOCCW,UCCW,EFTCCW =*,2X,4E14.7,/)
             C
             C
             C      FIND HX CAPACITY RATES ON HOT & COLD SIDES
             C
    45            CRHRH=FRHRH*1.0
    46            CRHRC=FRHRC*1.0
    47            CCCWH=FCCWH*1.0
    48            CCCWC=FCCWC*1.0
             C
    49            IF(CRHRH .LE. CRHRC)  CPMIN=CRHRH
    50            IF(CRHRH .GT. CRHRC)  CPMIN=CRHRC
    51            IF(CCCWH .LE. CCCWC)  CSMIN=CCCWH
    52            IF(CCCWH .GT. CCCWC)  CSMIN=CCCWC
             C
             C      START RHR-CCW HX LOOP CALCULATION
             C
    53            A1=(CPMIN*EFTRHR)/(CSMIN*EFTCCW)
    54            A2=QTOTMI/(CSMIN*EFTCCW)
    55            A3=TSCI*A2
    56            A4=-1.0*A1
    57            A5=A1*TPHI  + A3
    58            A6=(CSMIN*EFTCCW)/FCCWH
    59            A7=1.0/(1.- A6)
    60            A8=(A6*TSCI)/(1.0-A6)
             C
    61            TPCI=(A5+A9)/(A7-A4)
```

MAIN PROGRAM

```
62          TSHI=A4+TPCI +15
63          QRHR=CPMIN+EFTRHR+(TPHI-TPCI)
64          TPH3=TPHI- QRHR/FRHRH
65          TPCO=TPCI + QRHR/FRHRC
66          TM1=TPCI + QTOTM1/FTOTM1
   C
67          TSHI=(FRHRC+TPCO + FTOTM1+TM1)/FCCWH
68          QCCW=CSMIN+EFTCCW+(TSHI-TSCI)
69          TSC3=TSCI + QCCW/FCCWC
70          TSHO=TSHI - QCCW/FCCWH
71          TM2=TSCI + QTOTM2/FTOTM2
   C
   C        FIND COOLIN TOWER INLET WATER TEMP., TCLWIN, AND TOTAL
   C        SYSTEM HEAT LOAD, QTOTSY
   C
72          TCLWIN=(FCCWC+TSCO + FTOTM2+TM2 + FCOLER+TCOLER)/
   &        (FCCWC+FTOTM2+FCOLER)
73          QTOTSY=QCCW+QTOTM2+QCOLER
   C
   C        FIND TEMPERATURE CHANGE ACROSS EACH MISC. LOAD
   C
74          DO 50 I=1,2
75          TMIS1(I)=TPCI+(QMIS1(I)/FMIS1(I))
76      50  CONTINUE
77          DO 60 I=1,12
78          TMIS2(I)=TSCI+(QMIS2(I)/FMIS2(I))
79      60  CONTINUE
   C
   C        PRINT OUT THE RESULTS
   C
80          DO 70 I=1,2
81          TPCII(I)=TPCI
82      70  CONTINUE
83          DO 90 I=1,12
84          TSCII(I)=TSCI
85      80  CONTINUE
86          CALL MPRINT(QRHR,EFTRHR,TPHI,TPH3,TPCI,TPCO,FRHRH,FRHRC,URHR,
   &        QCCW,EFTCCW,TSHI,TSHO,TSCI,TSCO,FCCWH,FCCWC,UCCW,
   &        TCLWIN,QTOTSY,QTOTM1,FTOTM1,QTOTM2,FTOTM2)
   C
87          STOP
88          END
```

```
I N D E X                                                        PAGE  11

    1          SUBROUTINE HOCOEF(TAVGS,FMASS,DISHEL,NTUBE,NSHPAS,IBAFLE,DBAFLE,
          &    IPITCH,DPITCH,DO,HO)
          C
          C    THIS SUBROUTINE CALCULATES SHELL SIDE FILM COEFF., HO
          C
    2          REAL NTUBE,NSHPAS
          C
    3          P=PSL(TAVGS)
    4          VISCO=VISL(P,TAVGS)*3600.
    5          THCOND=CONDL(P,TAVGS)
    6          SPVOL=VCL(P,TAVGS)
    7          DEN=1./SPVOL
    8          PRANDL=VISCO/THCOND
    9          FVOL=FMASS/DEN
          C
          C    FIND EQUIVALENT DIAMETER FOR TRIANGULAR TUBE PITCH, DTRI
          C
   10          DTRI=4.*(0.43*DPITCH**2 - .125*3.1416*DO**2)/(.5*3.1416*DO)
          C
          C    FIND EQUIVALENT DIAMETER FOR SQUARE TUBE PITCH, DSQURE
          C
   11          DSQURE=4.0*(DPITCH**2 - 3.1416*D3**2)/(3.1416*DO)
          C
   12          IF(IPITCH .EQ. 1)  DE=DTRI
   13          IF(IPITCH .EQ. 2)  DE=DSQURE
          C
   14          AL=(3.1416/4.)*(DISHEL**2-(DO**2)*NTUBE)/144.
          C    FOR DOUBLE BAFFLE SEGMENT,THE LONGITUDINAL FLOW IS PREDOMINENT,
          C    AND ASSUME THE BAFFLE CUT IS 50%, SO
   15          ALA=0.3927*AL
          C
   16          AS=DISHEL*(DPITCH-DO)*DBAFLE/(144.*DPITCH)
   17          GLONG=FMASS/(ALA*NSHPAS)
   18          GCROSS=FMASS/AS
   19          G=(GLONG*GCROSS)**0.5
          C
   20          IF(IBAFLE .EQ. 1)  GMASS=GCROSS
   21          IF(IBAFLE .EQ. 2)  GMASS=GLONG
          C
   22          RENOLD=(DE/12.)*GMASS/VISCO
   23          A1=RENOLD**0.55
   24          A2=PRANDL**0.3333
   25          A3=THCOND/(DE/12.)
   26          HO=0.36*A1*A2*A3
          C
   27          WRITE(6,42) PRANDL,RENOLD,THCOND
   28   42     FORMAT(2X,'PR,RE,THCOND =',2X,3E15.7,/)
   29          WRITE(6,45) DE,AL,AS,GLONG,GCROSS,G

I N D E X                                                        PAGE  12
          SUBROUTINE HOCOEF(TAVGS,FMASS,DISHEL,NTUBE,NSHPAS,IBAFLE,DBAFLE,IP

   30   45     FORMAT(2X,'DE,AL,AS,GL,GS,G =',2X,6E14.6,/)
          C
   31          RETURN
   32          END
```

```
                              ***** INPUT DATA *****


        PRIMARY (RHR) HX DATA :


        TYPE OF HX =
        1 = SHELL & TUBE HX , 1 SHELL PASS,2,4,6... TUBE PASSES
        2 = COUNTER FLOW HX
        FLOW ARRANGEMENT =
        1 = TUBE SIDE - HOT FLUID, SHELL SIDE - COLD FLUID
        2 = TUBE SIDE - COLD FLUID, SHELL SIDE - HOT FLUID
        HOT SIDE FLUID INLET TEMP.. DEG F, =
        EFFECTIVE HEAT TRANSFER AREA, SQ FT, =
        HOT SIDE FLUID FLOW, LB/HR. =
        COLD SIDE FLUID FLOW, LB/HR. =
        TUBE OUTSIDE DIAMETER, IN., =
        TUBE INSIDE DIAMETER, IN., =
        SHELL INNER DIAMETER, IN., =



        TYPE OF TUBE PITCH, 1-TRIANGULAR, 2-SQUARE, =
        TUBE PITCH, IN. =
        NO. OF TOTAL TUBES =
        NO. OF TUBE PASS =
        NO. OF SHELL PASS =
        TYPE OF BAFFLE, 1-SINGLE SEGMENT, 2-DOUBLE SEGMENT. =
        BAFFLE SPACING, IN., =
        TUBE SIDE FOULING FACTOR, HR-SQ FT-F/BTU, =
        SHELL SIDE FOULINT FACTOR, HR- SQ FT-F/BTU. =
        TUBE WALL THERMAL CONDUCTIVITY, BTU/HR-FT-F, =
```

```
SECONDARY (CCW) HX  DATA :


TYPE OF HX =
1 = SHELL & TUBE HX . 1 SHELL PASS,2,4,6... TUBE PASSES
2 = COUNTER FLOW HX
FLOW ARRANGEMENT =
1 = TUBE SIDE - HOT FLUID, SHELL SIDE - COLD FLUID
2 = TUBE SIDE - COLD FLUID, SHELL SIDE - HOT FLUID
COLD SIDE FLUID INLET TEMP., DEG F =
** THIS IS THE SWITCH TO TURN ACCU SYSTEM ON
   IF INPUT 0., ACCW SYSTEM CALAULATION WILL BE PERFORMED
EFFECTIVE HEAT TRANSFER AREA. SQ FT. =
COLD SIDE FLUID FLOW. LB/HR, =
TUBE OUTSIDE DIAMETER, IN., =
TUBE INSIDE DIAMETER, IN., =
SHELL INNER DIAMETER, IN., =



TYPE OF TUBE PITCH, 1-TRIANGULAR, 2-SQUARE, =
TUBE PITCH. IN. =
NO. OF TOTAL TUBES =
NO. OF TUBE PASS =
NO. OF SHELL PASS =
TYPE OF BAFFLE, 1-SINGLE SEGMENT, 2-DOUBLE SEGMENT, =
BAFFLE SPACING, IN., =
TUBE SIDE FOULING FACTOR. HR-SQ FT-F/BTU, =
SHELL SIDE FOULINT FACTOR. HR- SQ FT-F/BTU, =
TUBE WALL THERMAL CONDUCTIVITY, BTU/HR-FT-F, =
```

MISCELLANEOUS LOADS ON CCW HX ARE LISTED AS FELLOWS :

ITEM IDENTIFICATION	HEAT LOAD (BTU/HR)	FLOW RATE (LB/HR)
1. SFP HEAT EXCH.	.86900+07	.19800+07
2. RHR PUMP SEAL COOLER	.24800+04	.76000+05

DO YOU WANT TO CHANGE THE ABOVE LOAD DATA ?
YES=1, NO=0

MISCELLANEOUS LOADS ON COOLING TOWER
ARE LISTED AS FELLOWS :

ITEM IDENTIFICATION	HEAT LOAD (BTU/HR)	FLOW RATE (LB/HR)
1. CTB AUX AIR COOLER	.45900+07	.70000+06
2. REACTOR CAVITY COOLER	.24000+06	.11000+06
3. SI PUMP LO COOLER	.10000+06	.20000+05
4. SI PUMP MTR COOLER	.10000+06	.20000+05
5. CNT CHARGING PUMP LO COOLER	.20000+06	.20000+05
6. CNT CHARGING PUMP MTR COOLER	.20000+06	.20000+05
7. CONTAINMENT SPRAY PUMP & MTR	.70000+05	.10000+05
8. RHR PUMP MOTOR	.37000+06	.99200+04
9. ESF CHILLER	.54800+07	.55000+06
10. CCW PUMP MTR COOLER	.14000+06	.44640+05
11. PIPE PENETRATION AREA COOLER	.19400+07	.12500+06
12. DIESEL GENERATOR	.21080+08	.75000+06

DO YOU WANT TO CHANGE THE ABOVE LOAD DATA ?
YES=1, NO=0

VERIFICATION AND VALIDITY OF COMPUTER PROGRAM

```
@PRT 912346*HTDATA.DATA
FURPUR 28R2 U1 S74T11 09/17/82 08:33:17

912346*HTDATA(1).DATA(57)
     1      @PRT 912346*HTDATA.DATA
     2      @XQT 912346*HM.ABSU
     3      1,1,182.3,6518.1480000.2480000,.75,.652,42.75
     4      1,.9375,1544.4,1,2,13.625,.0003,.0005,8.55
     5      2,2,93.2,34608,3950000,.75,.652,78
     6      1,.9375,4100,2,2,1,32.6,.0015,.0005,26
     7      0
     8      0
     9      2,2,88.5,14222,2697000,4850000,.75,.652,53.125
    10      1,.9375,1760,1,1,1,24.9,.0015,.0005,26
    11      2,2,88.5,14222,2697000,4850000,.75,.652,53.125
    12      1,.9372,1760,1,1,1,24.9,.0015,.0005,26
    13      57.794E+6
```

```
@XQT 912346*HM.ABSU


        ***** OUTPUT OF RHR-CCW HEAT EXCHANGER LOOP *****


    PRIMARY (RHR) HEAT EXCHANGER :
    ------------------------------


    HOT SIDE FLUID FLOW RATE =      .1480000+07  LB/HR

    HOT SIDE FLUID INLET TEMP. =   182.3000  DEG F

    HOT SIDE FLUID OUTLET TEMP. =  132.7935  DEG F

    COLD SIDE FLUID FLOW RATE =     .2480000+07  LB/HR

    COLD SIDE FLUID INLET TEMP. =  104.3160  DEG F

    COLD SIDE FLUID OUTLET TEMP. =    .1338602+03  DEG F

    OVER ALL HEAT TRANSFER COEFFICIENT =     .3790000+03  BTU/HR-SQ FT-F

    EFFECTIVENESS =      .6348

    HEAT  LOAD =    .7326955+08  BTU/HR


    SECONDARY (CCW) HEAT EXCHANGER :
    --------------------------------


    HOT SIDE FLUID FLOW RATE =     .4536000+07  LB/HR

    HOT SIDE FLUID INLET TEMP. =   122.3852  DEG F

    HOT SIDE FLUID OUTLET TEMP. =  104.3160  DEG F

    COLD SIDE FLUID FLOW RATE =     .3950000+07  LB/HR

    COLD SIDE FLUID INLET TEMP. =   93.2000  DEG F

    COLD SIDE FLUID OUTLET TEMP. =    .1139499+03  DEG F

    OVER ALL HEAT TRANSFER COEFFICIENT =     .2438000+03  BTU/HR-SQ FT-F

    EFFECTIVENESS =      .7110

    HEAT  LOAD =    .8196202+08  BTU/HR
```

MISCELLANEOUS LOADS ON CCW HEAT EXCHANGER :

ITEM IDENTIFICATION	HEAT LOAD (BTU/HR)	FLOW RATE (LB/HR)	INLET TEMP (DEG F)	OUTLET TEMP (DEG F)
1. SFP HEAT EXCHANGER	.86900+07	.19800+07	104.31599	108.70488
2. RHR PUMP SEAL COOLER	.24800+04	.76000+05	104.31599	104.34862
TOTAL	.86925+07	.20560+07		

MISCELLANEOUS LOADS ON COOLING TOWER :

ITEM IDENTIFICATION	HEAT LOAD (BTU/HR)	FLOW RATE (LB/HR)	INLET TEMP (DEG F)	OUTLET TEMP (DEG F)
1. CTB AUX AIR COOLER	.45900+07	.70000+06	93.20000	99.75714
2. REACTOR CAVITY COOLER	.24000+06	.11000+06	93.20000	95.38182
3. SI PUMP LO COOLER	.10000+06	.20000+05	93.20000	98.20000
4. SI PUMP MTR COOLER	.10000+06	.20000+05	93.20000	98.20000
5. CNT CHAR PUMP LO COOLER	.20000+06	.20000+05	93.20000	103.20000
6. CNT CHAR PUMP MTR COOLER	.20000+06	.20000+05	93.20000	103.20000
7. CONTAINMENT SPY PUMP & MTR	.70000+05	.10000+05	93.20000	100.20000
8. RHR PUMP MTR	.37000+06	.99200+04	93.20000	130.49839
9. ESF CHILLER	.54800+07	.55000+06	93.20000	103.16364
10. CCW PUMP MTR COOLER	.14000+06	.44640+05	93.20000	96.33620
11. PIPE PENETRATION AREA CLR	.19400+07	.12500+06	93.20000	108.72000
12. DIESEL GENERATOR	.21080+08	.75000+06	93.20000	121.30667
TOTAL	.34510+08	.23796+07		

TOTAL SYSTEM HEAT LOAD (EXCLUDING HVAC LOAD) ON COOLING TOWER = .11647202+09 BTU/HR

TEMP OF WATER RETURNING TO SAFETY INJECTION SYSTEM = 132.7935 DEG F

```
@PRT 912346*HTDATA.DATA
FURPUR 2BR2 U1 S74T11 09/17/82 08:41:46

912346*HTDATA(1).DATA(58)
     1      @PRT 912346*HTDATA.DATA
     2      @XQT 912346*HH.ABSU
     3      1,1,165.2,6518,1480000.2480000,.75,.652,42.75
     4      1,.9375,1544.4,1,2,13.625,.0003,.0005,8.55
     5      2,2,91.9,34608,3950000,.75,.652,78
     6      1,.9375,4100,2,2,1,32.6,.0015,.0005,26
     7      0
     8      0
     9      2,2,88.5,14222,2697000,4850000,.75,.652,53.125
     10     1,.9375,1760,1,1,1,24.9,.0015,.0005,26
     11     2,2,88.5,14222,2697000,4850000,.75,.652,53.125
     12     1,.9372,1760,1,1,1,24.9,.0015,.0005,26
     13     57.794E+6


@XQT 912346*HH.ABSU



     ***** OUTPUT OF RHR-CCW HEAT EXCHANGER LOOP *****



PRIMARY (RHR) HEAT EXCHANGER :
------------------------------



HOT SIDE FLUID FLOW RATE =      .1480000+07  LB/HR

HOT SIDE FLUID INLET TEMP. =    165.2000   DEG F

HOT SIDE FLUID OUTLET TEMP. =   124.5902   DEG F

COLD SIDE FLUID FLOW RATE =     .2480000+07  LB/HR

COLD SIDE FLUID INLET TEMP. =   101.2302   DEG F

COLD SIDE FLUID OUTLET TEMP. =     .1254651+03  DEG F

OVER ALL HEAT TRANSFER COEFFICIENT =    .3790000+03  BTU/HR-SQ FT-F

EFFECTIVENESS =     .6348

HEAT  LOAD =    .6010253+08  BTU/HR
```

Appendix IV: Chemical Reaction Fouling

MODELLING HYDROCARBON FOULING

By B. D. CRITTENDEN* (MEMBER), S. T. KOLACZKOWSKI* (MEMBER) and
S. A. HOUT**

*School of Chemical Engineering, University of Bath, Bath. **Procter and Gamble, Geneva, Switzerland

Existing models of hydrocarbon fouling are reviewed and a fundamental mathematical model is developed in order to predict the initial rate of fouling in a chemically reactive system of simple kinetics. The rate of deposition is based on the summation of resistances due to mass transfer of reactant to the surface and reaction kinetics at the surface. For deposits which are mobile, back convection of the foulant to the bulk fluid is subtracted to give the overall initial rate expression. The model is tested with the experimental results obtained from the polymerisation of 1% styrene in kerosene flowing at Reynolds numbers up to 5200 in a 0.02 m ID tube. Prediction of the polystyrene concentration at the solid/liquid interface, necessary for calculation of the back convection rate, is not possible. However, the values of this interfacial concentration which are necessary to guarantee agreement between predicted and experimental initial fouling rates are shown, as expected, to be independent of flow rate and to increase strongly with surface temperature. The model has applications throughout the oil and chemical industries, subject to the need to evaluate some model parameters from plant data.

INTRODUCTION

The formation of deposits on heat transfer surfaces in contact with hydrocarbon fluids can lead to substantial economic penalties arising out of increased energy consumption, loss of production and cleaning of equipment[1]. Fouling is fundamentally a dynamic phenomenon and yet the design of heat transfer equipment is generally based on the summation of time-independent resistances to heat transfer. Therefore, in order to provide an economically satisfactory surface area for an acceptable period of operation, it is necessary to be able to predict the dependence of fouling resistances not only on time but also on key design and operational parameters. An ability to carry out such modelling would also aid the determination of optimum cleaning cycles, the evaluation of anti-fouling treatments, and the identification of process control strategies for networks of heat exchangers which are prone to foul, and thereby to affect the operability of downstream process units.

MECHANISMS

The diversity of hydrocarbon feedstocks and thermal environments encountered in the process industries makes it difficult to generalise about the chemical mechanisms by which the deposits are formed. At high temperatures in the gas phase, coke can be formed via complex secondary or synthesis reactions of the products from primary or degradation reactions (cracking and dehydrogenation)[2]. Such a mechanism applies to cracking furnaces. Primary thermal cracking reactions do not proceed at temperatures much below 650 K, and it is generally accepted that deposition from hydrocarbons in the liquid phase is due to free-radical autooxidation reactions[3]. The presence of inorganic constituents in the deposits found in crude oil preheat exchangers[4] suggests that particulate, corrosion and crystallisation fouling mechanisms may be interrelated with chemical reaction fouling. Trace metals present in the feedstock may also act as catalysts.

When material of increasing molecular weight and structural complexity exceeds its solubility in the fluid, it forms a deposit which initially may not be rigid. This process occurs not necessarily at the heat transfer surface, but rather in a reaction zone where the local conditions are favourable. The foulant may have to be transported, perhaps in colloidal form, to be adsorbed on, or otherwise attached to, the surface. Reactants must be transferred by convective mechanisms to the reaction zone. Likewise reaction products, including the foulant if still mobile, may be transferred back to the fluid bulk to take part in further fouling reactions or to be deposited on cooler surfaces in downstream units. The possibility also exists for wholesale removal of deposits by the turbulent action of the fluid. Figure 1 summarises this overview of chemical reaction fouling.

Individual chemical reaction rates are strongly dependent upon temperature, pressure, composition and the presence of catalysts, but the overall rate of chemical reaction fouling may, in addition, be dependent upon other physicochemical mechanisms, such as mass transfer and surface phenomena. Thus, many parameters can affect, and in turn be affected by, the deposition process (Figure 2).

MATHEMATICAL MODELS

Chemical reaction fouling models are summarised chronologically in Table 1. Reflecting the general observation that organic deposits are often tenacious, some models do not include deposit removal or release terms. All models relate to in-tube fouling only. An assumption implicit in most of the models is that the fouling rate of a whole heat exchanger can be modelled in terms of a single set of parameters, such as temperature, Reynolds number, tube diameter, deposit properties, etc., thus implying that there are no significant changes in the properties of the fluid as a result of passing through the exchanger. Thus the results of calculations might be expected to be valid for the conditions at one particular position in a heat exchanger, but some form of averaging

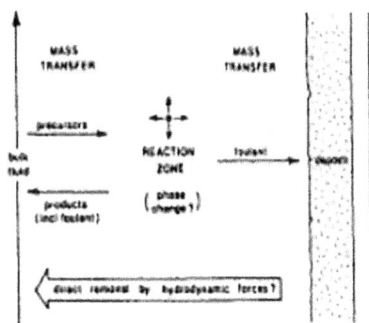

Figure 1. Overview of chemical reaction fouling.

will be necessary to obtain the mean R_f value that is usually quoted.

Atkins[8] published a guide to the selection of time-dependent fouling resistances for the petroleum industry. He postulated that as deposits build up on heater tubes, their structure is likely to change with further degradation. The total fouling resistance was assumed to be the sum of the resistance of a porous coke layer adjacent to the fluid film and the resistance of a growing consolidated hard coke layer adjacent to the tube wall. The former resistance could be obtained from standard sources, eg TEMA[11], whilst the latter, as a function of time, temperature and composition, could be obtained from a graph. Although such an approach would yield an exponential dependence of fouling rate on temperature, the analysis also leads to the suggestion of a linear dependence of fouling resistance on time.

Even for surface reactions of simple mechanisms and kinetics, solutions to the heat, mass and momentum transfer equations are difficult to obtain without invoking simplifying assumptions[4,5]. Nijsing[7] assumed that fouling from an organic nuclear reactor coolant was

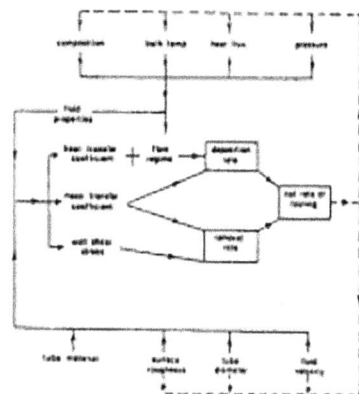

Figure 2. Effect of process parameters on chemical reaction fouling.

caused by the instantaneous, irreversible reaction of a precursor A to a product B which crystallised rapidly when compared with its diffusion rate to the surface. With the further assumptions, firstly, that all physical properties are independent of temperature, secondly, that the reaction is instantaneous at a temperature above a critical value T_c, and, thirdly, that the diffusivities of A and B were equal, the solution obtained was:

$$\text{Average rate of deposition} \propto [c_3 D (Re)^{8/13} (Sc)^{1/3}]/d \quad (1)$$

Thus, in effect, the rate limiting step in the deposition process was assumed to be the diffusion of reactant to the surface, the rate of which is increased by operation at higher fluid velocities.

Film theories of heat and mass transfer have been used to simplify the computational effort. Nelson[3] proposed that the rate of coking is directly dependent upon the volume of fluid in the film which is exposed to high temperatures, and which therefore can be reduced by increasing fluid velocities. Fernandez-Baujin and Solomon[11] developed a two-step, mass transfer and kinetics, model to account for the formation of coke in steam cracking furnaces. The mass transfer flux of reactant, or foulant precursor, to the reaction site is given by:

$$N_p = K_y (c_{pb} - c_{pe}) \quad (2)$$

Under steady conditions, this flux is balanced by the rate of deposition, assumed to be of first order, i.e

$$N_p = k c_{pe} \quad (3)$$

Hence the rate of fouling, which is proportional to N, is given by:

$$\frac{dR_f}{dt} = \frac{d}{dt}(x_0/k_f) = \frac{1}{\rho_f k_f}\left\{\frac{c_{pb}}{\frac{1}{K_y}+\frac{1}{k}}\right\} \quad (4)$$

Since tube wall temperatures in cracking furnaces are very high, Fernandez-Baujin and Solomon assumed that the reaction velocity constant would be very much greater than the mass transfer coefficient, thereby reducing equation (4) to:

$$\frac{dx_f}{dt} = \frac{K_y c_{pb}}{\rho_f k_f} \quad (5)$$

The mass transfer coefficient was expressed in terms of the average fluid velocity and fluid physical properties by the Chilton and Colburn analogy. Ultimately, the rate of increase in coke thickness was given by:

$$\frac{dx_f}{dt} = \frac{K^* G^{3/5}}{(d - 2x_f)^{3/5}} \quad (6)$$

The constant K^* was a function of feedstock, cracking severity and selectivity, dilution steam ratio and other system properties. The authors claimed very good agreement between actual furnace run lengths and those derived from equation (6). Chen and Maddock[18] have also reported that the coking rate in high temperature furnaces is dependent upon mass transfer and not upon surface kinetics, whilst the reverse is true at lower temperatures.

When reaction products are mobile, the possibility exists for their removal from the reaction zone by

Table 1. Chemical reaction fouling models.

Authors	Application	Deposition Term	Removal Term	Remarks
Nelson (1934)[a]	Oil refining	Rate is directly dependent upon thickness of thermal boundary layer	None considered	Fouling rate can be reduced by increasing fluid velocity
Atkins (1962)[b]	Fixed heaters in oil industry	Constant monthly increase in coke resistance for various refinery streams	None considered	Two layer concept—porous coke adjacent to fluid and hard coke adjacent to wall
Nijsing (1964)[c]	Organic coolant in nuclear reactors	Hydrodynamic boundary layer and diffusion partial differential equations (1) instantaneous first order reaction in zone close to wall (2) very rapid crystallisation at hot surface	Product diffusion back to the fluid bulk is an integral part of the differential equations	(1) Solution fits plant data. Control fits plant data. Fouling rate predicted to increase with velocity (2) Extended to consider colloidal transfer to the hot surface
Watkinson and Epstein (1970)[f]	Liquid phase fouling from gas oils	Mass transfer and adhesion of suspended particles (1) sticking probability proportional to $\exp(-E/RT)$ (2) sticking probability inversely proportional to hydrodynamic forces on particle as it reaches wall	First order Kern and Seaton[*] shear removal term	(1) Correct prediction of initial rate dependence on velocity (2) Incorrect prediction of asymptotic resistance on velocity
Jackman and Aris (1971)[g]	Vapour phase pyrolysis	Kinetics control—two reactions (1) first order dimerisation of A into products (2) zero order coke formation	None considered	(1) Quasi-steady state assumption (2) Untested
Fernandez-Baujin and Solomon (1976)[b]	Vapour phase pyrolysis	Kinetics and/or mass transfer control with first order reaction	None considered	Solution with mass transfer control fits plant run-time data, ie fouling rate increases with velocity
Sundaram and Froment (1979)[D]	Vapour phase pyrolysis of ethane	Kinetics control (1) at surface temperature (2) first order in propylene concentration, a parallel reactions	None considered	(1) Quasi-steady state assumption (2) Good agreement between industrial and individually simulated data
Crittenden and Kolaczkowski (1979)[m]	Hydrocarbons in general	Kinetics and/or mass transfer control with first order reaction	(1) Diffusion of foulant back into fluid bulk (2) First order Kern and Seaton[*] shear removal term	(1) Limited testing (2) Complex—many parameters (3) Extended to two layer concept proposed by Atkins[b]

convective or diffusive mechanisms". Crittenden and Kolaczkowski[11] extended the two step, mass transfer and kinetics, model to include convection of the foulant back into the fluid bulk, Figure 3. Thus,

$$\frac{dx_f}{dt} = \frac{1}{\rho_f}(N_p - N_i) \tag{7}$$

where

$$N_i = K_f(c_b - c_{fs}) \tag{8}$$

Assuming that the concentration of foulant in the fluid bulk is very small,

$$\frac{dR_f}{dt} = \frac{d}{dt}(x_f k_f) = \frac{1}{\rho_f k_f}\left\{\frac{c_{fb}}{\frac{1}{K_p} + \frac{1}{k}} - K_f c_b\right\} \tag{9}$$

Mass transfer coefficients were again expressed in terms of flow rate and physical properties by application of the Chilton and Colburn analogy, friction factors were assumed to vary only with Reynolds number, and the deposition reaction was assumed to be of first order, the dependency of the velocity constant on temperature being described by the Arrhenius equation. Thus,

$$\frac{dR_f}{dt} = \pi_1$$

$$= \frac{1}{\rho_f k_f}\left\{\frac{c_{fb}}{\frac{\rho(d - 2x_f)^{1.1}(Sc)_b^{0.67}}{0.607\lambda_b^{0.67}G^{0.8}} + \frac{1}{A\exp(-E/RT)}}\right.$$

$$\left. - \frac{0.607\lambda_b^{0.67}G^{0.8}c_b}{\rho(d - 2x_f)^{1.1}(Sc)_b^{0.67}}\right\} \tag{10}$$

Diffusion of foulant back into the fluid bulk was considered to be important when the deposit contains relatively mobile species, ie when $(Sc)_b$ is small. The concentration c_b of the foulant at the solid/fluid interface was considered to be the solubility of the deposit in the bulk fluid, and is likely to increase with temperature, but decrease as the molecular weight of the deposit increases.

Equation (10) indicates that the fouling rate can be critically dependent upon the mass flow rate G. Various scenarios for the deposition and removal were suggested:

Kinetics control ($k \ll K_p$; for relatively low temperatures, small tube diameters and high mass flow rates)

If the foulant back diffusion is important then the fouling rate decreases with time (and x_f) and asymptotic fouling is a possibility when

$$c_{fb}A\exp(-E/RT) = \frac{0.607\lambda_b^{0.67}G^{0.8}c_b}{\rho(d - 2x_f)^{1.1}(Sc)_b^{0.67}} \tag{11}$$

The initial rate of fouling ($x_f = 0$ in equation (10)) is predicted to decrease as the mass flow rate is increased.

If the foulant back diffusion is negligible, then the fouling resistance is predicted to increase linearly with time and thus asymptotic fouling can occur only if there is an alternative deposit removal or release mechanism which is dependent upon deposit thickness.

Precursor diffusion control ($K_p \ll k$; for relatively high temperatures, large tube diameters and low mass flow rates)

If foulant back diffusion is important, equation (10) becomes:

$$\frac{dR_f}{dt} = \frac{0.607\lambda_b^{0.67}G^{0.8}}{\rho(d - 2x_f)^{1.1}\rho_f k_f}\left\{\frac{c_{fb}}{(Sc)_b^{0.67}} - \frac{c_b}{(Sc)_b^{0.67}}\right\} \tag{12}$$

The terms within the bracket are independent of time and therefore the fouling resistance can only tend towards an asymptotic value ($dR_f/dt = 0$) if there is an alternative removal mechanism which is dependent upon deposit thickness. At very high temperatures, severe thermal degradation to hard coke is likely to occur and thus the interfacial foulant concentration c_b would tend towards zero. In such cases, equation (12) would become:

$$\frac{dR_f}{dt} = \frac{0.607\lambda_b^{0.67}G^{0.8}c_{fb}}{k_f\rho_f\rho(d - 2x)^{1.1}(Sc)_b^{0.67}} \tag{13}$$

which has the same dependency on G and d as the models proposed by Fernandez-Baujin and Solomon[12] and by Nijsing[7].

Deposit removal by fluid shear

A first order dependence of deposit removal rate on deposit thickness yields an asymptotic fouling resistance-time relationship, ie

when

$$\frac{dR_f}{dt} = \pi_1 - \pi_2 R_f \quad \text{for} \quad 2x_f \ll d \tag{14}$$

with $R_f = 0$, when $t = 0$

$$R_f = \frac{\pi_1}{\pi_2}(1 - \exp(-\pi_2 t)) \tag{15}$$

Kern and Seaton[5] originally proposed that a deposit may be removed in chunks from a wall by the shearing action of the fluid. Such a removal rate was assumed to be proportional to both the shear stress at the wall and to the deposit thickness. Thus,

$$\text{removal rate} = -\frac{\tau x_f}{\psi k_f} \tag{16}$$

in which ψ is a function of deposit structure. Equation (16) can be written in terms of R_f and fluid velocity for addition to equation (10).

$$\text{removal rate} = -\left\{\frac{0.607\lambda_b^{0.67}G^{1.8}}{\rho\psi k_f(d - 2x_f)^{1.1}}\right\}R_f = -\pi_2 R_f \tag{17}$$

Crittenden and Kolaczkowski[11] extended their model to consider the build-up of two layers on the tube wall, in the manner described by Atkins[6]. An overview of the model is presented in Figure 4. It was assumed that thermal degradation of a mobile tarry layer would result in the formation of coke, the thermal conductivity of which is greater. Much more experimentation would be required to characterise the feedstock and hence determinate the parameters for this model.

Watkinson and Epstein[1] attempted to rationalise their experimental results with both gas oils and sand-water slurries by proposing the following transfer-adhesion-release model:

deposition is caused by mass transfer of suspended

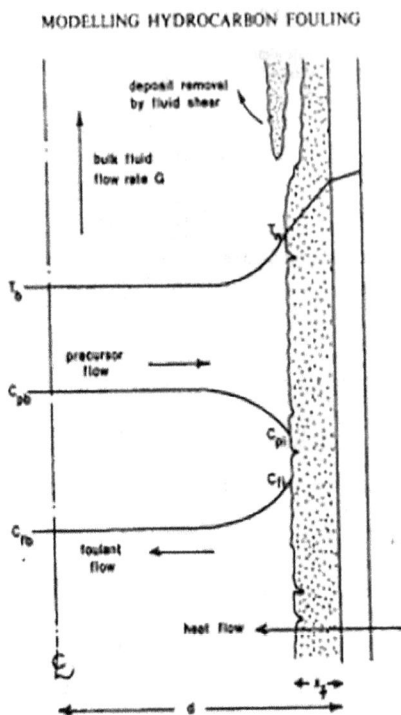

Figure 3 Mass transfer, kinetics and back convection model proposed by Crittenden and Kolaczkowski[10].

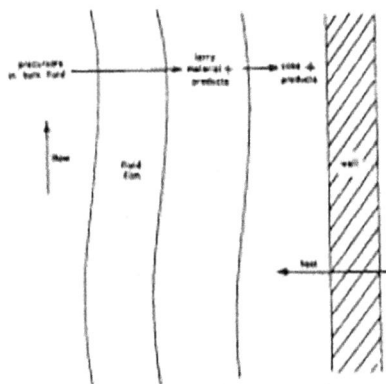

Figure 4 Overview of two layer model proposed by Crittenden and Kolaczkowski[10]

particles to the wall followed by adhesion of some particles to the wall;

—removal is a first order function of deposit thickness as proposed by Kern and Seaton.

The net rate of fouling was given by:

$$\frac{dR_f}{dt} = a_s SN - a_2 \tau x_f \tag{18}$$

The mass flux was assumed to be proportional to the difference in particulate concentrations in the bulk and at the surface. The sticking probability was assumed to be proportional to the physico-chemical adhesive forces binding a particle to the wall, dependent upon temperature according to the Arrhenius equation, and inversely proportional to the hydrodynamic forces on the particle as it reached the wall. Thus,

$$\frac{dR_f}{dt} = \frac{a_3(c_b - c_w)\exp(-E/RT)}{uf^{0.8}} - a_4 u^2 f x_f \tag{19}$$

Equation (19) correctly predicted the dependency of initial fouling rate on T and G as found by experiment. However, the model does not predict the correct dependency of asymptotic fouling resistance on flow rate.

Quasi-steady state

The vapour phase coking models of Jackman and Aris[13] and of Sundaram and Froment[12] contain neither mass transfer nor removal terms, but do invoke a quasi-steady state assumption, ie since the coking rate is very much less than the precursor throughput, the energy, continuity and pressure drop calculations need not contain explicit time dependencies but are updated periodically for changes in tube diameter as a result of deposition. Thus,

$$\Delta x_f = \frac{A \exp(-E/RT)c_{pb}\Delta t}{\rho_f} \tag{20}$$

By taking incremental steps along the tube, Sundaram and Froment numerically integrated the equations to give results which compared well with plant data.

TESTING THE CONCEPT OF FOULANT BACK CONVECTION

Crittenden and Kolaczkowski[14] assumed that deposition would be kinetically controlled at the relatively low surface temperatures, small tube diameter and high mass velocities used in Watkinson and Epstein's experiments[20] with gas oils. Two further assumptions were made, namely, that the deposit thickness remained much smaller than the tube radius, and that simple first order kinetics were applicable. Using the experimental initial fouling rates and asymptotic fouling resistances (when such resistances occurred), Crittenden and Kolaczkowski were then able to obtain an equation, derived from equations (10) and (17) which could be used to regenerate the fouling resistance-time data to very good agreement. Both back convection of foulant and shear removal of foulant were included in the equation.

The concepts embodied in the development of equation (10) are now applied to the model experiments of

chemical reaction fouling which used the polymerisation of 1% styrene in kerosene[21]. Since the order of reaction is 5/2 with respect to styrene:

$$N_p = k_r c_m^{5/2} \tag{21}$$

With back convection of foulant, the initial rate of fouling is given by:

$$R_f'(0) = \frac{1}{\rho_f k_f} \left\{ N_p - \frac{0.607 \lambda \mu^{0.2} G^{1.8} c_{fb}}{\rho d^{1.2} (Sc)_f^{0.67}} \right\} \tag{22}$$

From equations (2) and (21), the mass flux of styrene to the surface, N_p, for inclusion in equation (22), is obtained numerically from

$$c_{pb} = \left(\frac{N_p}{K_p}\right) + \left(\frac{N_p}{k}\right)^{2/5} \tag{23}$$

that is from

$$c_{pb} = \left\{ \frac{N_p \rho d^{1.2} (Sc)_p^{0.67}}{0.607 \lambda \mu^{0.2} G^{0.8}} \right\} + \left\{ \frac{N_p}{A \exp(-E/RT)} \right\}^{2/5} \tag{24}$$

For the experimental runs at the highest flow rate ($G = 649 \text{ kg m}^{-2} \text{ s}^{-1}$), which gives rise to the lowest surface temperatures for a given heat flux, it was concluded[20] that the overall deposition process tends to become kinetically controlled. For $G = 649 \text{ kg m}^{-1} \text{s}^{-1}$

$$R_f(0) = 659 \times 10^{-2} \exp(-39300/RT) \tag{25}$$

But

$$N_p = \rho_f k_f R_f'(0) = A \exp(-E/RT)c_{pb}^{5/2} \tag{26}$$

and[21]

$$\rho_c = 1050 \text{ kg m}^{-3}$$

$$k_f = 1.35 \times 10^{-4} \text{ kW m}^{-1} \text{ K}^{-1}$$

$$c_{pb} = 9.04 \text{ kg m}^{-3}$$

Hence

$$A = 3.801 \times 10^{-3} \text{ m}^{5.5} \text{ kg}^{-1.5} \text{ s}^{-1}.$$

From the friction factor-Reynolds number chart, λ was estimated to be 0.034 in the range of Reynolds numbers used in the experiments. Diffusion coefficients for the precursor styrene, D_p, and for the foulant polystyrene, D_f, were obtained as a function of temperature from the Wilke-Chang equation. The molecular weight of the kerosene was 142 and the molecular weight of the polystyrene taken to be 1000. The accuracy of determination of D_f and hence of $(Sc)_f$ could not be

Table 2. Physical properties of the styrene-polystyrene-kerosene system.

T °C	ρ kg m^{-3}	$\mu \times 10^3$ Nsm^{-2}	$D_p \times 10^9$ m^2 s^{-1}	$D_f \times 10^{11}$ m^2 s^{-1}	$(Sc)_p$	$(Sc)_f$
35	825	1.6	0.90	1.42	2155	136,600
55	808	1.2	1.29	2.06	1160	72,720
75	785	0.95	1.68	2.58	720	46,900
100	769	0.70	2.19	3.23	418	28,200
120	729	0.60	2.58	4.13	319	19,900
150	697	0.45	3.64	5.68	179	11,330
170	673	0.40	4.65	7.48	128	7959
190	657	0.35	5.94	9.29	90	5700
210	625	0.30	7.36	11.9	65	4000
230	609	0.22	9.03	14.6	40	2470

MODELLING HYDROCARBON FOULING

estimated with any reasonable degree of certainty. Physical property estimations are given in Table 2.

The mass flux of styrene, N_s, is calculated from equation (24) and converted to units of fouling rate by equation (26). The predicted data (with no allowance for back convection of foulant) and the experimental data are then compared in Figure 5. For corresponding surface temperatures and flow rates, the predicted initial fouling rates are consistently higher than the experimental values.

However for any given surface temperatures, the slopes of the predicted and experimental curves are broadly similar. As expected, agreement between the predicted and experimental values is best at low surface temperatures when the effect of flow rate is likely to be less important.

Equation (22) may be rearranged to give:

$$c_q = \left\{ \frac{N_s}{\rho_f k_t} - R'_f(0) \right\} \left\{ \frac{\rho d^{0.8} (Sc)_c^{\xi^{0.8}}}{0.607 \lambda \mu^{0.33} G^{0.8}} \right\} \qquad (27)$$

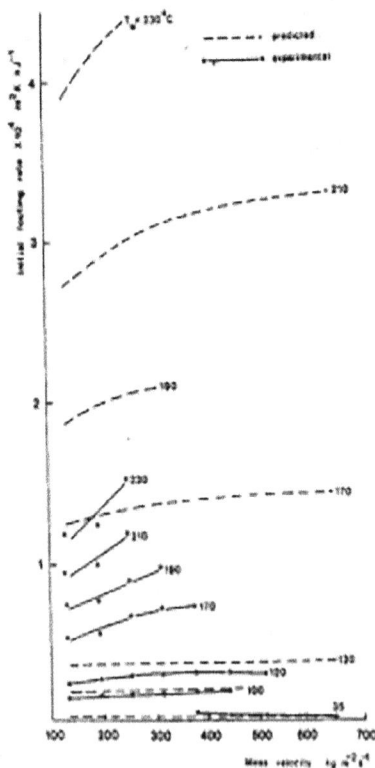

Figure 5. Comparison of predicted and experimental initial fouling rates for the system styrene-kerosene-polystyrene. Back convection and shear removal are excluded from the predictive model.

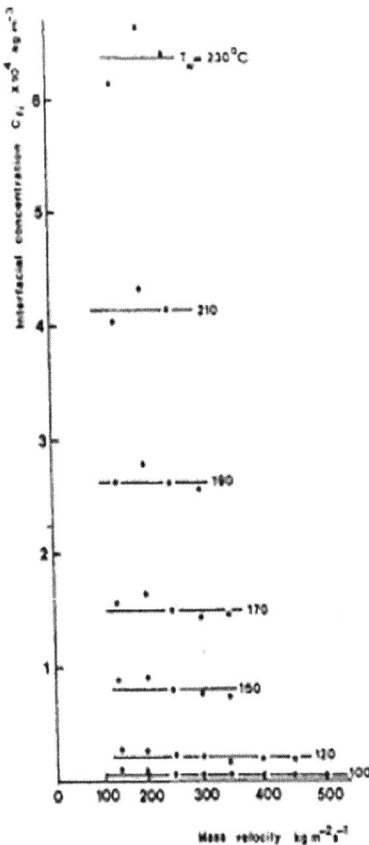

Figure 6. Concentration of foulant at the solid/fluid interface.

The predicted initial fouling rates were substituted for $N_s/\rho_f k_t$ and the experimental values for $R'_f(0)$. The values of c_q required to give equality of predicted and experimental fouling rates, as calculated from equation (27), are plotted as a function of flow rate and temperature in Figure 6. There is no fundamental reason to believe that the concentration of polystyrene at the deposit fluid interface should be dependent upon flow rate. The results shown in Figure 6 support this view. However, this interfacial concentration is expected to be strongly dependent upon temperature. Values of c_q, averaged from point values over the range of flow rates, are plotted against temperature in Figure 7 to demonstrate the strong temperature effect.

It is not possible to predict with any reasonable accuracy the solubility of polystyrene in styrene, as for this experimental system the solvent is complex. Never-

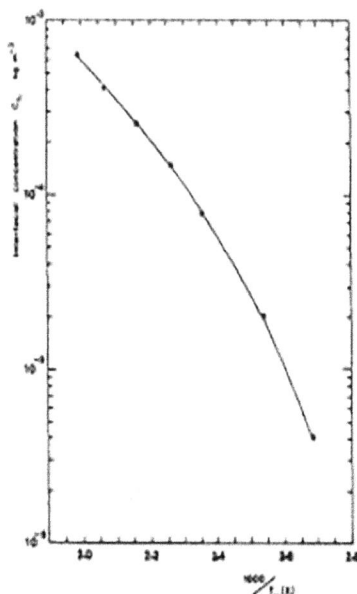

Figure 7. Effect of temperature on the concentration of foulant at the solid/fluid interface.

theless, the process of dissolving a polymer in a solvent is governed by the equation:

$$\Delta G = \Delta H_M - T\Delta S \qquad (28)$$

The dissolution of a polymer is always connected with a large increase in entropy[20] and therefore the magnitude of the heat of mixing is the deciding factor in determining the sign of the free energy change. The heat of mixing may be determined from

$$\Delta H_M = V_M \phi_1 \phi_2 (\delta_1 - \delta_2)^2 \qquad (29)$$

The solubility parameter for styrene[22] is 19,000 (Jm^{-3})$^{1/2}$, whilst that for polystyrene[22] is between 17,400 and 21,000 (Jm^{-3})$^{1/2}$. Thus ΔH_m is expected to be small for the styrene-polystyrene system, and hence ΔG is likely to be negative. Therefore, from equation (28) it is expected that the solubility of polystyrene in styrene will increase strongly with increasing temperature.

CONCLUSIONS

A three step chemical reaction fouling model, based on mass transfer of foulant precursor to the wall, reaction at the wall, and back convection of foulant to the fluid bulk, has been developed in order to predict the dependence of initial rates of fouling on key design and operating variables such as temperature, flow rate and tube diameter. The model predicts a complex dependency of fouling rate on flow rate. Whether the fouling rate increases or decreases with flow rate depends on the relative balance between mass transfer and kinetic

effects. For complex hydrocarbon streams, it is likely that some of the parameters in the model would need to be evaluated from plant data.

The model is tested with the formation of deposits from gas oils and the formation of deposits from a model chemical system comprising 1% styrene in kerosene. For the latter system, the only adjustable parameter is the concentration of polystyrene at the foulant-liquid interface. It is shown that, as expected, this interfacial concentration is independent of mass flow rate but increases strongly with interfacial temperature.

SYMBOLS USED

a_1, a_2, a_3, a_4	coefficients in models	various units
A	pre-exponential factor	m s^{-1}
		(first order)
		m$^{3.5}$ kg$^{-1.5}$ s^{-1}
		(5/2 order)
c	concentration	kg m^{-3}
d	tube diameter	m
D	diffusion coefficient	m^2 s^{-1}
E	activation energy	kJ kmol^{-1}
f	friction factor	
G	mass flow rate	kg s^{-1}
k	reaction rate constant	m s^{-1}
		(first order)
		m$^{5.5}$ kg$^{-1.5}$ s^{-1}
		(5/2 order)
k	thermal conductivity	kW m^{-1} K^{-1}
K	mass transfer coefficient	m s^{-1}
K^*	function defined by equation (6)	
N	mass flux	kg m^{-2} s^{-1}
R	universal gas constant	8.3143 kJ kmol^{-1} K^{-1}
Re	Reynolds number	
R_f	fouling resistance	m^{2}K kW^{-1}
S	sticking probability	
Sc	Schmidt number	
t	time	s
T	temperature	°C, K
u	bulk linear velocity	m s^{-1}
V_m	total volume of mixture	m^3
x	thickness	m
Greek symbols		
δ_1, δ_2	solubility parameter for components 1, 2	(Jm^{-3})$^{1/2}$
ΔG	Gibbs free energy	kJ kmol^{-1}
ΔH_m	heat of mixing	kJ kmol^{-1}
ΔS	entropy change	kJ kmol^{-1} K^{-1}
λ	defined from $f = \lambda Re^{-0.2}$	
μ	viscosity	Ns m^{-2}
π_1	function defined in equation (10)	
π_2	function defined in equation (14)	
ρ	density	kg m^{-3}
t	wall shear stress	Nm^{-1}
ϕ_1, ϕ_2	volume fraction of components 1 and 2 in mixture	
ψ	function of deposit structure	
Subscripts		
b	bulk	
c	critical	
f	foulant	
i	fluid-deposit interface	
p	precursor	
w	wall	

REFERENCES

1 Van Nostrand, W. L., Leach, S. H. and Haluska, J. L., 1981, *Fouling of heat transfer equipment*, Somerscales, E. F. C. and Knudsen, J. G., (Eds) (Hemisphere, Washington) 619-64)

MODELLING HYDROCARBON FOULING   179

2. Fitzer, E., Mueller, K. and Schaefer, W., 1971, *Chemistry and Physics of Carbon, Vol 7*, edited by Walker, P. L., (Marcel Dekker, New York) 237-383.

3. Crittenden, B. D. and Khater, E. M. H., 1984, *1st UK National Conf on Heat Transfer, IChemE Symp Ser No 86*, (IChemE, Rugby) 401-414.

4. Canapary, R. C., 1961, *Oil Gas J*, 59, (41): 114-118.

5. Nelson, W. L., 1934, *Refiner and Nat Gas Man*, 13, (7): 271-276, 13, (8): 292-298.

6. Atkins, G. T., 1962, *Petro/Chem Engr*, 34, (4): 20-25.

7. Nijsing, R., 1964, *Diffusional and kinetic phenomena associated with fouling, Euratom Report EUR 543.e*.

8. Watkinson, A. P. and Epstein, N., 1970, *Proc 4th Int Heat Transfer Conf, Paris, Vol 1*, Paper HE1.6, (Elsevier, Amsterdam).

9. Kern, D. Q. and Seaton, R. E., 1959, *Br Chem Engng*, 4: 258-262.

10. Jackman, A. P. and Aris, R., 1971, *Proc 4th European Symp on Chemical Reaction Engineering*, (Pergamon, Oxford) 411-419.

11. Fernandez-Baujin, J. M. and Solomon, S. M., 1976, *Industrial and Laboratory Pyrolyses*, edited by Albright, L. F. and Crynes, B. L., *ACS Symp Series No 32*, (American Chemical Society, Washington) 345-372.

12. Sundaram, K. M. and Froment, G. F., 1979, *Chem Engng Sci*, 34: 635-644.

13. Crittenden, B. D. and Kolaczkowski S. T., 1979, *Proc. Conf Fouling—Science or Art? Inst Corr Sci and Tech/IChemE, London*, 169-187.

14. Crittenden, B. D. and Kolaczkowski, S. T., 1979, *Energy for Industry*, edited by O'Callaghan, P. W., (Pergamon, Oxford) 257-266.

15. Tubular Exchangers Manufacturers' Association, 1978, *Standards of the Tubular Exchangers Manufacturers' Association, New York*, 138-142.

16. Randhava, S. S. and Wasan, D. T., 1972, *Chemical Reaction Engineering, ACS Symp Series No 109* (American Chemical Society, Washington) 564-568.

17. Wanker, E. H. and Schechter, R. S., 1962, *Chem Engng Sci*, 17, 937-948.

18. Chen, J. and Maddock, M. J., 1979, *Hydrocarb Process*, 52, (5): 147-150.

19. Denbigh, K., 1965, *Chemical Reactor Theory* (Cambridge University Press, London) 159-161.

20. Watkinson, A. P. and Epstein, N., 1969, *Chem Eng Prog Symp Ser*, 65, (92): 84-90.

21. Crittenden, B. D., Hout, S. A. and Alderman, N. J., 1987 *Chem Eng Res Des* 65(2): 165-170.

22. Brandrup, J. and Immergut, E. H. (eds), 1975, *Polymer Handbook, 2nd Ed.*, (Wiley, New York) IV. 337-IV 359.

ACKNOWLEDGEMENT

The authors express their thanks to the Science and Engineering Research Council for the award of a research grant to study aspects of chemical reaction fouling.

ADDRESSES

Correspondence in connection with this paper should be addressed to Dr B. D. Crittenden, School of Chemical Engineering, University of Bath, Bath BA2 7AY. Dr S. A. Hout is currently at Procter and Gamble AG, 1 rue du Pré-de-la-Bichette, 1211 Genève 2, Switzerland.

The manuscript was received 5 November 1986 and accepted for publication after revision 29 January 1987.

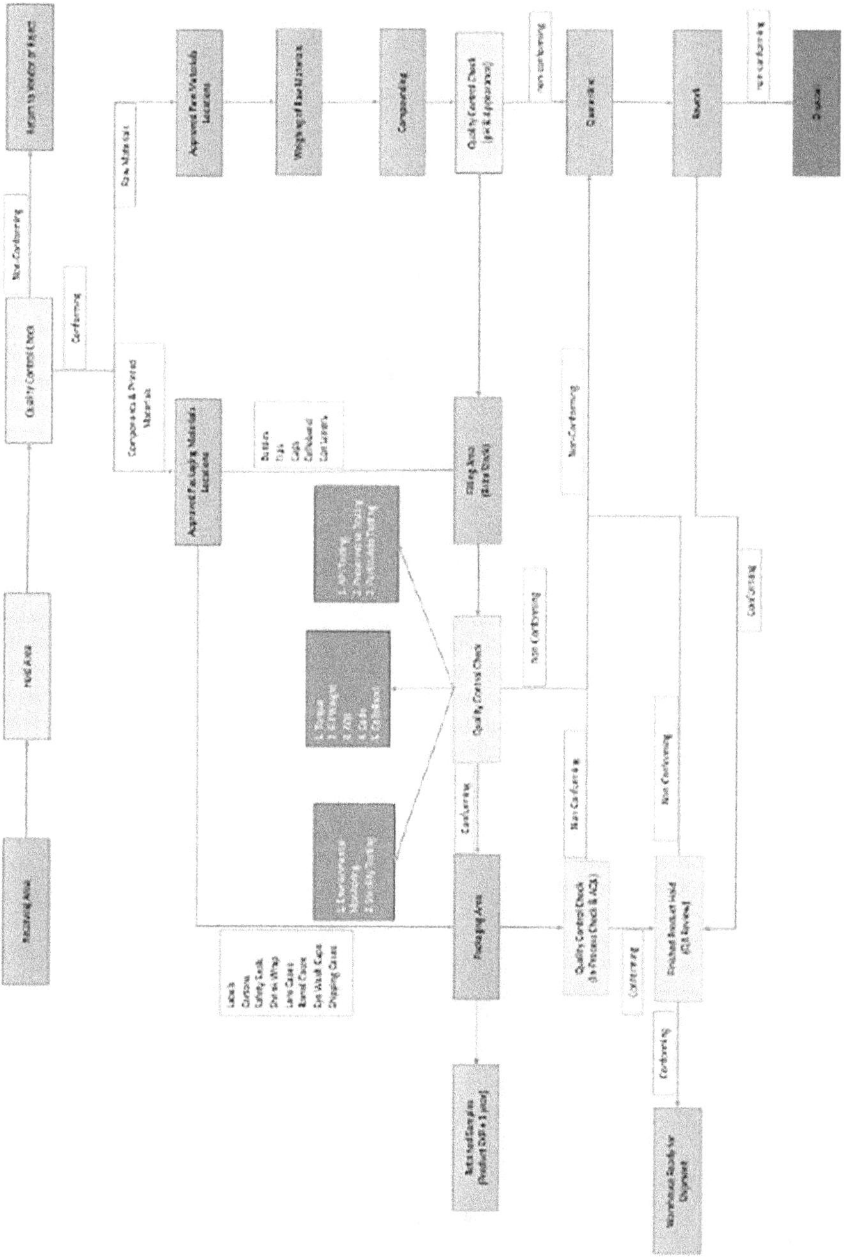

Appendix V: Proposal for New Plant Construction

DISCLAIMER

Contract Manufacturing Services of A, Inc. (A) has engaged B Consumer Healthcare (B) to act together in exploring a possible partnership in contract manufacturing. A and B are Partners.

This Confidential Information Presentation has been prepared from information obtained from the management of A and from other sources believed to be reliable. This Capital Investment Proposal (CIP) does not contain all of the information material to an evaluation of the business of the Company, does not contain or constitute a representation or warranty of any nature whatsoever, is provided solely for the purpose of assisting recipients in evaluating whether to pursue a business transaction and does not constitute an offer of any security by any person. The Partners expressly reserve the right, without giving reasons therefore, at any time and in any respect, to terminate discussions. Partners expressly disclaim any and all liability which may be based on the CIP or any of its contents, errors therein, or omissions therefrom. The recipient shall only be entitled to rely on the representations and warranties made to it in any final agreement.

The projections and forward-looking statements concerning the operation and financials included herein (the "projections") were prepared solely by A and were not prepared with a view toward public disclosure or complying with the published guidelines of the Securities and Exchange Commission. In addition, because such projections are based on a number of assumptions and are subject to significant uncertainties and contingencies, there is no assurance that the projections will be realized, and actual results may vary significantly from those shown. The inclusion of projections should not be regarded as a representation that the projections will be achieved.

THIS CIP IS BEING PROVIDED TO YOU UNDER A CONFIDENTIALITY AGREEMENT, THE TERMS OF WHICH WILL BE CLOSELY MONITORED AND STRICTLY ENFORCED. CONSEQUENTLY, YOU ARE URGED TO EXERCISE THE UTMOST DISCRETION IN MAKING THE ENCLOSED INFORMATION AVAILABLE TO YOUR EMPLOYEES, AFFILIATES, OR ADVISORS WHO MAY NEED TO ANALYZE SUCH MATERIAL IN THE COURSE OF YOUR EVALUATION. THIS CIP AND ANY SUBSEQUENTLY FURNISHED INFORMATION MUST BE DESTROYED OR RETURNED TO A, WITHOUT RETAINING COPIES OR EXCERPTS THEREOF, IF YOU DETERMINE NOT TO PROCEED WITH A TRANSACTION WITH A.

I. EXECUTIVE SUMMARY

A Is a Premier Sterile Private Label Development and Manufacturing Organization

- Leading reputation (OTC/NDC) and track record of performance in technically complex segments of aseptic pharmaceutical Ophthalmic drug/Liquid medical device (510k)
- Full range of process development and formulation through commercial fill/finish manufacturing services at strategically located cGMP facilities—Pomona, CA
- Only available North American sterile-focused fill/finish asset with proven commercial manufacturing capabilities; successful completion of FDA pre-approval inspections

Full Range of Capabilities with Proven Commercial Expertise

- Highly flexible, client-focused end-to-end solutions for aseptic drug and liquid-filled medical device products including complementary regulatory support, secondary packaging, and on-site analytical and microbiology laboratories
- Two automated liquid fill lines (high speed, auto fill, auto close). Focused on Ophthalmic category
- Currently manufacturing 12 commercial formulation products
- Deep expertise in complex formulations for small molecule products in a range of containers/closures

Superior Track Record of Quality and Regulatory Compliance

- Exemplary quality and regulatory track record with a long history of FDA cGMP compliance
- Successfully completed regulatory inspections since 2002
- Completed approximately 10 customer audits per year over the last 2 years

Strategically Located Modern Facilities with Substantial Capacity for Growth

- Modern cGMP facilities, spanning ~150,000 ft^2, offer a complete portfolio of industry-leading development and manufacturing capabilities
- Substantial capacity for growth; currently operating at ~50% capacity. Capable of operating at ~$75 million of revenue with minimal additional investment on the current footprint
- Strategically located close to two key U.S. pharmaceutical hubs (Los Angeles and Orange Counties). Facilities include big pharma-caliber redundancy and systems

II. INDUSTRY OVERVIEW

CDMOs PROVIDE INDEPENDENT DRUG MANUFACTURING SERVICES TO PHARMACEUTICAL COMPANIES

- The industry can generally be segmented into primary and secondary manufacturing; primary manufacturing of active pharmaceutical ingredients ("APIs") and secondary manufacturing of finished dosage forms.
- The finished dosage CDMO market is often divided into two primary categories: non-sterile and sterile.
  - Sterile forms are manufactured under highly stringent and technically advanced clean room conditions to prevent contamination and accommodate environmental sensitivities.
  - Sterile products may be synthetic drugs or biologics and include a variety of liquid and semi-solid forms.
  - Ophthalmic and Otic drugs are aseptically filled in Plastic Bottles/Tips/Caps.
  - Sterile formulations are typically filled into vials, ampoules, or syringes and may be lyophilized (freeze-dried). Outsourcing to CDMOs from pharmaceutical and biotechnology firms is accelerating due to:
- Lack of internal capabilities (most notably the smaller pharmaceutical firms)
- Enhanced flexibility to use company resources to focus on core competencies
- Increasing technical complexity and regulatory burdens
- Lower fixed costs and a reduction of capital expenditure needs

GROWING PHARMACEUTICAL SALES

- According to IMS, annual spending in the global pharmaceutical market is expected to grow from $887 billion in 2020 to approximately $1.4 trillion in 2030.
- An aging and growing population, the increased incidence of chronic diseases, and the continued development of new drugs are the primary drivers of expected growth.
- Pharmaceutical development between 2019 and 2021, global annual pharma R&D spending increased from $89 billion to $162 billion.

CDMO

- Pharmaceutical and biotechnology companies are increasingly deciding to outsource their manufacturing and development services to decrease costs and focus.
- The percentage of global finished dosage manufacturing currently outsourced is approximately 35% and is expected to increase by 5 percentage points annually.

III. COMPANY OVERVIEW

- Robust R&D is driving an increase in approvals of OTC/NDC Ophthalmic/Otic approvals, which increased as complex formulations of existing drugs have come to the forefront.
- CDMO segment gives a strong pipeline of specialty therapeutics and the reformulation of existing drugs into sterile formulations and non-sterile manufacturing capacity.
- A is a leading developer and manufacturer of complex formulations.
- The Company provides a full range of development, fill/finish services for Ophthalmic and Otic drugs, and medical devices.
- A is known throughout the industry for its exceptional quality and regulatory track record and deep formulation, process development, and sterile manufacturing experience combined with a flexible, client-focused business model.
- A is an innovative solution provider for difficult-to-formulate and -manufacture products.
- The Company's capabilities and track record of service have resulted in a diverse and loyal customer base and a well-balanced business mix of development projects and commercial products.
- It offers a full range of services from process development through commercial manufacturing.
- A's chemists refine formulations to optimize manufacturing time and yield.
- A employs a scientific approach and its extensive experience to develop and optimize the formulations of clients' drug substances.
- Utilizing proven tools and processes and by applying quality by design (QbD), A delivers leading process development solutions.

SCALE-UP

The Company's experienced development operations, quality, and project management teams work together to evaluate a client's project objectives. An optimal path forward is recommended to optimize regulatory submissions and minimize time to market.

TESTING CAPABILITIES

Research and development stability studies
Osmolality, reconstitution time, concentration, pH, density, HPLC, GC-MS, ion-chromatography
Hand filling in biosafety hood

DEVELOPMENT CAPABILITIES

Bulk and unit doses of small molecules

PROJECT MANAGEMENT

- Core Competency and Competitive Differentiator
- A's team of project managers provides a fully accountable, single point of contact for customers from quotation through ongoing commercial manufacturing.
- The Company's focused project management system optimizes the efficiency of functional departments in the organization—allowing each functional area to do what they do best.
- Project management also plays an important role in business development by qualifying and responding to all RFPs, generating new leads from current customers, and identifying and building alignment on value-added upselling for in-process projects.
- A's multi-disciplinary team approach ensures a strong client relationship and successful project completion.

SINGLE POINT OF CONTACT

Each project is assigned a Program Manager who is universally accessible throughout the project.

PROBLEM SOLVING

Addresses project issues in parallel to avoid delays.

COLLABORATION

Fosters partnership and teamwork among project members.

CLEARLY DEFINED OBJECTIVES

Ensures the client's needs and expectations are fully understood by the A team.

EXECUTION MANAGEMENT

Promotes focus and drives priorities to achieve project completion.

CUSTOMERS AND PRODUCTS

- Attractive Customer Base and Balanced Business Mix. The Company's expertise and track record of results has attracted a diverse set of longstanding customers
- Strong commercial base and rapidly growing pipeline; product mix continues to shift toward OTC drugs

IV. FINANCIAL OVERVIEW

CURRENT COMMERCIAL AND DEVELOPMENT PRODUCTS

Revenue forecast based on:

- Customer-provided demand forecasts by product, where available
- Management estimates
- Probability-weighted progression by product phase based on published research of drug progression

V. FUTURE EXPANSION OUTSIDE CALIFORNIA

STRATEGICALLY PLANNED WITH SUBSTANTIAL CAPACITY

- cGMP-compliant, purpose-built facilities located in the Nevada area enable cohesive operations and management
- Facilities offer a comprehensive suite of services, outfitted with premier development and manufacturing equipment to enable efficiency and flexibility

VI. VALUATION

ASSUMPTIONS

1. Land Acquisition is based on 5 acres for the proposed structure and another 5 acres for future expansion
2. Equipment valuation is based on relatively good conditions and lifetime not to exceed 10 years.
3. Asset values are for replacement or building cost.
4. There are no systems for automatic loading or unloading.
5. Equipment are not fitted with specific automation. Additional considerations for future expansion.
6. Aseptic filling equipment are CIP capable.
7. Automatic CIP is not installed.
8. Organic solvent handling exceeding 22% bulk concentration requires special Installation including effluent discharge or scrubbing.
9. Aseptic facilities are all integrated and connected to include two filling lines, four packaging lines, critical utilities, equipment mezzanine, chemistry and microbiology laboratories, and warehouse in a two-story building.
10. Aseptic manufacturing cascade from the aseptic core with a wraparound aseptic corridor.
11. Process flow from component preparation, weighing and compounding, filling, packaging, labeling bundling, and palletizing.
12. Warehouse will be operated with racks and two forklifts.
13. Lead time for construction is 9–11 months after permits. Lead time for filling equipment is 17–21 months from Kick off. Lead time for packaging equipment is 13–16 months from PO.

ANALYSIS

The following is a market value analysis for comparable equipment:

Buildings + Land

Description	Asset Type	Size, sq. ft	Asset Category	Location	Asset Replacement Value, $M	Note
Land	Industrial	10 Acres	5 acres for construction, 5 acres for future development	Industrial estate	0.75	
Facility	Building	24,000	MFG/R&D	Commerce Dr	4.80	Drug development and manufacturing; fill/finish (sterile)
Facility	Building	14,200	Office/lab	Commerce Dr	2.215	Office and laboratory space
Facility	Building	22,200	Warehouse	Commerce Dr	2.68	cGMP warehouse and storage
Facility	Building	8,400	Packaging	Commerce Dr	1.09	Packaging
Facility	Building	17000	Equipment Mezzanine	Commerce Dr	1.7	Manufacturing and critical utilities equipment
		85,800			**13.235M**	

Purified Water Production

City Water	Potable water	$35,000
Ozone Treatment	micro control	$157,000
Piping/plumbing/instrumentation	Connect to RO system	$100,000
Pretreatment	Sodium meta bisulfite + RO	$225,000
Piping/plumbing/instrumentation	Connect to RO tank	$110,000
RO tank		$75,000
Piping/plumbing/instrumentation	Connect to steam generator/stills	$100,000
3,008 still	Four-effect evaporator	$1,200,000
Piping/plumbing/instrumentation	Connect to PW	$100,000
Pumps	PW pumps	$125,000
PW tank		$300,000
RO/PW/CSM panels		$135,000

(*Continued*)

(Continued)

Purified Water Production

Industrial boiler	Steam	$225,000
CSM	Generator	$300,000
UV Equipment		$275,000
		$3,462,000

Manufacturing Equipment

Component preparation equipment	$250,000
Weighing	$50,000
Compounding tanks, mixer, and equipment	$750,000
Forklift x 2	$100,000
Fire protection	$175,000
Compressed clean gases	$300,000
Auxiliary equipment	$250,000
	$1,875,000

Testing Equipment

Description	Equipment type	Asset Category	Location	New Equipment Value, $	Note
Micro Lab	Micro Equipment	Micro	Commerce	$250,000	Hoods/BSC II
Micro Lab	Space	Micro	Commerce	$450,000	Micro labs
Chemistry					
HPLC	Chemical analysis	Lab Services		$720,000	HPLC units API/ Excipient testing
UPLC	Chemical analysis	Lab Services		$300,000	Method analysis
mDSC	Chemical analysis	Lab Services		$25,000	Modulated differential scanning calorimeter
Microscopy	Chemical analysis	Lab Services		$5,000	Refractive index
Particle analyzer	Chemical analysis	Lab Services		$35,000	Non-viable particle testing
KF titrator	Chemical analysis	Lab Services		$15,000	Karl Fischer titrator
Osmometer	Chemical analysis	Lab Services		$6,000	Osmolality
				$1,806,000	

Packaging lines

Description	Equipment Type	Size, sq. ft	Asset Category	Location	New Equipment Value, $
Pkg 1	Packaging	960	CNC/non-sterile	Commerce Dr	$1,894,857
Pkg 2	Packaging	960	CNC/non-sterile	Commerce Dr	$1,894,857
Pkg 3	Packaging	960	CNC/non-sterile	Commerce Dr	$1,894,857
Pkg 4	Packaging	960	CNC/non-sterile	Commerce Dr	$1,894,857
				TOTAL	**$7,579,428**
Total Filling					**$6,428,194**

Vial Filling Line

BTC filling line—200 bpm—**Line#1**

Description	Total
Rotary table	$35,900
Max load sensor	$3,354
Vial in-feed tray plastic	$6,970
plastic rotary disc	$11,100
Each Add. size change part for bottle	$3,780
Filler (filling-tip-capping)	$3,650,900
Servo drive—volumetric/peristaltic pump	$51,500
peristaltic pump	$18,300
Disposable product circuit	$4,540
Safety hood extension for STOPPER	$19,500
Safety hood extension for cap	$15,500
Tip loading system	$36,200
Cap loading system	$36,200
Single tray loading unit	$17,700
Installation—Startup	$60,830
Packing closed case for rotary table	$4,200
Packing closed case for filler	$8,300
Validation package	$55,700
Validation for check weighing system	**$25,100**
	$4,065,574

(Continued)

(Continued)

Vial Filling Line
Options

Nitrogen system with digital flow meter	$28,900
100% check weighing system single	$109,200
Isokinetic probe	$18,350
Petri plate	$14,720
Each additional transport size SINGLE step	$24,200
Each additional size for check weigher	$6,770
Additional tip size SINGLE step	$35,000
Additional capping size SINGLE step	$38,900
Additional capping roller	$4,540
Additional tip bowl	**$27,400**
	$307,980
Grand total—Line#1	$4,373,554

Filling #2	Unit price	Total price
	(USD)	(USD)
150 bpm	$1,884,648	$1,884,648
Integral filling/tipping/capping		
Capacity = 2,500-L batch		
Control system (Allen Bradley PLC + iFix)		
Factory acceptance test (FAT)		
SIP (w/ASME CODE VESSELS)	incl.	
STERILE INLET FILTER (x1)	incl.	
CIP	incl.	
HYDRAULIC STOPPERING	incl.	
STOPPERING PISTON BELLOWS	incl.	
MAIN VALVE BELLOWS	incl.	
OIL-SEALED VACUUM PUMP (x2)E2M28	incl.	
RECIPROCATING COMPRESSOR SYSTEM	incl.	
HTF CIRCULATION PUMPS (x2)	incl.	
MKS CAP. MANO. CHAMBER VACUUM MEASUREMENT	incl.	
DUAL DRAIN TEMPERATURE PROBES	incl.	
UPS, 10 MINUTES	incl.	
UL INSPECTED ELECTRICAL CABINETS	incl.	
SEISMIC ZONE COMPLIANCE	incl.	
MAINTENANCE SCREEN	incl.	
ALLEN BRADLEY CONTROL LOGIX PLC	incl.	
SCREEN PRINTER	incl.	
E-SIGNATURE	incl.	
GAMP5 COMPLIANT DOCUMENTATION	incl.	
IQ/OQ DOCUMENTATION PACKAGE	incl.	
MACHINE TOTAL		$1,884,648

Filling #2	Unit price	Total price
	(USD)	(USD)
INSTALLATION (2 MAN-WEEKS)	$33,800	$33,800
START-UP AND COMMISSIONING (2 MAN-WEEKS)	$33,800	$33,800
SITE ACCEPTANCE TEST (SAT) (1 MAN-WEEK)	$15,900	$15,900
DELIVERED AT Destination	$18,892	$18,892
SERVICES TOTAL		$102,392
FINAL TOTAL AMOUNT		$1,987,040
IQ/OQ ON-SITE EXECUTION (4 MAN-WEEKS)	$67,600	$67,600
Grand Total—**Line#2**		$2,054,640

	Unit price	Total price
Plant Critical Utilities	(USD)	(USD)
Integral distribution network	$884,648	
Control system (Allen Bradley PLC + iFix)	incl.	
Factory acceptance test (FAT)	incl.	
SIP (w/ASME CODE VESSELS)	incl.	
STERILE INLET FILTER (x1)	incl.	
CIP	incl.	
STOPPERING PISTON BELLOWS—auxiliary	incl.	
MAIN VALVE BELLOWS	incl.	
OIL-SEALED PUMP (x2)	incl.	
CIRCULATION PUMPS (x2)	incl.	
DUAL DRAIN TEMPERATURE PROBES	incl.	
UPS, 10 MINUTES	incl.	
UL INSPECTED ELECTRICAL CABINETS	incl.	
SEISMIC ZONE COMPLIANCE	incl.	
MAINTENANCE SCREEN	incl.	
ALLEN BRADLEY CONTROL LOGIX PLC	incl.	
SCREEN PRINTER	incl.	
E-SIGNATURE	incl.	
GAMP-5 COMPLIANT DOCUMENTATION	incl.	
IQ/OQ DOCUMENTATION PACKAGE	incl.	
System TOTAL		$884,648
INSTALLATION (2 MAN-WEEKS)	$23,800	$23,800
START-UP AND COMMISSIONING (2 MAN-WEEKS)	$23,800	$23,800
SITE ACCEPTANCE TEST (SAT) (1 MAN-WEEK)	$11,900	$11,900
DELIVERED AT Destination	$10,892	$10,892
SERVICES TOTAL		$70,392
IQ/OQ ON-SITE EXECUTION (4 MAN-WEEKS)	$47,600	$47,600
FINAL TOTAL AMOUNT		**$1,002,640**
Autoclave (2)	$657,000	$1,314,000

(*Continued*)

(Continued)

Plant Critical Utilities	Unit price (USD)	Total price (USD)
shelves	incl.	
Capacity = 346 L	incl.	
SIP with ASME code vessels	incl.	
Door for autoloading	incl.	
Variable frequency drive	incl.	
Integrity test	incl.	
Main valve bellows + integrity test	incl.	
Vacuum pumps + Qty 1 booster	incl.	
PC 21CFR Part 11 compliant	incl.	
Device mode SCADA screen	incl.	
Maintenance screen	incl.	
Batch report incl	incl.	
Screen printer incl	incl.	
Absolute vacuum pressure measurement chamber	incl.	
30 Point Yokogawa Paperless Recorder	incl.	
UPS PC and PLC (15 minutes)	incl.	
Redundant fluid inlet probe	incl.	
Sterile filters	incl.	
Cleaning in place (CIP)	incl.	
Water ring pump incl	incl.	
Jacket and door cooling	incl.	
IQ/OQ documentation package	incl.	
TOTAL		$1,314,000
GRAND TOTAL AMOUNT Plant Critical Utilities		**$2,316,640**

OVERALL VALUATION SUMMARY

ITEM	Description	Valuation
Plant critical utilities	Sterilization and distribution	**$2,316,640**
Purified water	PW-making plant	**$3,462,000**
Buildings + land	Site improvement	**$13,235,000**
Manufacturing equipment	Component prep., compounding tanks, weighing, CCA, FP	**$1,875,000**
Testing equipment	Micro and chemistry labs	**$1,806,000**
Packaging	Packaging lines	**$7,579,428**
Fill/finish	Filling lines	**$6,428,194**
	Total	**$36,702,262**

Manufacturing Capacity - Projected

Line	Operation	Product (format)	Qty/Shift	# Work Shifts	Effective Output Daily	Workdays/Week	Output Weekly	Output Monthly	Annual Capacity (total units)
#1	Aseptic Filling	0.5 oz, Eye Drops (round)	96,000	3	288,000	6	1,728,000	6,912,000	82,944,000
#2		0.5 oz, Eye Drops (round)	72,000	3	216,000	6	1,296,000	5,184,000	62,208,000
Pkg1	Packaging	0.5 oz, Eye Drops (round)	20,000	3	60,000	6	360,000	1,440,000	17,280,000
Pkg2		0.5 oz, Eye Drops (round)	20,000	3	60,000	6	360,000	1,440,000	17,280,000
Pkg3		0.5 oz, Eye Drops (round)	20,000	3	60,000	6	360,000	1,440,000	17,280,000
Pkg4		0.5 oz, Eye Drops (round)	20,000	3	60,000	6	360,000	1,440,000	17,280,000

Total Brite Stock (WIP) Output/Annually   145,152,000

Total Packaged Goods Output/Annually   69,120,000

Note1: "Work shift" defined as manufacturing working shifts (labor) directly generating throughput.

Manufacturing Labor Cost - Projected

Line	Operation	Product (Format)	Staff/Line	# Support Shifts	Total Staff	Daily	Weekly	Monthly	Annual	Labor Cost/Unit $	Run Strategy
A1	Aseptic Filling	0.5 oz. Eye Drops Solutions (round)	8	3	24	$ 3,774.72	$ 22,648.32	$ 90,593.28	$ 1,087,119.36	0.013	22.5 hours daily fill time, 6 days/week, 48 weeks/annually
A2		0.5 oz. Eye Drops Solutions (round)	8	3	24	$ 3,774.72	$ 22,648.32	$ 90,593.28	$ 1,087,119.36	0.017	22.5 hours daily fill time, 6 days/week, 48 weeks/annually
Pkg1	Packaging	0.5 oz. Eye Drops (cva)	3	3	9	$ 1,257.12	$ 7,542.72	$ 30,170.88	$ 362,050.56	0.021	24 hours daily (3-Shift), 6 days/week, 48 weeks/annually
Pkg2		0.5 oz. Eye Drops (cva)	3	3	9	$ 1,257.12	$ 7,542.72	$ 30,170.88	$ 362,050.56	0.021	24 hours daily (3-Shift), 6 days/week, 48 weeks/annually
Pkg3		0.5 oz. Eye Drops (round)	3	3	9	$ 1,257.12	$ 7,542.72	$ 30,170.88	$ 362,050.56	0.021	24 hours daily (3-Shift), 6 days/week, 48 weeks/annually
Pkg4		0.5 oz. Eye Drops (round)	3	3	9	$ 1,257.12	$ 7,542.72	$ 30,170.88	$ 362,050.56	0.021	24 hours daily (3 Shift), 6 days/week, 48 weeks/annually

Note 1: Labor rate/hour based on KE staff notes for filling at $29.56 and packaging at $17.28.

Note 2: Supporting staff (denoted as manufacturing support staff) specify generating throughput and staff scheduled by production/automation of classrooms/equipment.

	Production Staff	Production Supervisors	Production Support	Maintenance staff	Maintenance supervisors	Plant Staff	Total
	84	9	11	12	4	7	127
Labor Cost, $/hr	30	40	30	30	50	70	
Annual Labor Cost $	$5,201,600	$718,000	$686,070	$908,400	$416,000	$1,059,200	$9,110,400

Appendix VI: The Accident at Three Mile Island

SUMMARY OF EVENTS

The accident began about 4 a.m. on Wednesday, March 28, 1979, when the plant experienced a failure in the secondary, non-nuclear section of the plant. Either a mechanical or electrical failure prevented the main feedwater pumps from sending water to the steam generators that remove heat from the reactor core. This caused the plant's turbine generator, and then the reactor itself to automatically shut down. Immediately, the pressure in the primary system began to increase. In order to control that pressure, the pilot-operated relief valve opened, located at the top of the pressurizer. The valve should have closed when the pressure fell to proper levels, but it became stuck open. Instruments in the control room, however, indicated to the plant staff that the valve was closed. As a result, the plant staff was unaware that cooling water in the form of steam was pouring out of the stuck-open valve. As alarms rang and warning lights flashed, the operators did not realize that the plant was experiencing a loss-of-coolant accident. Other instruments available to plant staff provided inadequate or misleading information. During normal operations, the large pressure vessel that held the reactor core was always filled to the top with water. So there was no need for a water-level instrument to show whether water in the vessel covered the core. As a result, plant staff assumed that as long instruments showed that the pressurizer water level was high enough, the core was properly covered with water too. That was not the case. Unaware of the stuck-open relief valve and unable to tell if the core was covered with cooling water, the staff took a series of actions that uncovered the core. The stuck valve reduced primary system pressure so much that the reactor coolant pumps started to vibrate and were turned off. The emergency cooling water being pumped into the primary system threatened to fill up the pressurizer completely—an undesirable condition—and they cut back on the flow of water. Without the reactor coolant pumps circulating water and with the primary system starved of emergency cooling water, the water level in the pressure vessel dropped and the core overheated and was irreparably damaged.

TECHNICAL EVALUATION

On March 28, 1979, USA experienced the worst accident in the history of commercial nuclear power generation. This accounts for a summary of events and future recommendations. The accident was initiated by mechanical malfunctions in the plant combined with human error in responding to it. While equipment failure initiated the event, if the operators had kept the emergency cooling systems on through the early stages of the accident, Three Mile Island would have been limited to a relatively insignificant incident.

There was very extensive damage to the plant. While the reactor itself has been brought to a cold shutdown, there are vast amounts of radioactive materials trapped within containment and auxiliary buildings. Therefore, operators are faced with a massive cleanup process that carries its own potential dangers to public health.

The pilot operated relief valve (PORV) at the top of the pressurizer opened as expected when pressure rose, but failed to close when pressure decreased., thereby creating an opening in the primary coolant system. Loss of coolant accident (LOCA). The PORV indicator light in the control room showed only that a signal had been sent to close the PORV rather than the fact that the PORV remained open. The operators assumed that PORV was closed and were unaware of PORV failure. LOCA continued for 2 hours. The high-pressure injection system (HPI)(major safety system) came on automatically; however, operators concerned with the filled water level, cut back HPI from 1,000 gpm to less than 100 gpm, or even shut down on occasions. This led to the core being uncovered for extended periods and resulted in severe damage to the core. If the HPI had not been throttled, core damage would have been prevented in spite of stuck-open PORV.

TMI-2 (Three Mile Island reactor unit 2) had repeated problems with the condensate polishers demineralizers, 18 months before the accident. No effective steps were taken to correct these problems. It appears that the polishers initiated the 03/28/1979 sequence of events. To supply steam that runs the electric generating turbines, TMI relied on a pressurized water reactor steam generating system. The primary loop was kept under a high pressure of 2,155 psi. Nominal operations require that pressurized primary loop water is heated in the core, remains below saturation, and does not boil. In an accident, team is formed, and the reactor coolant system and containment building are breached. The first safety barrier is the fuel rods, which trap radioactive materials. The second barrier is the reactor vessel and closed reactor coolant system loop, which holds the reactor core and its control rods (40 ft high steel tank with 8–1/2 inches thick). The pressurizer tank maintains reactor water at a pressure high enough to prevent boiling. TMI-2 pressurizer held 800 c.ft of water and 700 c.ft. of steam above it. Steam pressure is controlled by heating or cooling the water in the pressurizer, which is used to control the pressure of water cooling the reactor. TMI-2 reactor coolant system (Primary loop) water is pumped through the reactor, pick up heat through convection and transfer to steam generators. The secondary water loop picks up heat from the primary loop. The feedwater system (secondary loop) is forced into the shell side of steam generators. Secondary loop water boils into steam, which turns steam turbines to produce electricity. Steam is condensed back to water. Condenser water is cooled in cooling towers. Neither water nor vapor or feedwater is radioactive under normal conditions. The primary water loop is radioactive as it runs through the reactor coolant system and it has been exposed to radioactive materials in the core.

In a LOCA, the Emergency Core Cooling System (ECCS) automatically ensures cooling water covers the core. At TMI-2 a vital part of ECCS is HPI is designed to automatically pour 1,000 gpm into the core to replace cooling water being lost through a stuck-open valve, broken pipe, or other leaks. ECCS is effective if operators do not interfere by stopping its function. TMI-2 operators did exactly that, against the procedure, and interfered in the ECCS operation.

A series of feedwater system pumps supply the system to TMI-2 steam generators tripped when the plant was operating at 97% power. When the pumps stopped, the flow of water to steam generators stopped. The plant safety system automatically shut down steam turbines. When the feedwater flow stopped, the temperature of the reactor coolant increased. The rapidly heating water expanded. Therefore the pressurizer level rose and steam in the pressurizer compressed. The pressure inside the pressurizer built to 2,255 psi, which caused the PORV valve to open and water began flowing out of the reactor coolant system through a drainpipe to a tank on the floor of the containment building. The control rods automatically dropped down into the reactor core to halt nuclear fission. The operators assumed that the PORV valve was closed, while it was stuck-opened, and did not close the backup valve to stem the flow of coolant water. In addition, they interfered with the plant automatic HPI system that would have kept the flow of emergency water to cover the reactor core, and the accident at TMI-2 would have been averted as a minor incident.

Two x 12 valve sets that were supposed to be opened in the feedwater system to steam generators were closed on March 26, 1979 and were not opened again. The valves were never opened after a routine test before the accident. The loss of emergency feedwater for 8 minutes had no significant effect on the outcome of the accident but added to confusion and human error.

Appendix VII: Water Hammer in Nuclear Power Plants

Water hammer includes transients involving steam flow and two-phase flow, e.g., water entrainment in steam lines, steam bubble collapse, and in addition classical water hammer transients such as those involving valve closing and pump start up in single-phase water systems. These events might result in damage to piping and restraints. In addition, severe cases might involve small cracks or ruptures in feed water piping systems. These occurrences of water hammer in fluid systems could have an impact on plant safety. Water hammer in vessels and pressure pulsations during steady operations, e.g., positive displacement pumps, cavitating valves, etc. are accounted for separately in fluid mechanics design assumptions and criteria. The main concerns are for lines that are larger than 1 inch in diameter.

Water hammer potential causes:

1. Pump startup with inadvertently voided discharge lines
2. Expected flow discharge into initially empty lines
3. Valve opening, closing, and instability
4. Check valve closure and delayed opening
5. Water entrainment in steam lines
6. Column separation
7. Steam bubble collapse and mixing of subcooled water and steam from interconnected systems
8. Slug impact due to rapid condensation
9. Pump startup, stopping, and seizure

CORRECTIVE MEASURES

Pump startup with inadvertently voided discharge lines. In water systems designed for operation with full discharge lines, inadvertent voiding of the lines due to air entrapment or draining may result in excessive dynamic loads following pump startup should be prevented through a gradual increase in flow rate by proportional valve opening and controls. The effects of the resulting water hammer are particularly pertinent to the residual heat removal (RHR), emergency core cooling (ECC), component cooling water (CCW), and service water systems because of their safety significance. In many systems, e.g., RHR and ECC systems for PWRs and some reactor core isolation cooling (RCIC) and high-pressure coolant injection (HPCI) systems for BWRs, the relative elevation of the water supply, injection points, and isolation valves are such that the piping which initially vented tends to remain full. For such systems, the principal protection is obtained from administrative controls and proper vent locations and filling and venting procedures.

Expected flow discharge into initially empty lines. Discharge lines in water systems that are normally empty and discharge lines from various pressure relief valves should be designed to withstand the expected dynamic loads.

Valve opening, closing, and instability. Rapid valve opening and closing in both water and steam systems and instability of control valves in water systems may cause excessive dynamic loads that should be compensated for in design criteria.

Check valve closure and sudden delayed opening. Normal check valve closure following pump stopping is not expected to result in large dynamic loads but should be considered in system design. For certain checks, valves that perform a safety function and associated piping should be designed to withstand large dynamic loads resulting from postulated rupture upstream of the valve. The sudden opening of a stuck check valve after pump startup can also produce damaging pressure pulses.

Water entrainment in steam lines. Water slugs driven by steam may cause excessive dynamic loads while being swept through bends in the lines and from the impact on tees or closed, or partially closed valves.

Transient cavitation (column separation). The subsequent collapse of voids formed in the water system by low-pressure transients resulting from pump stopping or seizure and change in valve settings or check valve closure may produce excessive dynamic loads.

Steam bubble collapse and mixing of subcooled water and steam from interconnected systems. Damaging water hammer may result from the collapse of steam bubbles in water systems due to pressurization and condensation following pump startup or valve opening from the mixing of steam and subcooled water from interconnected systems.

Slug impact due to rapid condensation. The impact of water slugs, formed and driven by forces resulting from rapid condensation of steam on subcooled water in feedwater rings and adjacent piping, has been identified as the cause of damaging water hammer in certain PWR steam generators with top feed.

Pump startup, stopping, and seizure with full lines. Dynamic loads resulting from pump startup with full lines are expected to be relatively small but should be considered in system design. Pump stopping in some systems may produce column separation. Postulated pump seizure can result in large dynamic loads and may cause column separation.

Appendix VIII: Selection and Design of Industrial Furnace Heaters

FURNACE DESIGN

1. Select fuel—gas or oil.
2. Calculate the heat duty of the process.
3. Decide proportions of heat duty.
4. Select the flue gas temperature.
5. Calculate the flow rate of fuel and air.
6. Calculate the enthalpy input rate.
7. Calculate the temperature of combustion products.
8. Calculate the adiabatic flame temperature.
9. Calculate reduced furnace efficiency.
10. Calculate the effective gas temperature in the furnace.
11. Select furnace geometry. Calculate the view factor between refractory and sink.
12. Use Hading's correlation for gas emissivity.
13. Calculate the ratio of sink area to diverted area.
14. Calculate the convective coefficient to sink.
15. Calculate the ratio of sink temperature to gas temperature.
16. Calculate heat flux across the heat transfer area. Check flux distribution.
17. Consider five furnace zones (imaginary position). Neglect radiation between zones.
18. Use heat balance to calculate the heat transfer coefficient for each zone.
19. Calculate the heat transfer parameters for each zone.
20. Compute the heat flux to sink in each zone.
21. From the process side (temperature, heat duty, heat transfer coefficient), calculate the sink temperature for each zone from the heat flux obtained data.
22. Repeat heat transfer calculations with sink temperature for each zone. Check flux to the critical sink area.
23. Determine flow, i.e., furnace area dynamics.
24. Correlate gas emissivity more accurately.
25. Divide gas space into gas zones. Select gas zone slope. Furnace flow is either normal to or parallel to their boundaries.

26. Divide the sink area into surface zones each with a fixed temperature.
27. Assign temperature to each gas zone. Assign temperature to refractory surface zones.
28. Calculate net heat input and output in real time for each gas and surface zone.
29. Divide net heat in or out by the size of a parcel.

Appendix IX: Chemical Reaction Fouling in Unit Operations

This study is concerned with the chemical reaction fouling from a liquid feedstock flowing inside a heated round tube. The aim of the project is to understand further the manner in which key operating parameters affect fouling rates and fouling resistances. This is to be done by simulating the dynamic coking phenomenon by the addition of a reactive monomer (styrene) to a non-fouling hydrocarbon feedstock (kerosene). The principal variables to be studied are surface temperature and mass flow rate. The initial rates of deposition are to be compared with those predicted by a general fouling model.

CATEGORIES OF FOULING

Type	Definition
Precipitation	The precipitation of dissolved substances on the heat transfer surface. Where the dissolved substances have inverse rather than normal solubility versus temperature characteristics, the precipitation occurs on superheated rather than subcooled surfaces, and the process is often referred to as scaling.
Particulate	The accumulation of finely divided solids suspended in the process fluid onto the heat transfer surface. In a minority of instances settling by gravity prevails, and the process may then be referred to as sedimentation fouling.
Chemical reaction	Deposits are formed at the heat transfer surface by chemical reactions in which the surface material itself is not a reactant.
Corrosion	The heat transfer surface itself reacts to produce corrosion products which foul the surface and may promote the attachment of other foulants.
Biological	The attachment of macro-organisms (macro-biofouling) and/or microorganisms (micro-biofouling or microbial fouling) to a heat transfer surface, along with the adherent slimes often generated by the latter.
Freezing	Solidification of a liquid or some of its higher melting constituents onto a subcooled heat transfer surface.

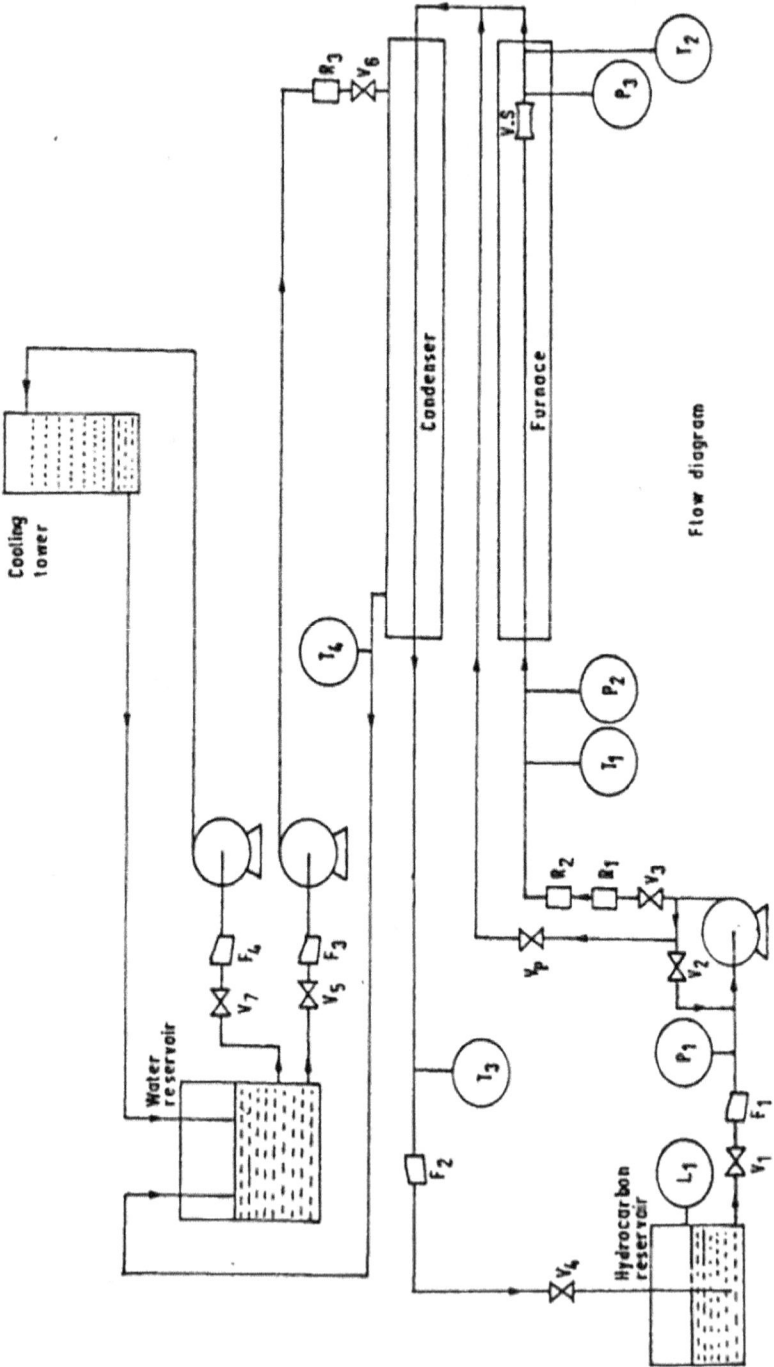

Flow diagram

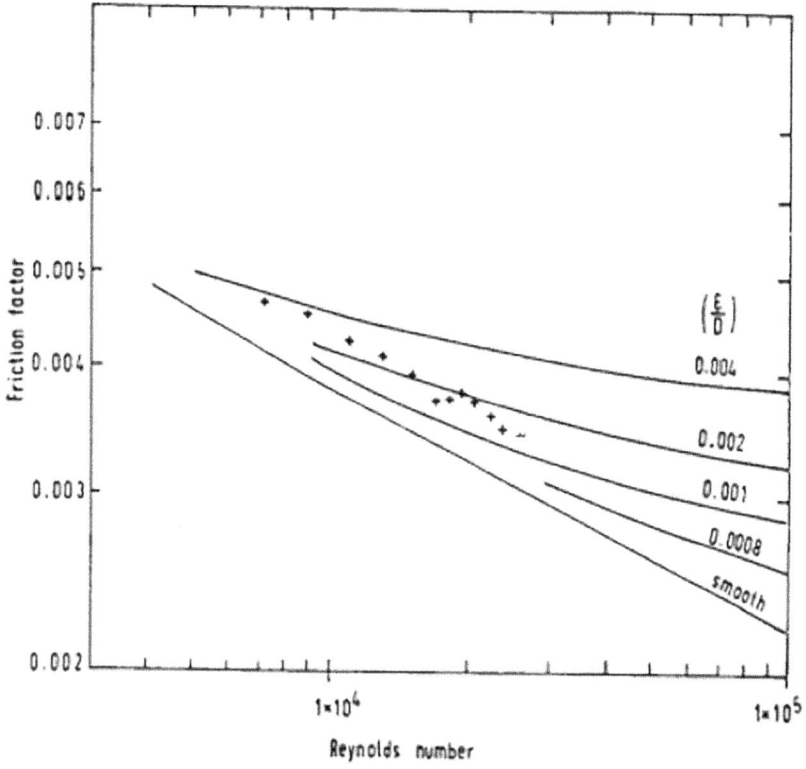

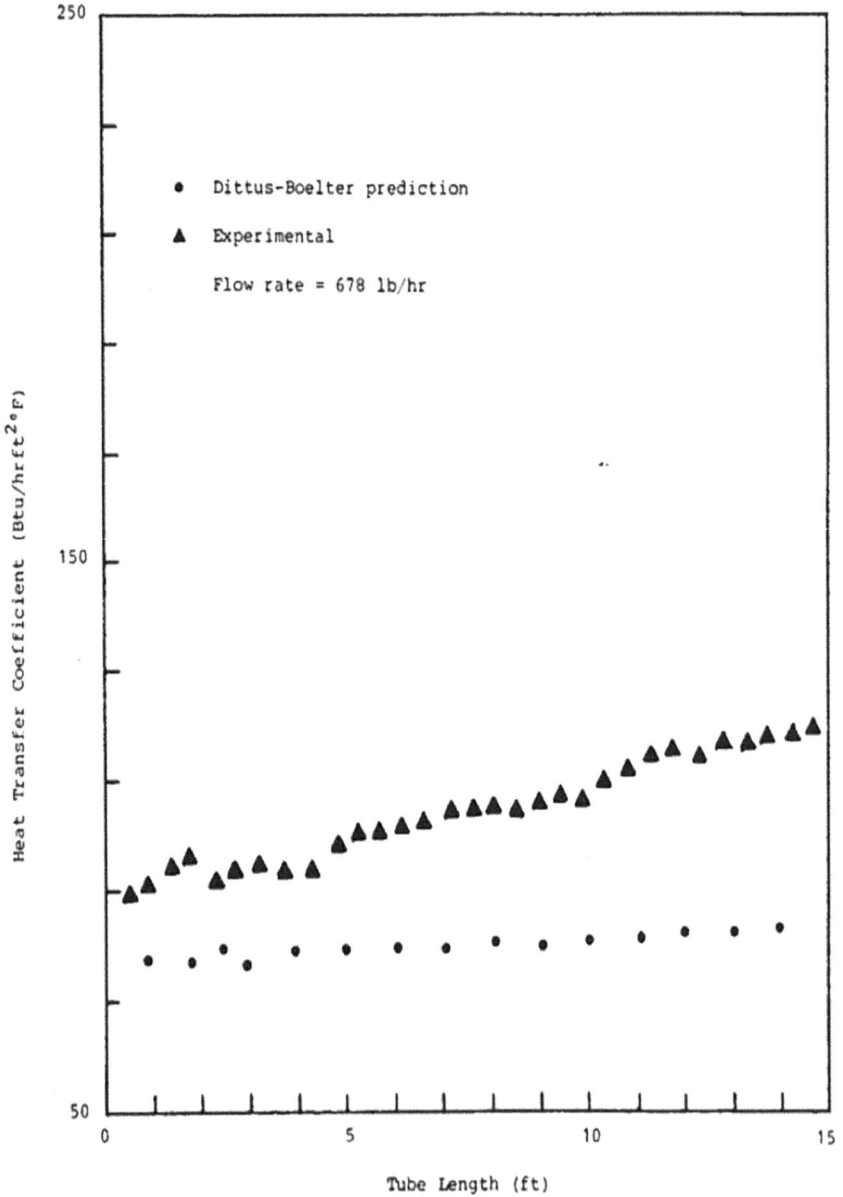

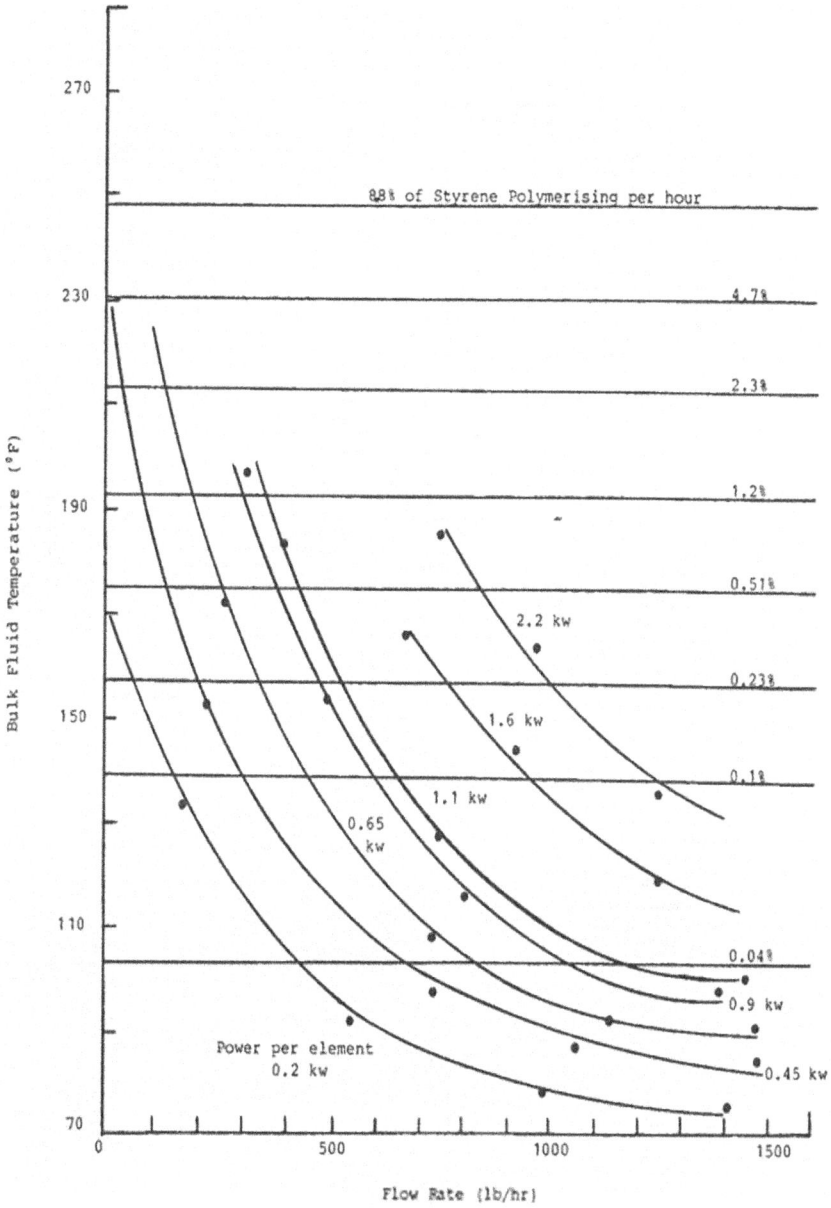

Flow Rate (lb/hr)

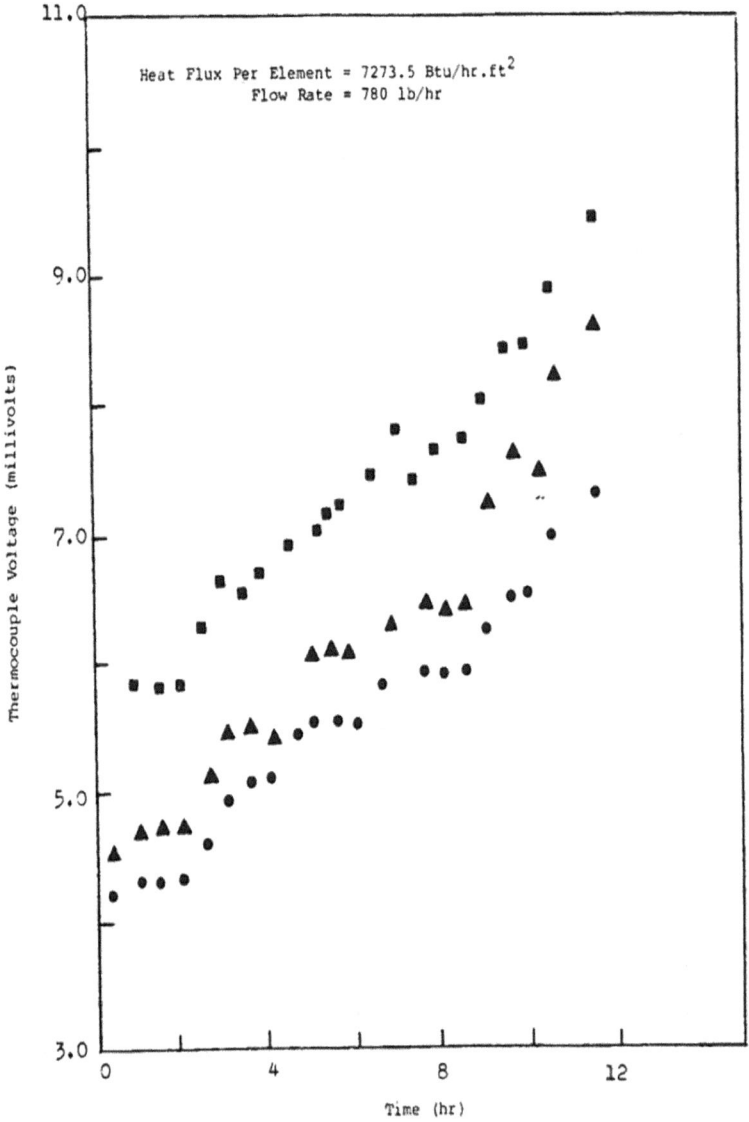

Heat Flux Per Element = 7273.5 Btu/hr.ft^2
Flow Rate = 780 lb/hr

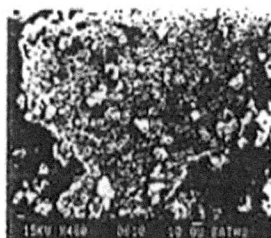

POLYSTYRENE DEPOSITS

POLYSTYRENE DEPOSITS

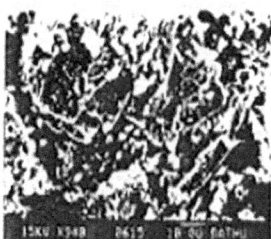

POLYSTYRENE DEPOSITS

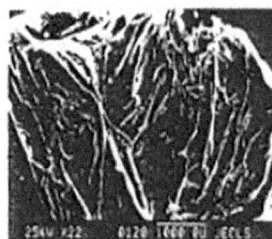

POLYSTYRENE DEPOSIT COLLECTED ON
THE DOWNSTREAM FILTER AFTER COOLING

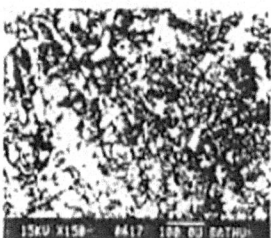

TUBE SURFACE

TUBE SURFACE

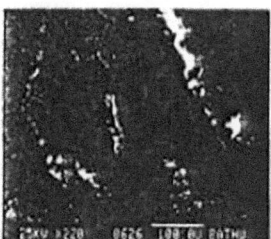

TUBE SURFACE

TUBE SURFACE

PLATE J2 POLYSTYRENE DEPOSIT SCRAPED OFF THE FURNACE TUBE SURFACE

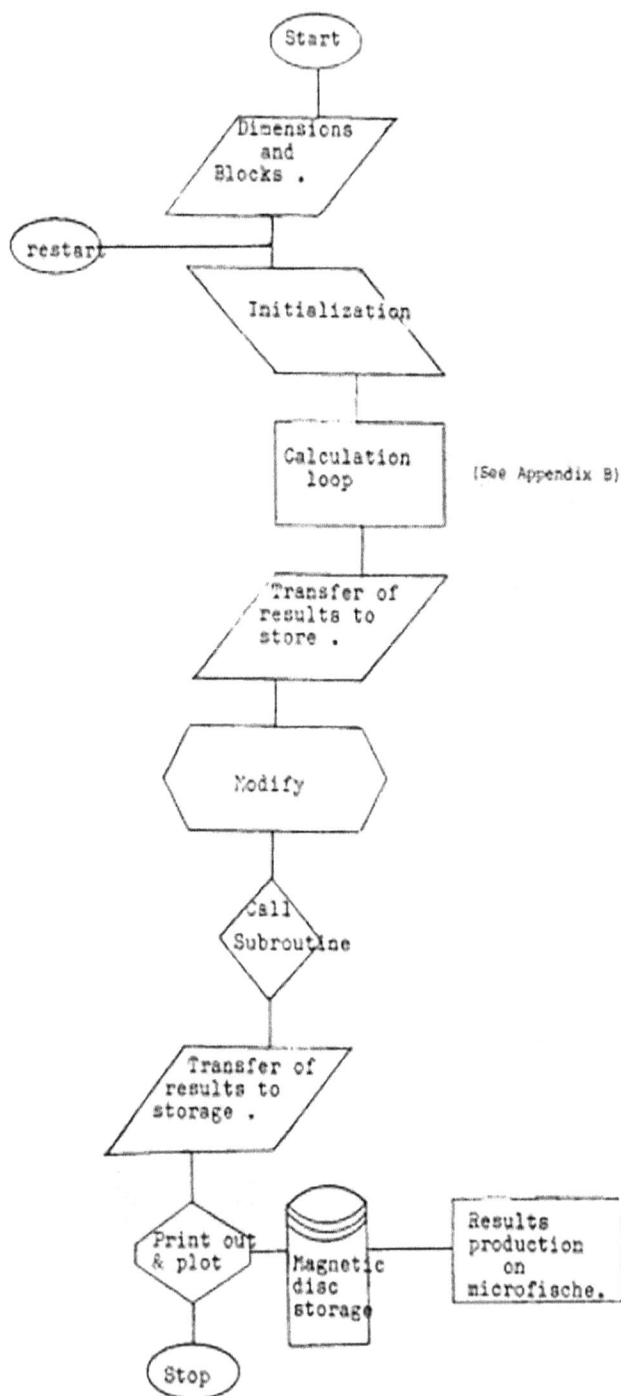

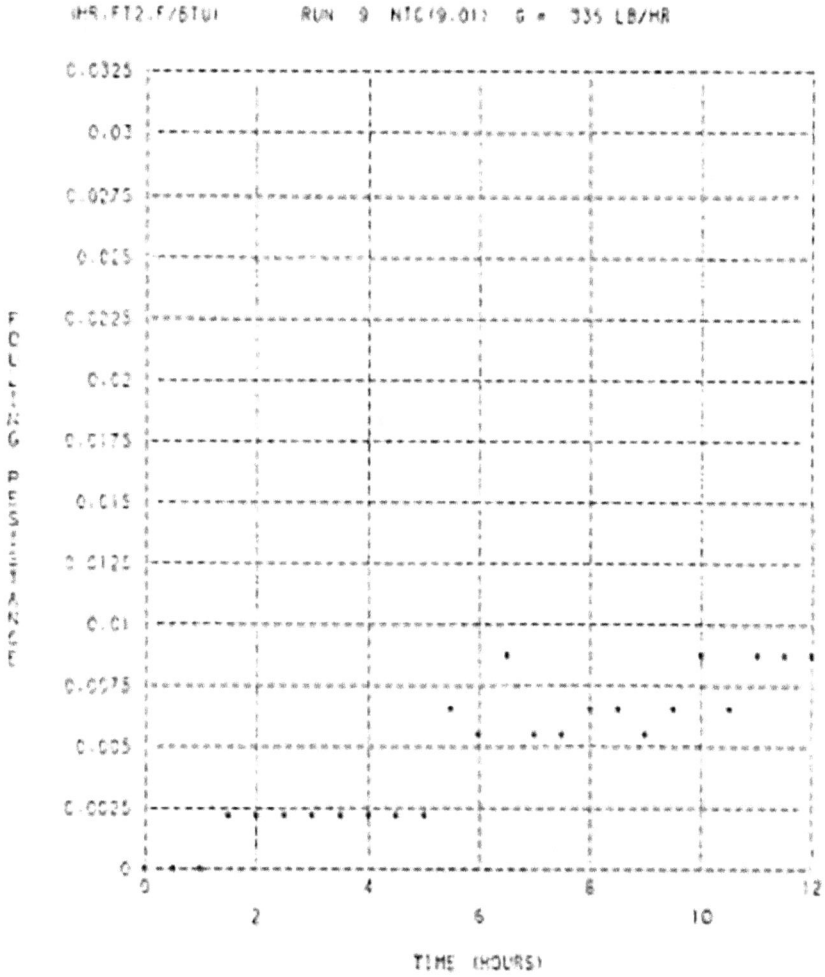

Figure 5.32 Fouling Resistance vs. Time.

Heat flux per element = 1851.0 BTU/HR.FT^2.

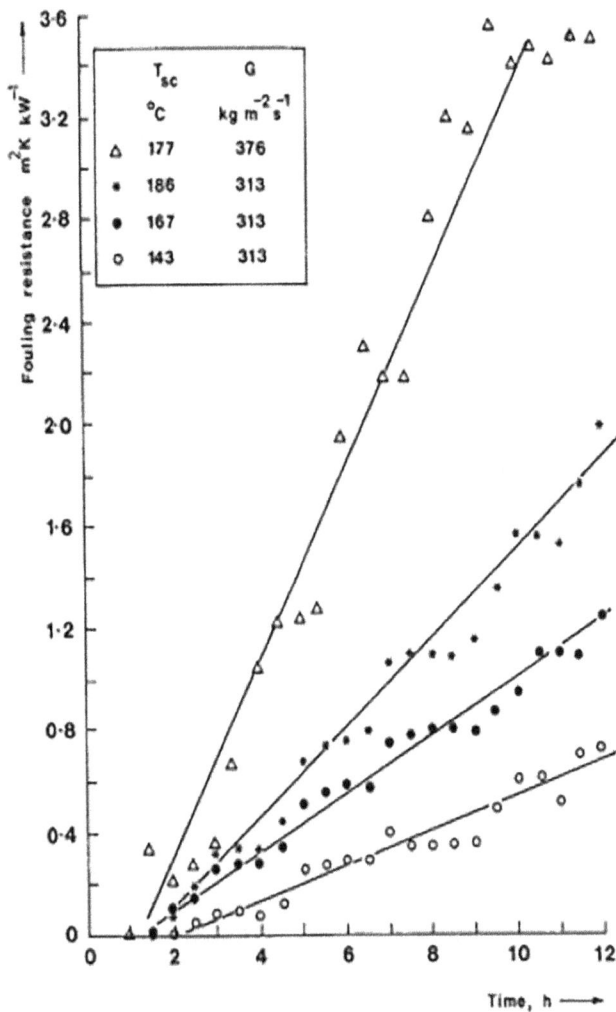

Typical fouling transients with 1% v/v styrene in kerosene

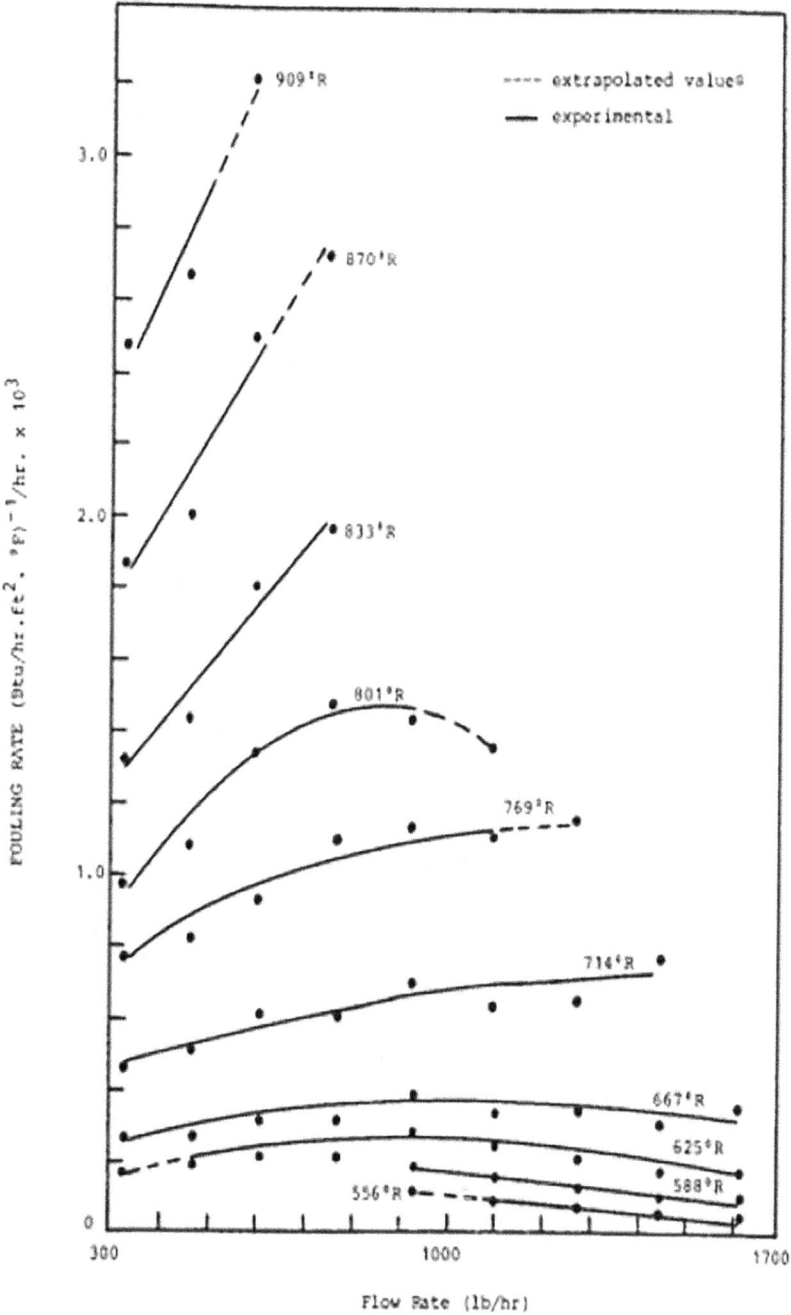

Flow Rate (lb/hr)

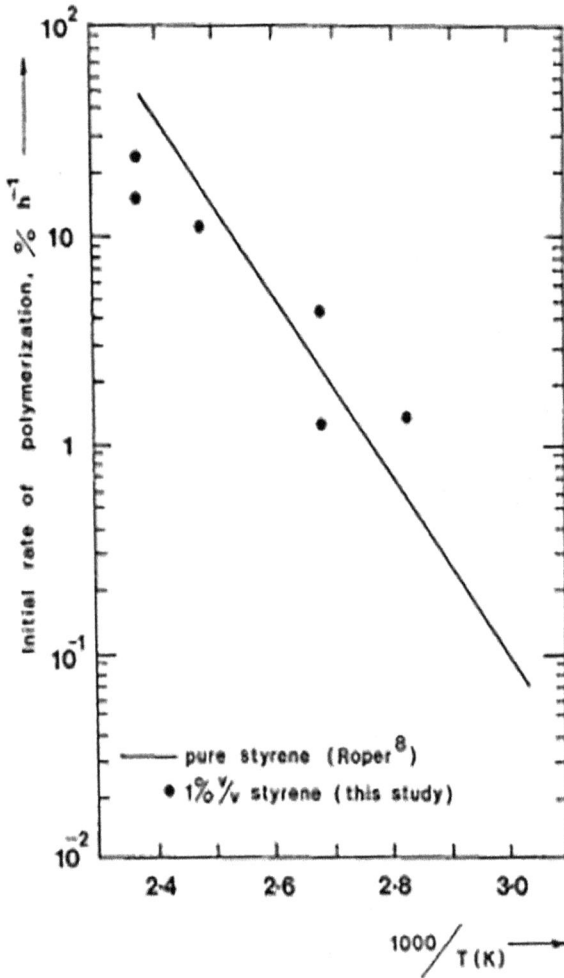

Initial rate of polymerisation of 1% v/v styrene in kerosene

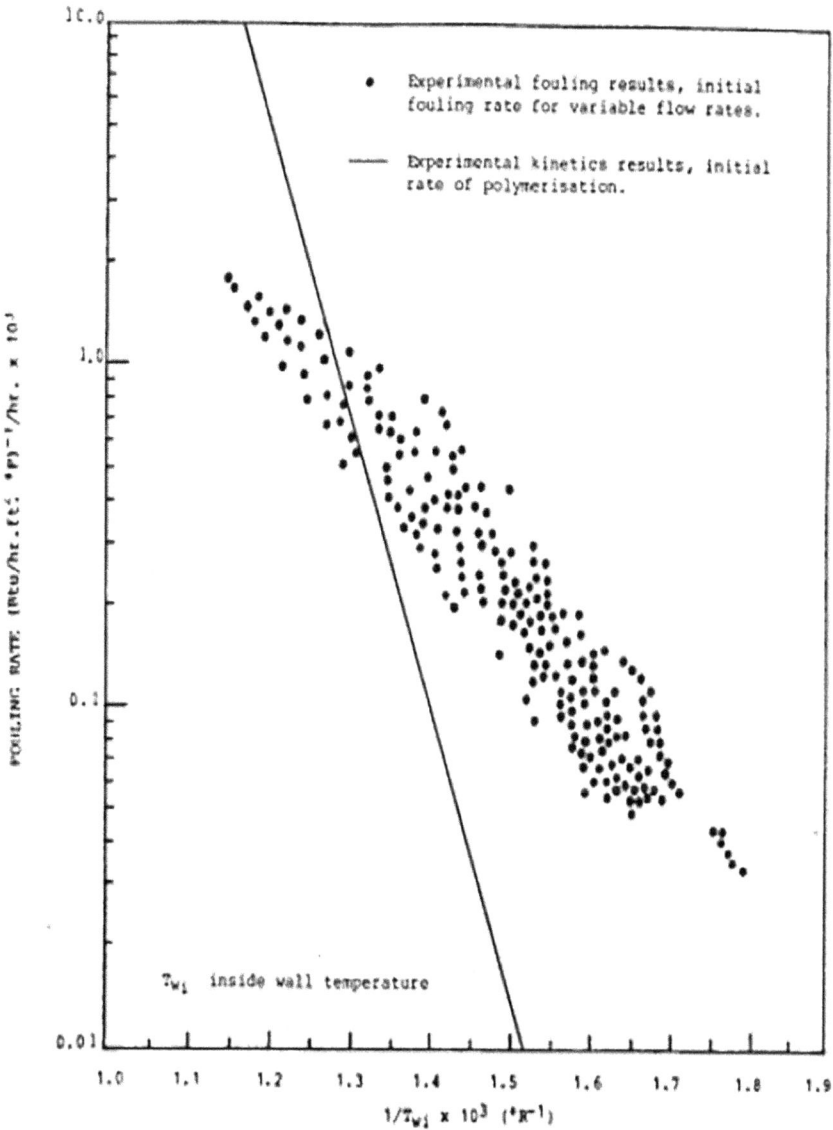

Graph axes and legend:

FOULING RATE (Btu/hr.ft² °F)⁻¹/hr. × 10³ (y-axis, logarithmic from 0.01 to 10.0)

1/T_{w_i} × 10³ (°R⁻¹) (x-axis, from 1.0 to 1.9)

• Experimental fouling results, initial fouling rate for variable flow rates.

—— Experimental kinetics results, initial rate of polymerisation.

T_{w_i} inside wall temperature

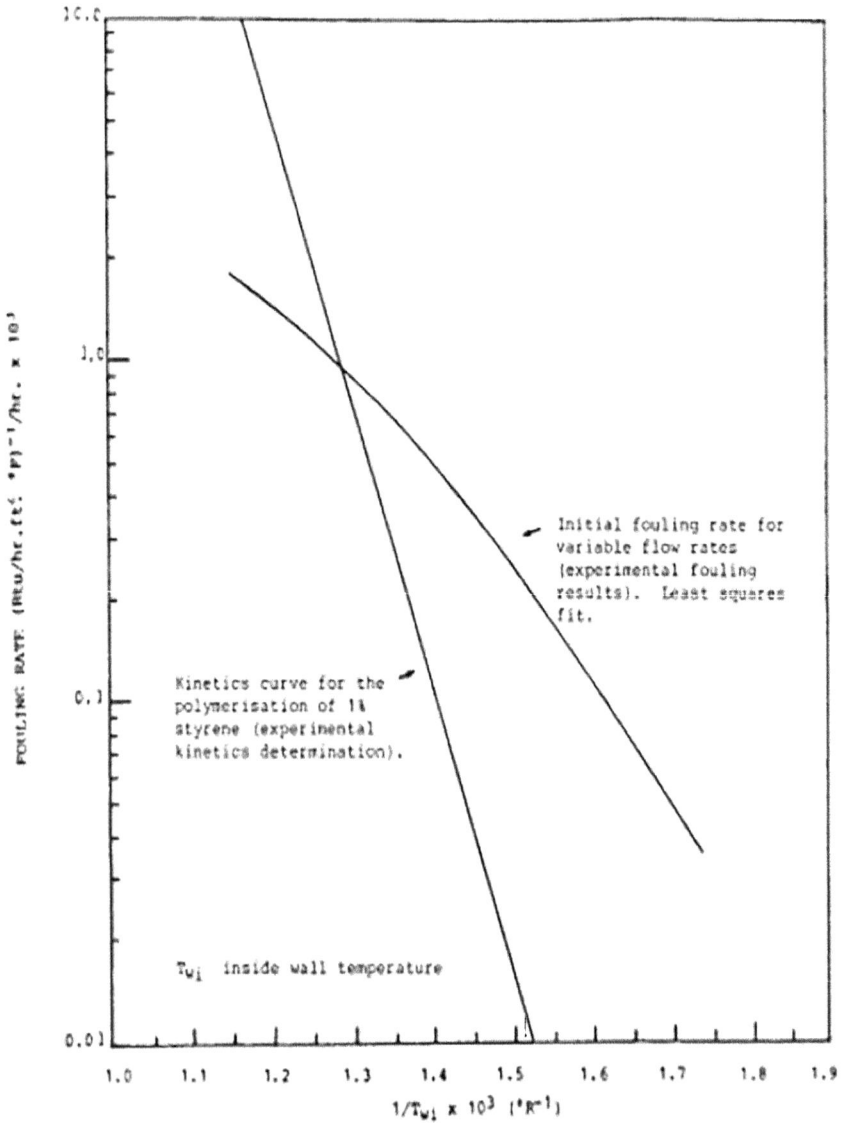

$$\dot{R}_f(\theta) = \frac{1}{\rho_f T \kappa_f}$$ (transport and reaction of − diffusion of foulant)
precursors at the transfer back to bulk fluid
surface to create foulant

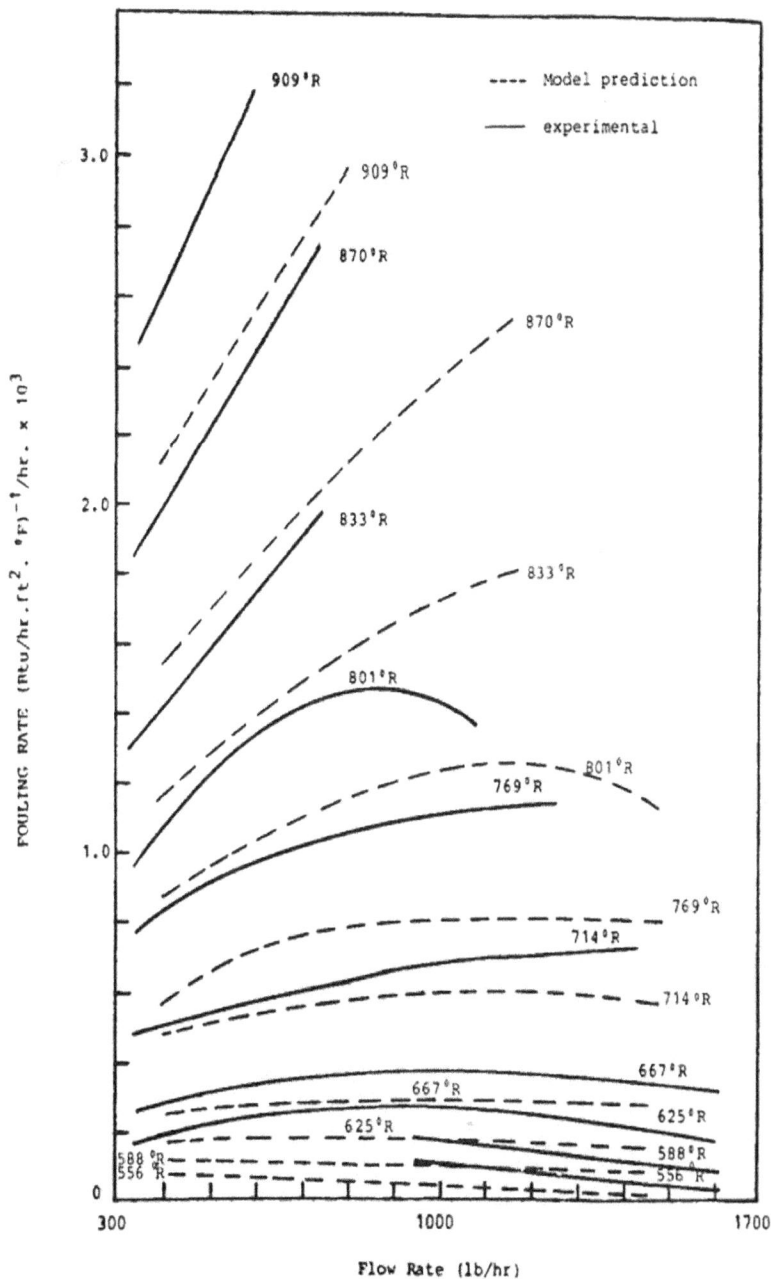

Flow Rate (lb/hr)

Appendix X: Analysis of Single-Phase Flow in Power Plant Drain Systems

Analysis of Single-Phase Cascade Flow in Power Plant Drain Systems

S. A. Hout[1]

Process Engineer.
Orange, CA

The performance of a typical drain system was checked through a pressure drop analysis and the determination of valve sizing coefficients. The operation checks that two-phase flow does not occur upstream of the control valve. Erosion checks are performed at any elbow installed downstream of the control valve where flashing may take place. The analysis does not cover the performance of the feedwater heater or moisture separator and reheater drain systems under transient conditions such as load rejection, i.e., turbine trip. It is believed that accurate valve sizing and appropriate drain cascade are vital to proper drainage.

Introduction

Two-phase pressure drop depends on flow regime and system geometry and is not a simple application in power plant complex drain systems. Benjamin and Miller [2] proposed an analytic approach for sizing lines carrying flashing mixtures. Isbin et al. [7] and later Fauske [5] proposed experimental data that apply to steam-water critical flow under fixed conditions.

Lalo [8] considered that the drain systems calculations must be performed for all possible operating conditions to assure its adequacy; any system in a power plant must be operated with varying process conditions.

Proper design of the drain system from turbine periphery facilities requires correct line layout and control valve sizing [1]. From a purely economic viewpoint, an undersized valve cannot do the job and must be replaced; too large a control valve costs more initially and has high maintenance costs since the seating surfaces wear rapidly. Considering the operation, an oversized valve provides poor control and can cause system instability [4]. The most expensive, sensitive, and accurate controller is of little value if the control valve cannot correct the flow properly to maintain the control point.

Normally, the drain system is designed to avoid flashing upstream of the control valve. This has been accomplished by maintaining sufficient static head up to the control valve to suppress flashing [8]. This proved to be a limitation and some drain systems are designed for flashing mixtures.

Process

Power plant drain systems may generally operate under conditions of saturated or slightly subcooled liquid, which has a flashing potential. To maintain a flow in the drain system between the upstream heater and the downstream vessel, the pressure relationship should be satisfied as follows:

$$P_u + pH - \Delta p = P_b \qquad (1)$$

[1] Currently at Procter & Gamble, Geneva, Switzerland

Contributed by the Power Division for publication in the JOURNAL OF ENGINEERING FOR GAS TURBINES AND POWER. Manuscript received by the Power Division 1983.

(pH may be negligible, e.g., drains to deaerator).

For single-phase liquid flow the differential pressure between pipe end critical pressure and back pressure is usually negligible [8].

The static head should be satisfied to avoid two-phase flashing flow before the control valve assuming that heat losses are negligible, then

$$H \gtrsim \frac{1}{\rho} (\Delta P_u - (P_u - P_s)) \qquad (2)$$

If the drain system is handling saturated liquid, then

$$P_u = P_s$$

and

$$H \gtrsim \frac{\Delta P_a}{\rho} \qquad (3)$$

If the liquid is subcooled, then

$$\Delta P_a > P_s - P_t \qquad (4)$$

and system A is above system B (see Fig. 1), or

$$\Delta P_a < P_u - P_t \qquad (5)$$

and system B is above system A.

If the heaters are at the same elevation, the drain line should still contain the control valve loop as the upstream heater is usually operated at a higher pressure.

Darcy's equation gives the friction loss

$$\Delta P_f = \frac{3.36 \times 10^{-6} \times f \times L \times W^2}{\rho d^5} \qquad (6)$$

The elevation changes are given by

$$\Delta P_b = \frac{\rho H}{144} \qquad (7)$$

The velocity head is accounted for with respect to the velocity at previous section

$$\Delta P_v = \frac{(V^2 - V_1^2)^\rho}{144 \times 64.4} \qquad (8)$$

If the fluid entering the control valve is saturated liquid,

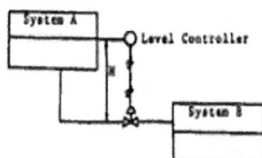

Fig. 1 Typical drain system

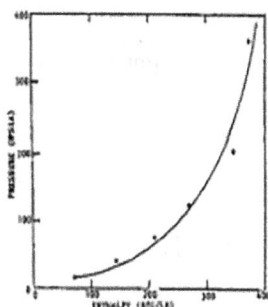

Fig. 2 Pressure-enthalpy plot of feedwater heater shell pressure and enthalpy of the feedwater heater drain

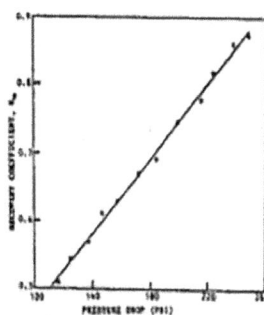

Fig. 3 A plot of the pressure drop across the valve against the recovery coefficient; single-ported, multiple-seat globe control valve

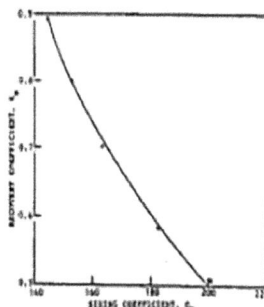

Fig. 4 The variation of the recovery coefficient with the valve sizing coefficient; single-ported multiple-seat globe control valve

then sonic flow exists downstream of the valve orifice; the presence of the liquid phase shifts the sonic point downstream. The critical pressure ratio is defined by

$$r_c = \frac{P_c}{P_i} \tag{9}$$

The maximum allowable pressure drop that is effective in producing flow is not necessarily the maximum ΔP that may be handled by the valve. The maximum allowable differential pressure for sizing purposes is then given by [6]

$$\Delta P = K_m(P_i - r_c P_v) \tag{10}$$

K_m is determined from empirical tests for liquid flow, and it is a function of the physical geometry of the valve.

After ΔP is determined, then it is used in the liquid sizing equation (6)

$$W = C_v \left(\frac{\Delta P}{G}\right)^{\frac{1}{2}} \tag{11}$$

to determine the flowrate W or the valve sizing coefficient C_v. Equation (11) may be used to calculate the body pressure drop at which significant cavitation can occur. It is recognized that minor cavitation may occur at a pressure drop slightly less than the one predicted by equation (11).

Results

Forty-six drain lines were individually analyzed. Single ported with multiple seats globe control valves were used. The feedwater heaters had an integral drain cooler for the purpose of reducing the flashing potential. Figure 2 is a pressure-enthalpy plot of feedwater heater shell pressure against enthalpy of the feedwater heater drain.

The differential pressure across the valve correlated well with the valve recovery coefficient (see Fig. 3). Figure 4 is a plot of the recovery coefficient against the sizing coefficient. The flow upstream of the valves was single-phase liquid over the range of data. Some flashing occurred downstream of the valve. This was accepted and not considered crucial as flashing upstream was completely suppressed and served the purpose of design.

Nomenclature

C_v = sizing coefficient
d = inside diameter
f = friction factor
G = specific gravity
H = static head
K_m = recovery coefficient
L = length

M = Mach number = $\frac{V}{V_s}$

P = pressure

P_v = vapor pressure of liquid at body inlet
r = pressure ratio
Re = Reynolds number = $\frac{\rho d V}{\mu}$

V = fluid velocity
V_s = speed of sound in the fluid
W = mass flowrate
Δ = differential
ϵ = absolute roughness
ρ = fluid density
μ = fluid viscosity

Subscripts

b = back
c = critical
f = friction
h = elevation
i = inlet
L = section
s = saturated
u = upstream
v = velocity
v = vapor

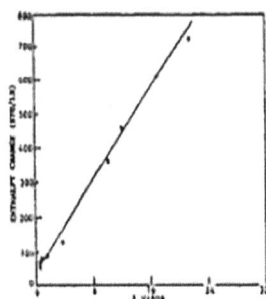

Fig. 5 A plot of the percentage of vapor downstream of the control valve against enthalpy change

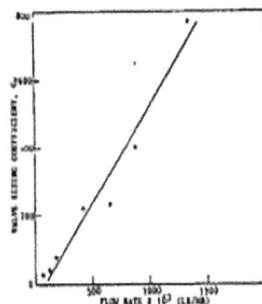

Fig. 6 Valve sizing coefficient against flow rate; single-ported, multiple-seat globe control valve

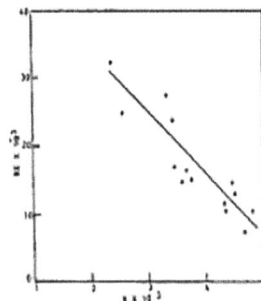

Fig. 7 A plot of Reynolds number against Mach number; single phase flow

Figure 5 is a plot of the percentage of vapor downstream of the control valve line against the existing enthalpy change of the drain system. A direct linear relationship is established over the range of data gathered.

Table 1 SI units conversion

Quantity	Multiply by	SI (metric)
Pressure, psia	703.1	kg/m²
Enthalpy, Btu/lb	2326	J/kg
Flowrate, lb/hr	0.454	kg/hr

The valve sizing coefficients showed a direct proportionality to flowrate as predicted by equation (11) (see Fig. 6).

The Reynolds and Mach numbers were evaluated for the purpose of determining the kind of flow existing and relevant turbulence. Figure 7 shows a plot of Re against M. The data showed some scatter as the flow showed dependence on the variations in viscosity and density of the fluid. The scatter is attributed to geometry, pressure drop, and two-phase variations which affect the thermodynamic equilibrium of the system.

Discussion

The sizing criterion for a drain line carrying a flashing flow downstream of the control valve is to maintain the pressure at the control valve outlet less than the pressure at the valve orifice, in order that the mixture flow will not limit the flow through the control valve. Therefore, the control valve will have full control of the flowrate of the drain system.

The procedure used to size control valves for liquid service should consider the possibility of cavitation and flashing since they can limit the capacity and produce physical damage to the valve.

Data were obtained from the results of a computer analysis. The program used upstream heater conditions; upstream heater pressure, pressure drop to heater outlet, and temperature at heater outlet to find the saturation pressure and water data (specific gravity and viscosity). The geometric configuration of the upstream line was analyzed to account for the fittings in the pressure drop calculation. Downstream conditions and line configuration were also used.

Conclusion

Over the range of data, the design criteria for power plant drain systems have been established. Single-phase flow in the upstream cascade of the control valve in the drain line should be common practice while two-phase flashing downstream is acceptable. The sizing coefficient of the control valve and the prevention of flashing in the drain system are greatly dependent on the flowrate and cascade layout.

References

1 "Recommended Practices for the Prevention of Water Damage to Steam Turbines Used for Electric Power Generation," No. TWDPS-1, ASME Turbine Water Damage Prevention Committee, July 1972.
2 Benjamin, M. W., and Miller, J. G., Trans. ASME, Oct. 1942, pp. 657–669.
3 Crane Co., "Flow of Fluids," Technical Paper No. 410, 20th printing, 1981.
4 Driskell, L., "Control Valve Sizing with ISA Formulas," Instrumentation Technology, July 1974.
5 Fauske, H., "Critical 2-Phase, Steam-Water Flows," Proc. Heat Transfer and Fluid Mechanics Institute, Stanford University Press, 1961.
6 Sizing and Selection Data, Catalog 10, Fisher Controls, IA, 1974.
7 Isbin, H. S., et al., J. AIChE, Vol. 3, No. 2, Sept. 1957, pp. 361–365.
8 Liao, G. S., and Larson, J. K., ASME Winter Annual Meeting, New York, Dec. 1976.
9 Liao, G. S., ASME Winter Annual Meeting, Houston, TX, Nov. 1975.

Appendix XI: Case Studies

CASE STUDY 1—COMMERCIAL PROCESS FOR CATALYTIC DEHYDROGENATION OF LIGHT HYDROCARBON TO PRODUCE MONO-OLEFINS

An adiabatic process incorporates fixed-bed catalysis to dehydrogenate light hydrocarbons. Feedstock can be propane, n-butane, isobutane, n-pentane, or isopentane. Products are propylene, n-butylene, isobutylene, n-amylene, and isoamylene, respectively. A continuous process employing a multi-reactor system that can be on stream processing hydrocarbon or being regenerated and purged. A catalyst in the form of cylindrical pellets consists of activated alumina impregnated with chromic oxide (nominal lifespan is approximately 2 years dependent on the severity of operation). Feedstock including recycle stream is heated to design inlet temperature before entering a fixed bed reactor loaded with the specified catalyst. Following compression, absorption, stripping, and stabilization, the product is separated from unconverted hydrocarbon. This thermally balanced operation involves heat absorbed in the endothermic dehydrogenation equivalent to the heat evolved in the subsequent catalyst regeneration periods, where preheated air burns the coke deposits of the catalyst surfaces. Proper adjustment of air and hydrocarbon feed input temperatures gives reactor temperatures in the range of 1,000–1,200 °F and maintains reactor heat balance. Low pressures allow for optimized process equilibrium and selectivity for the production of the desired olefin.

CASE STUDY 2—COMMERCIAL PROCESS FOR CATALYTIC DEHYDROGENATION OF LIGHT HYDROCARBON TO PRODUCE DI-OLEFINS

An adiabatic process incorporates fixed-bed catalysis to dehydrogenate light hydrocarbons. Feedstock can be n-butene, or isopentene. Products are di-olefins. A continuous process employing a multi-reactor system that can be on stream processing hydrocarbon or being regenerated and purged. A catalyst in the form of cylindrical pellets consists of activated alumina impregnated with chromic oxide (nominal lifespan is approximately 2 years dependent on the severity of operation). Feedstock including recycle stream is heated to design inlet temperature before entering a fixed bed reactor loaded with the specified catalyst. Following compression, absorption, stripping, and stabilization, the product is separated from unconverted hydrocarbon. This thermally balanced operation involves heat absorbed in the endothermic dehydrogenation equivalent to the heat evolved in the subsequent catalyst regeneration periods, where preheated air burns the coke deposits of the catalyst surfaces. Proper adjustment of air and hydrocarbon feed input temperatures gives reactor temperatures in the range of 1,000–1,200°F and maintains reactor heat balance. Low pressures allow for optimized process equilibrium and selectivity for the production of the desired diolefin.

CASE STUDY 3—OXYGEN DIFFUSION IN AIR

Kinetic diffusion of gases equations allows for the calculation of diffusion coefficient, D, which is a function of temperature and pressure. Diffusion increases with increased temperature and decreases with increasing pressure.

Diffusion refers to the process by which molecules intermingle because of their kinetic energy of random motion, e.g., consider two containers of gas A and B separated by a partition. The molecules of both gases are in constant motion and make numerous collisions within the partition. If the partition is removed, the gases will mix because of the random velocities of their molecules. In time, a uniform mixture of A and B will reach equilibrium in the container. The tendency toward diffusion is strong even at room temperature because of molecular velocities associated with thermal energy of particles. Since the average kinetic energy of different types of molecules (different masses) which are at thermal equilibrium is the same, then their average velocities are different. Their average diffusion rate is expected to depend upon their average velocity.

By comparison, if two solutions of different concentrations are separated by a semi-permeable membrane, which is permeable to the smaller solvent molecules but not to the larger solute molecules, then the solvent will tend to diffuse across the membrane from the less concentrated solution to the more concentrated solution defined as osmosis. Thus, osmosis is a selective diffusion process driven by the internal energy of solvent molecules. It is customary to express this tendency toward solvent transport in pressure units relative to pure solvent defined as osmotic pressure. One approach to the measurement of osmotic pressure is to measure the amount of hydrostatic pressure necessary to prevent the fluid transfer by osmosis. The transport of water and other types of molecules across membranes is the key to many processes in living organisms. Many of these transport processes proceed by diffusion

TABLE AI.1
Effect of Temperature and Gas Mixture on Oxygen Diffusion Coefficients

Temperature, C	RH, %	Oxygen Diffusion coefficient, 2% (cm²/s)	Oxygen Diffusion coefficient, 15% (cm²/s)
20	50	0.203	0.214
20	100	0.203	0.214
60	50	0.259	0.273
60	100	0.264	0.278

through membranes that are selectively permeable, allowing small molecules to pass but blocking larger ones.

When a gas is in contact with the surface of a liquid, the amount of gas that will go into the solution is proportional to the partial pressure of the gas. A simple rationale will imply that if the partial pressure of this gas is twice as much, then on average twice as much of this gas will be captured in the solution. However, for a gas mixture, different gases have different solubilities, which will affect the rate of diffusion for each gas. Hence, gases are dissolved in liquids, and the relative rate of diffusion is proportional to its solubility in the liquid. The net diffusion rate of a gas across a fluid membrane is proportional to the difference in partial pressure, proportional to the area of the membrane, and inversely proportional to the thickness of the membrane.

Appendix XII: Heavy Metals Testing

EQUIPMENT

Reagents: Use only reagents of recognized analytical grade.

a) Hydrochloric acid solution, 0.25% by weight (0.07 N)
b) Hydrochloric acid solution, 0.50% by weight (0.14 N)
c) Hydrochloric acid solution, 7.3% by weight (2.0 N)
d) Hydrochloric acid solution, 21.9% by weight (6.0 N)
e) Concentrated nitric acid
f) Concentrated sulfuric acid
g) Acid-free 1,1,1-trichloroethane or other suitable solvents
h) 0.1% potassium chloride (KCl) solution
i) 2% ammonium chloride (NH_4Cl) solution
j) Aqua regia
k) Alumina or calcium oxide
l) Water that has been filtered and deionized through a cation and anion mixed bed resin
   Metal sieve of 0.5 mm aperture
   pH meter having an accuracy of ±0.1 pH unit
   Membrane filter having a pore size of 0.45 μm
   Whatman #541 filter (or equivalent textured, rapid filtering paper having a retention rating of 20–25 μm)

TEST METHODS

Sample Preparation:
   Preparation for the Determination of Total Heavy Elements in Surface Coatings
   NOTE—If the total heavy elements are determined to be greater than the allowable soluble levels in coating materials, then the soluble levels must be determined.

A) Stir the test material thoroughly and place 10 grams of the wet sample in an aluminum laboratory weighing dish measuring approximately 57 mm in diameter.
B) Dry the sample on a hot plate. Maintain the hot plate surface temperature between 150 and 200°F.

Note: To ensure the sample has been dried thoroughly, perform the following steps:

1) Dry until the sample is presumed to be "completely" dry.
2) Cool and measure the weight of the sample and dish.
3) Dry for 1 additional hour, cool and weigh the sample again.
4) If the weight change after the 1 hour drying time is less than 1% of the original sample weight determined in step 2, no further drying is necessary. If the weight change is greater than 1%, repeat step 3.

C) Remove the entire sample from the weighing dish and comminute the sample until it is capable of passing through a metal sieve having an aperture of 0.0555 inch (U.S. sieve No. 14).
D) Weigh about 1 gram of material (representative of the entire dried sample) to the nearest 0.001 gram and place it into a 50-mL capacity porcelain crucible.
E) Add a thin layer (approximately 1/2 gram) of alumina or calcium oxide on top of the sample, just enough to cover the exposed surface.
F) Start with a cold muffle furnace, insert the sample, raise the temperature to 300°C, and ash the sample for 8 hours.
G) Allow the furnace to cool, remove the sample from the furnace, and place in a desiccator for a period of at least 2 hours.
H) Immediately after removing from the desiccator, dissolve the ash in 5 mL of Aqua Regia, warming the solution if necessary to dissolve the residue.
I) Cool, centrifuge, and filter the solution through a Whatman #541 filter (or equivalent) into a 50-mL volumetric flask.
J) With the filter paper still over the flask, wash the filter paper with deionized water to ensure the entire solution has been rinsed into the flask.
K) If barium and chromium are to be determined, add 5 mL of each of 0.1% KCl and 2% NH_4Cl solution per 50 mL dilution.
L) Dilute the solution to volume with deionized water and mix well.
M) Examine the resulting solution within 6 hours of preparation of the extract.

Preparation for the Determination of Soluble Heavy Elements:

Test portions are to be taken from accessible parts of one or more laboratory samples. Alternatively, test portions may be taken from materials in a form such that it is representative of the relevant material. For example, test portions of coating samples scraped from a laboratory sample may be substituted by wet paint samples that have been dried prior to testing.

Note: It has been shown that the extraction of soluble cadmium can show a twofold to fivefold increase when extraction is carried out in the light rather than dark. Therefore, this process is to be carried out in a dark environment.

A) Coatings of Paints, Varnishes, Lacquers, Printing Ink, and Similar Materials:
   1) Obtain a portion of coating from the laboratory sample by scraping off the coating.

2) Comminute the coating so the material is capable of passing through a metal sieve having an aperture of 0.5 mm (500 μm).

3) If the coating on the laboratory sample is not uniform (e.g., differing in color or substance) obtain a test portion from each different coating covering an area greater than 500 mm^2. Coatings covering an area less than 500 mm^2 shall form part of the test portion obtained from the main coating. Coatings covering an area greater than 500 mm^2 with a width less than 5 mm shall form part of the test portion obtained from the adjacent coating.

4) Mix the test portion with 50 times its mass of an aqueous solution (at a temperature of 37°C ± 2°C) of 0.07 N hydrochloric acid and stir for 1 minute.

5) Check the acidity of the mixture. If the pH is greater than 1.5, add drop by drop, while stirring continuously, an aqueous solution of 2 N hydrochloric acid until the pH is 1.5 or less.

6) Stir the mixture for 1 hour and then allow the mixture to stand for 1 hour at 37°C ± 2°C. Both operations are to be carried out in the dark.

7) If necessary, centrifuge the mixture and separate the solids from the mixture by filtration through a membrane filter having a pore size of 0.45 μm. Examine the resulting solution within 6 hours of preparation of the extract to determine the presence and quantity of the appropriate elements.

B) Non-Textile Plastic and Similar Materials:

1) Obtain a test portion of the material from the laboratory sample by cutting out test pieces. Each of the test pieces must be of a size that can be contained in a cube having the dimensions of approximately 6 mm × 6 mm × 6 mm. The test portion is to be taken from the sample in the area(s) having the thinnest material thickness. (This will assure the test portion has a surface area as large as possible in proportion to its mass and will allow for the greatest possible metal migration.)

2) If the laboratory sample is not uniform in its material, a test portion is to be obtained from each different material forming a mass greater than 100 mg. Material forming a mass less than 100 mg is to form part of the test portion obtained from the main material.

3) If the material to be tested is coated with paint, varnish, lacquer, printing ink, or similar material and the coating can be scraped off and it covers an area greater than 100 test portions are to be obtained separately from the coating by scraping the coating away from the base material.
If the coating cannot be scraped off or the coating covers an area of less than approximately 100 test portions are to be taken from the base material in a way so that they also include parts of the coated area.

4) Treat the test portion as specified in paragraphs (A) (4) through (A) (7) of this section.

C) Paper, Cardboard, Chipboard, and Similar Materials:

1) Obtain and prepare test portions as specified

2) Macerate the test portion with 25 times its mass of water at 37°C ± 2°C so that the resulting mixture is uniform in color and texture.

3) Transfer the mixture to a conical flask. Add 25 times the mass of the test portion an aqueous solution of 0.14 N hydrochloric acid at 37°C ± 2°C to the mixture in the conical flask. Stir for 1 minute.

4) Check the acidity of the mixture. If the pH is greater than 1.5, add drop by drop, while stirring continuously, an aqueous solution of 2 N hydrochloric acid until the pH is 1.5 or less.

5) Stir the mixture for 1 hour and the mixture to stand for 1 hour °C. Both operations are out in the dark.

6) If necessary, centrifuge the mixture and separate the solids from the mixture by filtration through a membrane filter having a pore size of 0.45 μm. Examine the resulting solution within 6 hours of preparation of the extract to determine the presence and quantity of the appropriate elements in accordance with Section 5.0.

D) Textiles

1) Obtain and prepare test portions as specified.

2) Treat the test portion as specified.
NOTE: Observe whether the test portion is completely wetted after stirring for one minute as specified. If this is not the case, continue to stir until the test portion is completely wetted.

E) Mass Colored Materials

1) Obtain and prepare test portions as specified.

2) Treat the test portion as specified.

F) Solid Materials Intended to Leave a Trace (e.g., chalk, crayons, and pencil leads): Material or cutting the material into pieces. A test portion is to be obtained for each material, which will leave a trace, present in the laboratory sample.

2) Comminute the test capable of passing having an aperture portion so the material is through a metal sieve of 0.5 mm (500 μm).

3) If the material contains any grease, oil, wax or similar substance, remove such substances with acid-free 1,1,1-trichloroethane or other suitable solvent by using Soxhlet extraction. At least 10 cycles are to be carried out.
NOTE: Trace amounts of hydrochloric acid can be produced in chlorinating solvents under certain conditions which may have an adverse effect on subsequent extractions.

4) Treat the test portion as specified.

G) Liquid Materials Intended to Leave a Trace (e.g., the ink used in a pen):

1) Obtain a test portion of the material from the laboratory sample. A test portion is to be obtained for each material, which will leave a trace, present in the laboratory sample.

2) If the material contains any grease, oil, wax, or similar substance, remove such substances with acid-free 1,1,1-trichloroethane or other suitable solvent by using Soxhlet extraction. At least 10 cycles are to be carried out.

3) Treat the test portion as specified.

NOTE 1: If the test portion contains large quantities of alkaline materials, generally in the form of calcium carbonate, adjust the pH to 1.5 or less with an aqueous solution of 6 N hydrochloric acid in order to avoid over dilution.

NOTE 2: The volume of the 0.07 N hydrochloric acid solution is calculated on the mass of the test portion prior to dewaxing, if necessary.

H) Pliable Modeling Materials or Gels:

1) Obtain and prepare test portions as specified in paragraphs (H)(1) and (H)(2) of this section.

2) Treat the test portion as specified in paragraphs (A)(4) through (A) (7) of this section.

I) Oil and Water-Based Paints, Varnishes, Lacquers, and Similar Materials Intended to be Used and Applied by the Consumer (e.g., as used in an arts and crafts set):

1) Solids—Obtain and prepare test portions as specified in paragraphs (G)(l) through (G)(3) of this section.

2) Liquids—Obtain and prepare test portions as specified in paragraphs (H)(l) and (H) (2) of this section.

3) Treat the test portion (of both solids and liquids) as specified in paragraphs (A) (4) through (A) (7) of this section.

Preparation for the Determination of Water-Soluble Barium in Coatings of Paints, Varnishes, Lacquers, Printing Ink, and Similar Materials:

Note: This method is not suitable for use with iron blue pigment. Use the conductivity method given in ASTM Methods D 1135, for chemical analysis of blue pigments.

A) Obtain a portion of coating from the laboratory sample by scraping off the coating.

B) Comminute the coating so the material is capable of passing through a metal sieve having an aperture of 0.5 mm (500 μm).

C) Weigh about 10 grams of the sample to the nearest 0.001 gram and place it into a 500-mL beaker.

D) Add 100 mL 5 minutes, cool, to a 250 mL of distilled water, boil for, and transfer quantitatively volumetric flask.

Note: If the pigment is found to be strongly water-repellant, wet the sample with a small amount of alcohol or carry out a preliminary washing with chloroform.

E) Dilute with distilled water to 250 mL.

F) Stir the mixture for 1 hour and then allow the mixture to stand for 1 hour.

G) If necessary, centrifuge the mixture and separate the solids from the mixture by filtration through a membrane filter having a pore size of 0.45 μm.

H) Examine the resulting solution within 6 hours of preparation of the extract. Use the method stated in Section 4.3 to determine the presence and quantity of water-soluble barium.

TEST ENVIRONMENT—VERIFICATION
OF LABORATORY ACCURACY

Testing is to be performed by a laboratory whose accuracy and reliability have been verified. The verification must be performed the first time a laboratory is employed and at regular intervals not to exceed six months. Verification may be accomplished by the analysis of prepared samples with known concentrations of heavy elements. The known concentrations should be in the same order of magnitude as expected to be found in the test sample. It may also be accomplished by the analysis of duplicate samples which were previously analyzed by a lab whose accuracy and reliability has been established. An alternative would be to utilize laboratories which have been certified by an established program of an appropriate group or agency. An example of such certification in the United States would be the National Bureau of Standards' National Voluntary Laboratory Accreditation Program.

PROCEDURE

For the determination of the concentration of elements in question, methods having a detection limit not greater than 1/10 of the values to be determined must be applied. As an example, for determination of antimony in the order of 250 mg/kg (5 mg/L in the resulting solution), the detection limit must be no greater than 0.5 mg/L; flame atomic absorption spectrophotometry employing background correction may be used in this case. The report of such determinations must include at least the following information:

a) Type and identification of the product tested
b) A reference to this procedure
c) The methods used for determining the quantity of each element present and detection limits of that method
d) Applicable limitations for heavy elements per this procedure
e) The results of each test expressed as: milligrams, to the nearest 0.1 mg (of element present) per kilogram (of material)
f) Any deviation, by agreement or otherwise, from this test procedure
g) Date of the test

All new products should have applicable materials tested during the development stage for conformance to this procedure. The following actions are the minimum required to ensure conformance to heavy elements requirements for all production materials covered by this procedure:

A) The first delivery of every material formulation/color from a given vendor must be accompanied by an analysis for heavy elements. The analysis must have been performed by a laboratory whose accuracy and reliability has been verified (see Section 4.2).
B) Samples of that delivery must also be analyzed independently, either in house or by an outside laboratory, to verify the accuracy of the vendor's analysis.

C) If the two analyses are in substantial agreement, each subsequent shipment of the same formulation/color from that vendor need only be accompanied by a statement of conformance. (See attached last page for example of the required certification form.)

D) If the analyses are not in substantial agreement or if either analysis indicates a nonconformance, steps A and B must be repeated. In the case of a nonconformance, the lot must immediately be rejected, isolated, and returned to the vendor. Any substantial differences in the analyses must be investigated.

E) If the same formulation/color from a given vendor is to be received on a continuing basis, steps A and B must be repeated at least every six months.

F) Any change in formulation/color or vendor requires repeating steps A and B.

EVALUATION

Limits for total heavy elements, in paints, varnishes, or similar surface coatings, as a percent of total solids in the test sample:

The sample must not contain lead compounds of which the lead content (calculated as Pb) is in excess of 0.01% of the total weight of the solids.

The sample must not contain compounds of antimony, arsenic, cadmium, mercury or selenium, individually or in total (calculated as Sb, As, Cd, Hg, and Se, respectively), in excess of 0.05% by weight of the total weight of the solids.

Coatings of paints, varnishes, lacquers, printing ink, and similar materials must not contain compounds of barium of which the water-soluble barium (calculated as Ba) is in excess of 1% of the total barium in such materials.

Limits for soluble levels of heavy elements as a percent of total solids in the test sample:

Coatings of paint, varnishes, lacquers, and similar substances, textiles (dyed and undyed), and plastics used as textiles—the sample must not contain compounds of antimony, arsenic, barium, cadmium, chromium, or mercury (calculated as Sb, As, Ba, Cd, Cr, and Hg, respectively), in excess of the following values by weight of the total weight of the solids.

Coatings of paint, varnishes, lacquers, and similar substances on drawing and painting instruments, substances intended to leave trace (i.e., chalk, crayon, pen, pencils, etc.), plastics, paper, and cardboard—The sample must not contain compounds of antimony, arsenic, barium, cadmium, chromium, lead, or mercury (calculated as Sb, As, Ba, Cd, Cr, Pb, and Hg, respectively), in excess of the following values by weight of the total weight of the solids:

Antimony	0.025%	(250 mg/kg)
Arsenic	0.010%	(100 mg/kg)
Barium	0.050%	(500 mg/kg)
Cadmium	0.010%	(100 mg/kg)
Chromium	0.010%	(100 mg/kg)
Lead	0.010%	(100 mg/kg)
Mercury	0.010%	(100 mg/kg)

Oil- and water-based paints, varnishes, lacquers, and similar materials intended to be used and applied by the consumer, modeling clays and similar materials.

The sample must not contain compounds of antimony, arsenic, barium, cadmium, chromium, lead, or mercury, (calculated as Sb, As, Ba, Cd, Cr, Pb, and Hg, respectively), in excess of the following values by weight of the total weight of the solids:

Antimony	0.025%	(250 mg/kg)
Arsenic	0.005%	(50 mg/kg)
Barium	0.025%	(250 mg/kg)
Cadmium	0.005%	(50 mg/kg)
Chromium	0.0025%	(25 mg/kg)
Lead	0.010%	(100 mg/kg)
Mercury	0.0025%	(25 mg/kg)

Bibliography

ASME (1967) *Steam Tables*, ASME.

Bennett, C.O., Myers, J.E. (1974) *Momentum, Heat and Mass Transfer*, McGraw-Hill.

Bett, K.E., et al. (1975) *Thermodynamics for Chemical Engineers*, Athlone Press.

Bird, R.B., et al. (1960) *Transport Phenomena*, Wiley.

Brinkworth, B.J. (1978) *An Introduction to Experimentation*, Bodder & Stoughton.

Butterworth, D., Hewitt, G.F. (1977) *Two-Phase Flow and Heat Transfer*, Oxford University Press.

Carrol, G.C. (1962) *Measuring Instruments*, McGraw-Hill.

Collier, J.G. (1972) *Convective Boiling and Condensation*, McGraw-Hill.

Comings, E.W. (1956) *High Pressure Technology*, McGraw-Hill.

Cooper, A.P., Jeffreys, G.V. (1973) *Chemical Kinetics and Reactor Design*, Prentice-Hall.

Coulson, J.M., Richardson, J.F. (1977) *Chemical Engineering*, Pergamon.

Crane Co. (1981) *Flow of Fluids*, Crane Co.

Denbigh, K.G., Turner, J.C.R. (1971) *Chemical Reactor Theory*, Cambridge University Press.

Eckert, E.R.G., Drake, R.M. (1972) *Analysis of Heat and Mass Transfer*, McGraw-Hill.

Evans, U.R., et al. (1975) *An Introduction to Metallic Corrosion*, Edward Arnold.

Foust, A.S., et al. (1960) *Principles of Unit Operations*, Wiley.

Foust, A.S., et al. (1979) *Principles of Unit Operations*, Wiley.

Gerald, C.F. (1973) *Applied Numerical Analysis*, Addison-Wesley.

Gerhartz, W. (1990) *Enzymes in Industry*, Weinheim.

Himmelblau, D.M. (1974) *Basic Principles and Calculations in Chemical Engineering*, Prentice-Hall.

Holland, C.D. (1975) *Fundamentals and Modeling of Separation Processes*, Prentice-Hall.

Hout, S.A. (1983) *Chemical Reaction Fouling*, Ph.D. Thesis, University of Bath.

Hout, S.A. (2022a) *Manufacturing of Quality Oral Drug Products*, Taylor & Francis, CRC Press.

Hout, S.A. (2022b) *Sterile Manufacturing*, Taylor & Francis, CRC Press.

Hout, S.A. (2022c) *Sterile Processing of Pharmaceutical Products*, John Wiley & Sons.

Jones, E.B. (1965) *Instrument Technology*, vol. 1–3, Butterworths.

Kay, J.M., Nedderman, R.M. (1974) *An Introduction to Fluid Mechanics and Heat Transfer*, Cambridge University Press.

Kern, D.Q. (1950) *Process Heat Transfer*, McGraw-Hill.

Kick, F.W., Rimoi, N.R. (1975) *Instrumentation*, American Technion Society.

King, C.J. (1971) *Separation Processes*, McGraw-Hill.

Marks, L.S. (1958) *Mechanical Engineers Handbook*, McGraw-Hill.

McAdams, W.H. (1954) *Heat Transmission*, McGraw-Hill.

McCabe, W.L., Smith, J.C. (1976) *Unit Operations of Chemical Engineering*, McGraw-Hill.

McCraken, D.D. (1978) *Guide to Programming for Micro Computer Applications*, Addison-Wesley.

Miller, J.T. (1964) *The Revised Course in Industrial Instrument Technology*, United Trade Press.

Mohr, C.M., et al. (1988) *Membrane Applications and Research in Food Processing*, USDOE.

Nelson, W.L. (1949) *Petroleum Refinery Engineering*, McGraw-Hill.

Nelson, W.L. (1969) *Petroleum Refinery Engineering*, McGraw-Hill.

Perry, R.H., Chilton, C.H. (1973) *Chemical Engineers Handbook*, McGraw-Hill.

Ryder, G.H. (1975) *Strength of Materials*, Macmillan.

Scully, J.C. (1966) *The Fundamentals of Corrosion*, Pergamon.

Shapiro, A.H. (1977) *Shape and Flow*, Heineman.

Sherwood, T.K., et al. (1975) *Mass Transfer*, McGraw-Hill.

Smith, G.D. (1969) *Numerical Solutions of Partial Differential Equations*, Oxford University Press.

Smith, J.M. (1970) *Chemical Engineering Kinetics*, McGraw-Hill.

Smith, J.M., Van Negs, H.C. (1975) *Introduction to Chemical Engineering Thermodynamics*, McGraw-Hill.

Spalding, D.B. (1977) *Genmix—A General Computer Program for 2-D Parabolic Phenomena*, Pergamon.

Thibaut Brain, P.L. (1972) *Staged Cascades in Chemical Engineering*, Prentice-Hall.

Treybal, R.E. (1968) *Mass Transfer Operations*, McGraw-Hill.

Vilbrandt, F.C., Dryden, C.E. (1959) *Chemical Engineering Plant Design*, McGraw-Hill.

Westway, C.R., Loomis, A.W. (1981) *Cameron Hydraulic Data*, Ingersoll-Rand.

Glossary of Terms

Accelerated Aging: Artificial aging of a product by subjecting it to higher-than-normal temperature conditions (usually 120–140°F) for an extended length of time.

Adhered: Material to be bonded using an adhesive-containing substance. More specific than the substrate to an adhesive application.

Adhesion: The state in which two surfaces are held together by interfacial forces which may consist of valence forces or interlocking action, or both.

Adhesion, Specific: Adhesion between surfaces which are held together by valence forces of the same type as those which give rise to cohesion.

Adhesive: Material used to combine two or more materials in interfacial contact, forming a bond, which may be temporary or permanent.

Adhesive, Hot Melt: An adhesive that is applied in a molten state and forms a bond on cooling to a solid state.

Adhesive, Pressure Sensitive: A viscoelastic material which is solvent-free form remains permanently tacky. Such material will adhere instantaneously to most solid surfaces with the application of very slight pressure.

Adhesive, Water Based: An adhesive which has the adhesive components dissolved or dispersed in water.

Adhesive Failure: Bond failure characterized by a separation at one or both of the adhesive-to-substrate interfaces. This occurs when the bond strength is less than the cohesive strength of the adhesive and substrate.

Adhesive Mass: The quantity of adhesive coating, most often indicated in a weight per area unit. A measurement of thickness may not be accurate due to fine lines in coating and due to possible penetration of adhesive into the stock.

Adhesive Transfer Coating: Coating the release liner and laminating to the face stock. The adhesive will transfer when stripped. This can be one way to make a tighter release and may be the only way to laminate hot melt adhesives to films.

Ammonia: An alkaline material with a sharp odor. When dissolved in water, it helps stabilize rubber emulsions and dissolve alkali soluble of many types of adhesives. Leaves a gas when adhesive dries.

Anchorage: The ability of an adhesive to penetrate the surface of the Usually refers to Pressure Sensitive label or tape face stock.

Anti-Foam: Additive that reduces the tendency of an adhesive or coating to form long-lasting air bubbles when agitated

Arcing: Grounding of a radio frequency (RF) gluer due to too much conductivity between top and bottom platen.

Assembly Time: Time between the first glue application until pressure is applied to the composite. Particularly important in cold press and edge gluing.

Back Off: Defect in finger jointing when the jointed pieces separate slightly from each other yielding a weak joint.

Balanced: Refers to equal number and dimension of plies on either side of core material when veneering or laminating to prevent warpage.

Barrier Coating: Coating which creates an impermeable surface. Used to prevent migration of material across the interface.

Barrier Sheet: Nonporous sheet or film in a multiwall bag that keeps moisture out.

Baume: A method of calculating concentrations using degree (°) as units of measure; based on specific gravities.

Blade Assembly: Unit typically consisting of Doctor Blade and Backer Blade.

Blade Holder; part of the Gravure Metering System.

Blanks: Pre-cut paperboard or corrugated sheets which are used to form the finished case or carton. May be coated or uncoated.

Bleached Board: Kraft paper board which has been chemically lightened.

Bleached Paper: Paper made from a hardwood and softwood mixture and subjected to liquors in the paper-making process which remove color from the fibers.

Bleed Through: Unwanted penetration of the substrate which is either visible or noticeable (tacky) from the opposite side. Also known as strikethrough.

Blister: An elevation of the surface of an adhered, somewhat resembling in shape a blister on human skin: its boundaries may be indefinitely outlined, and it may have burst and become flattened.

Brushes: Stiff bristles which have been placed on the labeling machine to "wipe down" label and provide good contact between label and container.

Carton: Box which contains an individual selling unit. Usually made of some type of chip board which may be coated or uncoated.

Case: Box which contains a group of selling units. Usually made of corrugated Kraft paper which may be coated or uncoated. Also known as a shipper.

Catalyst: Additive material which causes a chemical reaction and cross inking in adhesives.

Cellulose Acetate: Thermoplastic resin which in fiber form is used as cigarette filter material.

Chalk: Bond defect due to the lumber temperature being below the coalescence temperature of the adhesive (film-forming temperature). This causes weak bonds and a white appearance to the glue line.

Char: Advanced stage of oxidation process of organic material. Characterized by a very dark or black appearance, like charcoal.

Chip Board: Paperboard material made from course ground pulp. Usually the base material for carton stock. Commonly known as cardboard.

Clamp Carrier: Device which enables an edge gluing operation to be semi-automatic. Edge-glued material is cycled around the carrier during the cold press cycle.

Cobb Test: Used to measure the relative speed of water absorbency of various paper stocks.

Cohesion: The ability of an adhesive to resist splitting. The state in which the particles of a single substance are held together by primary or secondary valence forces.

Cohesive: An adhesive which (after drying) forms a bond when brought in contact with itself or a like adhesive material under pressure. Similar to a cement having basically an infinite open time with an uncontaminated surface.

Cohesive Failure: Bond failure characterized by a splitting of the adhesive film leaving adhesive on both substrates. This occurs when the internal strength of the adhesive is less than the bond strength to the substrates and the internal strength of the substrates.

Cohesive Strength: property of the adhesive to resist internal bond failure or splitting. See shear—Cohesion.

Cold Drop Test: A performance test in which cooled, filled bottles are dropped from a distance of six feet. It is a measure of shock and impact resistance of PET base cup adhesives at low temperatures.

Cold Flow: The property of a thermoplastic or thermo-elastomeric adhesive to act like a heavy viscous liquid over long periods of time.

Cold Press: Method of laminating by making a stack of identically sized laminations and compressing in a room temperature press usually for 30–60 minutes.

Cold Seal: Adhesive type which forms a permanent or peelable bond to many substrates when coated dried and combined with another cold seal coated surface with bonding pressure only (no heat).

Colorimeter: An apparatus to compare the brightness of a stock and 1s generally used to measure the whiteness of a coated face paper.

Compression: The act of holding together of substrates in the bonded configuration under pressure.

Compression Rollers: A series of rollers which press the substrates in contact with the adhesive.

Compression Section: Portion of labeling machine where brushes, compression rollers, or pressure pads are located.

Compression Time: The period of time which the bonding area is under compression.

Consolidated Seal: Top sealing of an SOS style bag by folding the top over several times and gluing it to the side of the bag.

Contamination Free: Multiwall bag in which the inner ply is the barrier product in the bag and cannot be exposed to paper fibers or other contaminants. Abbreviated "CF."

Corona Treatment: A method used to increase the polarity of a plastic film by oxidizing the surface with electrical discharge. Used to improve adhesion to polyethylene and polypropylene.

Corrugated: Board made by bonding two Kraft paper liners to a corrugated Kraft paper medium using a starch or modified starch adhesive. Used to produce most shipping cases.

Creep: The dimensional change with time of a material under load, following the initial instantaneous elastic or rapid deformation. Creep at room temperature is sometimes called Cold Flow.

Curl: Caused by the same conditions as tunneling. Storage of laminated product or atmospheric changes will not correct this problem.

Cycle Time: Time to complete the gluing process. Usually the time in radio frequency (RF) gluer with power on.

Cylinder: Usually refers to the rollers which merge or transfer the adhesive or coating through tab coating process. Many times refers to the gravure roller.

Defoamer: Additive which helps break and release trapped air bubbles from an adhesive or coating which has been agitated.

Delta Seal: Top sealing for a POM style bag where one side is folded over, adhesive is applied to it, the other side is folded over and mated to the first side, and the triangular ends are folded over the inner seal and glued together.

Density Reduction: The amount of adhesive in a given volume that is displaced by gas. Expressed in percent.

Desert Test: Extended storage test run at 100°F and low relative humidity.

Die Cutting: Process whereby the converter cuts the roll label stock into a series of individual labels.

Dilution: Extension of material by addition of carrier material. Commonly used to decrease viscosity or improve penetration. Specific to cigarette manufacture, refers to the reduction of nicotine, tar, and noxious gasses in smoke stream by means of perforated filter papers.

Doctor-bar or Blade: A scraper mechanism that regulates the amount of adhesive on the spreader rolls or on the surface being coated.

Doctor-roll: A roller mechanism that is revolving at a different surface speed, or in an opposite direction, resulting in a wiping action for regulating the adhesive supplied to the spreader roll.

Dry: To change the physical state of an adhesive on an adhered by the loss of carrier material to a 100% non-volatile state.

Dwell Time: Time which adhesive or coating is under a certain process usually heat and or pressure.

Edge Glue: Operation to assemble lumber to form a wider surface dimension as in tabletops, butcher block or hardwood post construction.

Edge Ooze: A "squeezing out" of the adhesive from between the liner and face stock. When labels or tapes are in roll form, the edges of these rolls will be sticky

Elastic Attachment: Adhesive used for attaching elastic to poly in the manufacture of diapers.

Elastomer: A polymeric substance that has elastic rubber-like properties.

Elongation: The length that a tape will stretch before breaking.

End Seal: Adhesive used to seal the end of Feminine Pads, hospital under pads, and diapers. Usually non-woven to polyethylene film.

Exterior: Refers to ATM D-3110–72 wet use requirement having boiling water resistance. For adhesive applications which have expected contact with water and weather. All are two-part systems.

FSK: Refers to foil/scrim/kraft lamination.

Face or Label Stock: The label, tag, or tape surface.

Failure, Adhesive: See Adhesive Failure.

Failure, Cohesive: See Cohesive Failure.

Failure, Substrate: See Substrate Failure.

Fiberboard: Board stock consisting of compressed wood fibers held together with Urea-Formaldehyde binder.

Fiber Tear Bond: See Substrate Failure.

Film Laminating: Bonding polyolefin film to a kraft ply in a multiwall bag with latex or pressure-sensitive liquid or hot melt adhesive.

Finger Joint: Type of complex joint where lumber has the appearance of the fingers meshing together. This type of joint is very strong and used to join pieces for dimensional lumber.

Flagging: When label has partially released from the container at corners, most often occurs during the drying cycle.

Flame Treatment: A method of increasing the polarity by passing an open flame near the surface causing oxidation. Used to promote adhesion to polyethylene and polypropylene.

Flax Fibers: Fibers taken from plant *Linum usitatissimum* and used in the manufacture of cigarette papers. Chosen because of pleasant taste and good burn rates.

Flexible Packaging: An industry whose business is to provide non-rigid packaging materials.

Flexographic: A printing process that uses resilient pads or "stencils" to pattern apply the adhesive or coating. Similar to a rubber stamp mounted on a rotating cylinder.

Fluff Stabilization: Adhesive which is spray applied to fluff in diapers to help hold fluff together and keep from shifting in use.

Foam (aeration): Encapsulation of small air bubblethroughout the adhesive mass. Increases viscosity and reduces tack.

Foam Half Life: The time required for a foamed material to decrease to one-half of its original volume.

Foamability: The ability of a material to foam when gas is injected into the solution.

Foil Label: Label stock which has an outer layer of aluminum foil laminated to the paper substrate.

Fugitive Glue: Adhesive used to temporarily bond two materials. Fugitive bond allows cigarette cartons to be opened later for the application of tax stamp.

Gasket: A cushioning material used to seal a mechanically fastened joint.

Gauge: Refers to the thickness of film or paper stock.

Gel: Stage of partial cross linking of a material. Can refer to thickening in an over pot life exterior wood glue mix or partially oxidized hot melt adhesive which is solid at normal application temperature.

Glue Line (bond line): The layer of adhesive which attaches two adherends.

Glued Lap: Tab on a corrugated case, when bonded forms the body of the box. Also known as the manufactures joint.

Grain Raise: Phenomenon occurring on particle board which is caused by water swelling wood particles at the surface.

Gravure: Short for rotogravure. A method of coating using fine engraved cells of specific size. These cells carry the coating onto the substrate in a very uniform pattern. Used for precise application of coating or adhesive.

Green Strength: Property ofadhesive which allows handling of the bonded material before adhesive has completely cured.

Ground Wood Paper: A variety of paper usually very porous, wood fibers easily seen, used widely with aluminum foil for foil labels.

Gusset: Area on each side of an SOS bag which folds inward.

HDPE: High-density polyethylene—a thermosetting thermoplastic white solid characterized by linear and closely aligned molecular chains.

Hardboard: Special type of fiberboard with a Melamine binder which is dense and has a smooth, non-porous surface.

Heat Seal: Adhesive type which forms a permanent or peelable substrate when coated dried and combined with another heat surface with heat and bonding pressure.

Homogenous: The same composition or construction throughout.

Hot Melt Adhesive: See Adhesive, hot melt.

Hot Press: Lamination process where heat accelerates the cure. Usually requires a catalyzed system.

Humectant: Adhesive ingredient which delays water release, aiding in the lay flat of paper/paperboard laminations.

ICS Tray: A multi-chambered rigid plastic tray used for stacking PET soft drink bottles. Places much stress on the base cup bond.

Ice Proof Adhesive: An adhesive which when a labeled container is immersed in ice water (32–34°F) will maintain its bond and not allow the label to come off (usually 24–72-hour time period).

Inert Gas: A group of gasses that exhibit great stability and extremely low reaction rates.

Instron Tensile Tester: A mechanical device which pulls apart two adherends and measures the force required to separate them.

Interior: Refers to adhesives which pass ASTM D-3110–72 shear and fiber failure requirements. Not to be used in contact with water.

Internal Construction: Adhesives used in the internal construction of Feminine Pads.

Joint, Lap: A joint made by placing one adherend partly over another and bonding together the overlapped portions.

Jungle Test: Extended storage test run at 90°F and 90% relative humidity.

Kaymich Nozzle: Gravity feed system used to apply plug wrap hot melts and rod seam adhesives for cigarette manufacture.

Kill: Refers to adhesive being driven into the paper by draw rolls or pinch drums, leaving too little adhesive on the surface to obtain a bond.

Kraft Paper—Natural—paper made from softwood fibers, primarily from pine and fir trees.

Label Basket—The part of the labeling machine which holds the labels prior to adhesive application (also magazine).

Label Curl: A condition whereby the labels do not lay flat but curl up on the edges. Caused by lack of moisture or other storage conditions.

Laminate: A product made by bonding together face to face two or more layers of material or materials.

Lamination: Bonding process where two or more materials are combined by their largest dimension. As in lay flat lamination, panel lamination, bag lamination, and so on.

Land Area: The sections of the (gravure) cylinder surface which are not engraved.

Landing Strip: Hot melt adhesive applied to polyfilm which is applied at point of adhesion for diaper tape tab.

Lap Shear: The testing of an overlap joint shear mode (parallel to glue in line) of a specific area. Dynamic shear is a test of increasing strain to failure. Static shear is the constant stress of a weight in shear where time to failure is recorded.

Laquer: Usually refers to a solvent based over print varnish used as a protective coating on printed labels and carton materials.

Layflat: Absence of wrinkling, blisters, curl, or warping of laminated paper structures.

Legging: The drawing of filaments or strings when adhesive-bonded substrates are separated.

Litho: Highly calendared printing paper designed for use with offset press inks.

Lithography (Litho): A process using a metal plate to transfer an ink or coating to the substrate to be printed.

Longitudinal Seam: Adhesive used in the seam of Feminine Pads.

Loop Tack Test: Measures the force to remove a tape sample that has been in contact with a surface for just one or two seconds. There is also a probe test to measure tack, and it can also be measured with any one of the several tensile testers.

MDF: Medium-density fiberboard.

Machine Glazed: Paper run through a series of heated rollers at the mill leaving one side very smooth and semi-shiny.

Maker: Machine where tobacco rod is formed and encircled with cigarette paper.

Manufactures Joint: See Glued Lap.

Mastic: A viscous pasty substance used as a coating or adhesive.

Matrix: The disposable part of the DIE CUT roll label stock that is stripped and removed from the release liner.

Mechanical Adhesion: Adhesion between surfaces in which the adhesive holds the parts together by interlocking action.

Metalized Label: Label produced by the deposition of an aluminum surface to the paper. Appearance is similar to foil laminated to label, but running and bonding characteristics are different.

Michelmann Coating (IM): Water-based coating in the manufacture of corrugated boxes to increase water and oil resistance, angle of slide, etc. Generally requires a special glued lap and case seal adhesive.

Mil: Unit of measurement for paper and film thickness: 1 mil =.001 inch.

Mileage: The number of units which can be bonded with a specified weight or volume of adhesive.

Multi-bead: Many lines of adhesive used for laminating poly to nonwoven in the manufacture of diapers.

Newtonian: A fluid whose viscosity is generally unaffected by shear. A hot melt adhesive is a non-Newtonian fluid and shear does affect its viscosity and must be considered in the coating parameters for slot coating nozzles.

Offset: Printing process using a rubber roller to transfer the ink or coating to the substrate.

Open Time: The time lapse between the application of adhesive until the mating of the second surface. Also, open assembly time.

Oriented Polypropylene Film (OPP): Film used in many packaging applications having excellent strength.

Over Print Varnish (OPV): Protective coating to increase gloss and abrasion resistance of printed materials. Also, prevents surface contamination by inks in coating applications.

pH: Measurement of the relative acidity (low pH) or alkalinity (high pH) of water base materials. Ranges from O to 14 with 7 being neutral.

PET: Polyethylene Terephthalate—A thermoplastic polyester used in the manufacture of soda bottles, recording tapes, packaging films, and textiles.

Pad Attachment: A pressure-sensitive adhesive used to attach Feminine Pad to undergarments.

Pallet: On a rotary-style labeler, the portion of the machine that transfers the adhesive from the glue roller to the label.

Particleboard: Board stock consisting of compressed wood chips held together with a urea formaldehyde binder.

Pasted Open Mouth (POM): Same type of bag as the SOS except that it does not have a square bottom; it also is used in bulk packaging.

Peel Bond: Adhesive failure bond with no fiber or substrate destruction. For pressure-sensitive adhesives, also called removable bond.

Penetrometer: An apparatus for lab determination of the softness of an adhesive. In the test, a needle with a weight is pressed into the surface of the adhesive for five seconds. The distance of penetration is a relative value.

Picker Fingers: On a labeling machine such as a Super CM the picker fingers transfer the adhesive from the rubber transfer roll to the labels in the label basket.

Pick-Up Roll: A spreading device where the roll for picking up the adhesive turns in a reservoir of adhesive.

Pinch Bottom (PIN): Open mouth bag with its bottom closed by folding it over and sealing with a liquid or, more commonly, a hot melt adhesive; open end will have pre-applied hot melt for heat reactivation and closure by the end user.

Plasticizer: A material incorporated in an adhesive to increase its flexibility, workability, or distensibility. The addition of the plasticizer may cause a reduction in melt viscosity, lower the temperature of the second-order transition, or lower the elastic modulus of the solidified adhesive.

Plasticizer Migration: Movement of plasticizer from vinyl film into adhesive, weakening the bond.

Platen: Metal plate used to apply pressure in cold and hot press or plates which radio frequency energy passes between for RF glue.

Plug Wrap: A highly treated, hard-to-bond paper which surrounds the filter cigarette media to form filter plugs.

Point: Measurement of paperboard thickness; 1 point= 1 mil.

Pop Open: Bond failure usually in case and carton sealing where compression time is less than the adhesive set time. This causes end flaps to release out of compression.

Press Time: Length of time glued material is under pressure. In hot press, cold press, or RF gluing operation.

Radio Frequency (RF): Method of curing using microwaves to heat the water in the glue line for fast set. Uses catalyzed system.

Recirculation: Adhesive or coating delivery process which returns excess material to a reservoir for later return and use.

Release Paper: A sheet, serving as a protectant and/or carrier for an adhesive film or mass, which is easily removed from the film or mass prior to use. Also, Release Liner.

Release Liner: A silicone-coated substrate, usually kraft that is coated with silicone. Available in various weights and also available with lamination of poly, or as a total film. Various levels generally available from a range of 10 grams to 250 grams.

Repositionable: Materials can be combined and recombined more than once such as a diaper tape tab or label.

Ridging: Adhesive does not level properly and leaves lines which show through paper or film.

Ring and Ball: An apparatus for lab determination of the softening point and melting point of adhesives. An indication of end use performance or high temperature requirements.

Roll Blocking: When Nipweld coating adheres to the other side of the roll and does not release.

Roll Fed Stock: Some labelers will use a continuous roll of labels as opposed to a bundle of individual labels. Label is cut from the roll by the labeling machine.

Rolling Ball Test: This is a relatively inexpensive test and probably one of the oldest. It is an indicator of relative tack within a family of adhesives. Results may not be consistent with other types of adhesives.

SBS Board: Clay coated box board having a synthetic rubber binder (SBS) often used in beer and soft drink carriers for water and moisture resistance.

Scavenger Blades: Etching or engravings on the (gravure) cylinder face, outside the coating area; used to improve machinability.

Sealant: A substance used for caulking and sealing joints. Fills gaps and irregularities between parts.

Set: To convert an adhesive into a fixed or hardened state by chemical or physical action, such as condensation, polymerization, oxidation, vulcanization, gelation, hydration, or evaporation of volatile constituents, such as water.

Set Time: Time lapse between adhesive application and bond strength build-up in excess of required amount. Usually to substrate destruct bond.

Shear—Adhesion: The separation of two substrates when subjected to forces in opposite directions parallel to the glue lines.

Shear Strength: See Lap Shear.

Shipper: Usually refers to the corrugated case which is the basic shipping unit for a product.

Shortness: A qualitative term that describes an adhesive that does not string cotton, or otherwise form filaments or threads during application.

Side Seam: Lap joint on a carton, when bonded forms the body of the box. Similar to the glued lap on a case.

Sizing: The process of applying a material on a surface in order to fill pores and thus reduce the absorption of the subsequently applied adhesive or coating or to otherwise modify the surface properties of the substrate to improve the adhesion. Also, the material used for this purpose. The latter is sometimes called Size.

Skip Tipping: The pattern-application of cigarette tipping adhesive to tipping paper. Came into vogue with the advent of perforated tipping paper.

Sleeve Stock: Polyolefin film to kraft laminated tube placed in the pasted valve of a poly valve bag for filling; it is fed from a roll into a designated corner as the top of the bag is formed and glued in with latex on the film side; it will have an extension on the inside of the bag so that after the bag is filled, the product will seal it off.

Slip Agents: Material added to films to help in the manufacture and conversion of the films.

Self-Opening Sack (SOS): Open-mouth bag with a square, glued bottom like a common grocery bag; its most common use is for baler bags used in bulk packaging of smaller bags; e.g., 5-pound sugar bags packed 24 to a bale.

Sewn Open Mouth (SOM): Bag sewn closed at one end during manufacture; end user fills from the open end and sews shut.

Sewn Valve (SV): Bag sewn at both ends which has a valve in an upper corner for customer filing; the valve is folded into an inside pocket after filling.

Snap Back: Delamination of the film when it is the inner ply of a multiwall bag as it goes into the forming section of the multiwall tuber.

Solids Content: The percentage by weight of the nonvolatile matter in an adhesive. For comparison, standard test methods must be used.

Solubility: The maximum amount of a substance that can be dissolved into a given amount or volume of another substance. Salt dissolved in water is an example of solubility.

Spot Paste: Dots of adhesive applied at each end of a multiwall bag tube across the ply webs to tack the plies together.

Spray: Atomized adhesive is applied to the substrate by mechanical means.

Spread: The quantity of adhesive per unit area applied to an adherend, usually expressed in points of adhesive per thousand square feet of area.

Squeeze Out: Excess adhesive which oozes out around the glue line that indicates adequate adhesive amount and pressure.

Stain Test: A penetrating die solution that is brushed on a silicone liner. When wiped dry, any pin-holloing or streaks in the silicone coating will be indicated as dark spots.

Storage Life: The period of time during which a packaged adhesive can be stored under specified temperature conditions and remain suitable for use. Sometimes called Shelf Life.

Streaking: Coating defect characterized by liner adhesive or coating voids perpendicular to the coating head. Usually caused by material trapped in or behind the adhesive metering device.

Stringiness: The property of an adhesive that results in the formation of filaments or threads when adhesive transfer surfaces are separated. Also in hot melt applications where excessive cooling occurs between nozzle and substrate.

Strip: Adhesive is extruded on the film or wheel applied on the kraft in continuous or broken lines for bag manufacture. Strip laminating.

Substrate: A material upon the surface of which an adhesive-containing substance is spread for any purpose, such as bonding or coating. A broader term than adherend.

Surface Energy: A measure of the polarity of a substrate, the greater the polarity the easter to adhere. Normally measured in dynes/centimeter; 42 dynes or greater is preferred for ease of adhesion.

Surface Preparation: A physical and/or chemical preparation of an adherent to render it suitable for adhesive joining.

Surface Tension: The tension at the surface of a liquid likened to a stretched elastic skin. Normally measured in dynes/cm. Related to surface wetting and adhesion.

Surfactant: An additive that lowers the surface tension of a substrate.

Syneresis: The exudation of small amounts of liquid by gels on standing.

Systaltic Pump: Low shear pumping device which uses bladders to transfer adhesive similar to a heart-pumping motion. Preferred for most adhesive and coatings.

T&O Evaluation: Taste and Odor evaluation by a panel of experts, usually long-time employees. Can be very subjective and inconsistent.

Tack: The property of an adhesive that enables it to form a bond of measurable strength immediately after adhesive and adherend are brought into contact under low pressure.

Tack Range: The period of time in which an adhesive will remain in tacky-dry condition after application to an adherent, under specified conditions of temperature and humidity.

Tamperproof: A label that cannot be removed from an end-use substrate without destroying the label. Such a label is relative both to the peel value of the adhesive and the tear strength of the paper.

Tape Tab: Pressure-sensitive tab used for securing diapers around the waist. May be used in conjunction with tab landing strip.

Telegraphing: Board surface imperfections which show through paper or film after laminating process.

Telescoping: A laminated or coated roll that forms a one-like shape after some interval. A result of too much tension when winding a coating that is not uniform.

Tensile: The strength of a material when subjected to opposite forces parallel to the plane of the sample as measured by the force to cause it to break.

Thermoplastic: A material that will repeatedly soften when heated and harden when cooled.

Thixotropy: A property of adhesive systems to thin upon isothermal agitation and to thicken upon subsequent rest.

Throwing or Spitting: Small droplets of adhesive will be thrown from the fast-moving adhesive transfer rolls. Problem on rotary style machines.

Tipper: Machine where cigarette rods are combined with filter plugs and wrapped with tipping paper. Adhesive is applied by pot and wheel set-up.

Toughness: The total energy absorbed by a tape before it breaks. Represented as the area under a curve when plotting the elongation against force in tensile tests.

Tow: Generic name for cellulose acetate fibers which make up filter media.

Transfer Roller: A roll used to transfer adhesive from the pick-up roll located in the glue pot to picker fingers as on a Super CM machine.

Treatment Level: Polyethylene and polypropylene bottles can be flame or electrical discharge treated to provide a more receptive bonding surface. Measured in Dyne level.

Triacetin: A resinous plasticizer. Used to harden cigarette to material into semi-rigid filter plugs.

Try-Ply: Foil/kraft/foil lamination used for skins with foamed in place Urethane insulation.

Tunneling: Separation of liner and face paper either across the web, or along the web direction. When it is in the web direction, it is caused by a difference in moisture content between the two stocks when laminated. When it is across the web, then it is the result of improper tension between the two webs at the point of lamination.

Varnish: See over print varnish.

Veneer: Thin wood cut by slicing or rotary cutting lumber for plywood or flat stock.

Viscometer: An apparatus for lab determination of the viscosity of an adhesive at a given temperature. For one adhesive, the consistency of this viscosity is an indication of its degradation.

Viscosity: The ratio of the shear stress existing between laminae of moving fluid and the rate of shear between these laminae. A measurement of fluidity, usually measured in centipoise and determined by a Brookfield viscometer.

Warping: Curling or bowing of lamination caused by uneven adhesive drying or moisture content on the dissimilar surfaces.

Water Drop Test: A drop of water placed on the adherend side of the label to provide a measure of receptivity to water-based adhesives.

Webbing: See Stringing.

Web-Coat: Laminating process in which adhesive is applied to paper or vinyl film, which is then combined with board stock.

Well-Dressed Blade: Doctor Blade which has been mounted evenly and is free of nicks.

Wet Strength: The strength of an adhesive joint determined immediately after removal from a liquid in which it has been immersed under specified conditions of time, temperature, and pressure. Also used to describe the joint strength of adherends with the adhesive still in a wet state.

Wing-Up: The tendency of a label edge to lift or flag from the substrate to which it is adhered.

Index

A

adhesives, 133
  ethylene vinyl acetate (EVA), 133
  foamed hot melt, 136
  laminating adhesives, 139
  polyethylene (PE)—hot melt, 133
  pressure-sensitive adhesives, 138
  trouble shooting problems—adhesives,
    134–136

B

bio-chemical engineering, 115
  catalytic enzymes, 115
  fermentation, 115
  NADH, 115

C

chemical plant design, 107
  materials creep, 108
  Poisson's ratio, 108
  strain, 107
  stress, 107
  Young's modulus, 108
chemical plant—engineering materials, 13
  alloying metals, 15
  materials of construction, 14
  materials' cost, 18
  Newtonian liquids, 19
  rheological behavior, 18
  stiffness of materials, 16
  strength of materials, 15
  strength properties, 16
  visco-elastic liquids, 19
chemical reaction engineering, 51
  batch reactor, 52
  catalyst poisoning, 53
  catalytic processes, 54
  catalytic reactors, 54
  chemisorption, 53
  intraparticle diffusion, 54
  pellet equations, 53
  plug - ow (tubular reactor), 51
  purge, 51
  reactors, 52
  reactor types, 55
  space time yield, 54
  space velocity, 54

chemical reaction fouling, 237, 273
coating, 131
computations for nuclear reactor heat removal
    systems, 211

E

electrochemistry/corrosion, 67
  cathodic reaction, 67
  corrosion number, 72
  corrosion testing, 71
  Pourbaix diagram, 67, 68
  protective coatings, 71
  relative molar volume, 67, 68

H

heat, mass, and momentum transfer, 1
  Avogadro number, 1
  De Broglie duality, 1
  electrons, 1
  equivalent proportions, 1
  laws of de definite, multiple, and, 1
  Schrodinger, 1
heat transmission—natural convection, forced
    convection, 37
  A.S.M.E, 41
  boiling heat transfer, 40
  circulation rate, 39
  circulation ratio, 39
  composition effects, 45
  fluid velocity, 41
  fouling, 37
  heat capacity, 38
  heat flux, 45
  heat transfer coefficient, 37
  horizontal surface, 43
  immiscible mixture, 38
  LMTD, 40
  overall heat transfer coifficient,
    40, 42
  pitch, 39
  prandtl number, 37
  reboilers, 39
  sensible heat, 38
  submergence, 38
  T.E.M.A, 41
  viscosity ratio, 40
heavy metals testing, 297

I

industrial inks, dyes, and pigments, 117
    additives, 120, 125
    cadmium pigments, 121
    colorants, 121
    composition, 117
    detection limits of heavy metals, 120
    dispersion, 125
    dissolution, 119
    flame retardants, 120
    fluorescent pigments, 122
    heat stability, 125
    heat stabilizers, 120
    ink manufacturing, 128
    lightfastness, 125
    preparation of the specimens, 119
    printing inks, 117
    printing inks—manufacturing process,
        127, 128
    synthetic organic pigments, 118
    toxicity, 126
    UV curing offset inks, 126
    UV inhibitors, 120
    water extraction for the barium
        determination, 119
instrumentation, 3
    chromatography, 8
    closed-loop control, 10
    compressible - fluids, 4
    connection of pressure gauges in chemical
        plants, 3
    differential pressure, 4
    electronic transducer devices, 5
    flow measurement devices, 7
    gas chromatography (GC), 7
    Geiger-Muller, 7
    instrumentation and control of distillation
        columns, 8
    instrumentation of heat exchangers, 10
    ionization gauge, 7
    level measurements, 7
    liquid chromatography (HPLC), 7
    liquid column-type measurements, 6
    oxygen meters, 7
    pressure drop, 4
    pressure measurement, 3
    pulsation dampers, 4
    rotameters, 7
    scintillation counter, 7
    static head, 4
    stress hysteresis, 4
    temperature measurements, 4, 8
    thermocouple, 4, 5
    vacuum distillation, 9
    variable area meters, 7

M

master validation plan, 181
    final report template, 205, 207
    protocol template, 205, 206
    site validation requirement, 195
mathematics, 95
    differential equations, 95
multi-phase flow, 113

N

new plant construction, 249
nuclear power, 153
    ACCW, 153
    CCW, 153
    computations for nuclear reactor, 211
    cooldown, 155
    NSCW, 153
    power plant drain systems, 289
    process variables, 161
    RCS, 157
    RCS thermal capacity, 163
    RHR, 153
    SFP, 158
    system design, 157
    water hammer in nuclear power plants, 269
numerical methods—computation, 97
    computer program flow chart, 103
    DO-Loop, 97
    least square regression, 99
    mathematical models, 97
    numerical integration, 100
    numerical method calculations, 97
    numerical solution, 104
    operational validity, 98

P

pharmaceutical manufacturing, 169
    adjuvants, 173
    API transfers, 176
    CAPAs, 176
    cGMP, 176
    charging, 170
    clean room classicalisms, 174
    Clean Rooms GMP Classicalisms, 177
    clinical trials, 172
    compression, 170
    controlled-release drug, 169
    equipment qualifications cation, 176
    FDA aseptic processing guide, 176
    granulation, 170
    laser drilling, 170
    membrane coating, 169
    microbial limits, 175

milling, 170
NDA, 176
oral solid dosage (OSD), 169
osmotic pump system, 170
PAI, 176
permeability, 170
PFS, 174
PPE, 176
printing, 171
process flow - liquid aseptic operation, 179
QbD, 174
RABS, 176
SOPs, 176
sorting, 171
sterile injectables, 171
sterile products, 171
sub-coating, 170
vaccine, 172
WFI trending, 176
plastic materials, 129
power plant drain systems, 289
pressure vessels, 21
stress, 21
valve type selection, 23
process dynamics, 89
feedback control, 92
flow sheets, 89
laplace transforms, 93
Nyquist criterion, 92
solvent extraction, 90
steady state, 89
system stability, 92
unsteady state, 89

T

thermodynamics—mass, enthalpy, entropy, free
        energy, 25
adiabatic, 25
extensive property, 25
first law of thermodynamics, 26
heat engines, 29
ideal gas, 25, 27

the ideal gas model, 27
intensive property, 25
isentropic, 29
polytropic, 29
second law of td, 26
thermodynamic properties of
        fluids, 26
vapor–liquid equilibria, 28
three miles island, 265
transport phenomena, 47
Buckingham Pi method, 48
del operator, 49
dimensionless groups, 47
equations of motion and energy, 50
Fick's law, 48
Fourier's, 48
mass balance, 47
mass balance (steady state), 47
mass flow, 48
mass flux, 47, 53
Nabla, 49
Newton's, 48
rate of accumulation of internal and kinetic
        energy, 50
rate of accumulation of mass, 50
rate of accumulation of
        momentum, 50
Rayleigh's method, 47
vector, 49
vector analysis, 49

U

unit operations, 75
crystallization, 77
discharge flux, 72
distillation, 78, 81, 83
fluidization, 77
leaching, 78
McCabe-Thiele model, 86
Murphree efficiency, 86
sedimentation, 77
solvent extraction, 75

For Product Safety Concerns and Information please contact our EU
representative GPSR@taylorandfrancis.com
Taylor & Francis Verlag GmbH, Kaufingerstraße 24, 80331 München, Germany

www.ingramcontent.com/pod-product-compliance
Lightning Source LLC
Chambersburg PA
CBHW060811220326
41598CB00022B/2592

*9 7 8 1 0 3 2 4 7 0 1 1 5 *